STATISTICS AS PROOF

FUNDAMENTALS OF QUANTITATIVE EVIDENCE

DAVID W. BARNES
Assistant Professor of Law and Economics
Syracuse University College of Law
Maxwell School of Citizenship and Public Affairs

Little, Brown and Company
Boston and Toronto

Copyright © 1983 by David W. Barnes
All rights reserved. No part of this book may be reproduced in any form or by any electronic or mechanical means including information storage and retrieval systems without permission in writing from the publisher, except by a reviewer who may quote brief passages in a review.
Library of Congress Catalog Card Number 82-082846
ISBN 0-316-08149-3

B-B

Published simultaneously in Canada
by Little, Brown and Company (Canada) Limited
Printed in the United States of America

STATISTICS AS PROOF

EDITORIAL ADVISORY BOARD

LITTLE, BROWN AND COMPANY
LAW BOOK DIVISION

A. James Casner, *Chairman*
Austin Wakeman Scott Professor of Law, Emeritus
Harvard University

Francis A. Allen
Edson A. Sunderland Professor of Law
University of Michigan

Clark Byse
Byrne Professor of Administrative Law
Harvard University

Thomas Ehrlich
Provost and Professor of Law
University of Pennsylvania

Geoffrey C. Hazard, Jr.
John A. Garver Professor of Law
Yale University

Willis L.M. Reese
Charles Evans Hughes Professor of Law, Emeritus
Columbia University

Bernard Wolfman
Fessenden Professor of Law
Harvard University

To my parents,
whose support I value

SUMMARY OF CONTENTS

Contents		ix
Preface		xv
Acknowledgments		xvii
Chapter I	Introduction	1
Chapter II	Descriptive Statistics	5
Chapter III	Introduction to Inferential Statistics	31
Chapter IV	Standard Deviations	75
Chapter V	Significance Testing	143
Chapter VI	Inferring from a Sample	231
Chapter VII	Regression Analysis	283
Chapter VIII	Healthy Skepticism	379
Appendix A	Statistical Tables	397
Appendix B	Answers to Selected Problems	415
Table of Cases		427
Substantive Law Index		429
Index		433

CONTENTS

Preface	xv
Acknowledgments	xvii

CHAPTER I
INTRODUCTION   1

A.	How We Will Proceed	2
B.	Two Types of Statistics	3

CHAPTER II
DESCRIPTIVE STATISTICS   5

A.	Introduction	5
	Alfred F. Conard, The Quantitative Analysis of Justice	5
B.	Histograms, Tables, and Graphs	13
C.	Flexible Figures	18
	Notes and Problems	21
	Pennsylvania v. Local 542, International Union of Operating Engineers	24

CHAPTER III
INTRODUCTION TO INFERENTIAL STATISTICS   31

A.	Inferential Statistics as Proof		31
	1.	The Role of the Statistician	32
	2.	Examples of Inferential Statistical Proof	33
	3.	Two Illustrations of Statistical Inference	34

ix

B.	"Simple" Statistical Evidence	36
	Craig v. Boren	36
	Notes and Problems	43
C.	Using Statistical Studies	46
	Hans Zeisel, . . . And Then There Were None: The Diminution of the Federal Jury	47
	Ballew v. Georgia	58
D.	Binomial Probabilities	65
	Notes and Problems	71

CHAPTER IV

STANDARD DEVIATIONS — 75

A.	Group Averages	75
B.	Statistical Notation	76
C.	Variability	77
D.	Calculating a Standard Deviation	79
E.	A Special Case: The Binomial Distribution	81
F.	Standard Deviations as Statistical Proof	82
	Castaneda v. Partida	82
	Notes and Problems	89
	United States v. Goff	93
	Notes and Problems	95
	Inmates of the Nebraska Penal & Correctional Complex v. Greenholtz	97
	Notes and Problems	107
	Board of Education v. Califano	108
	Notes and Problems	115
	EEOC v. United Virginia Bank/Seaboard National	117
	Notes and Problems	123
	United States v. Maskeny	124
	Notes and Problems	127
	Note: In re Coca-Cola Co.	129
G.	The Coefficient of Variation	131
	In re Forte-Fairbairn, Inc.	131
	Notes and Problems	137
	Note: B.F. Goodrich Co. v. Department of Transportation	138
	Note: Taylor v. Weaver Oil & Gas Corp.	138
H.	The Normal Distribution	139

CHAPTER V

SIGNIFICANCE TESTING — 143

- A. Statistical versus Practical Significance — 143
- B. Type 1 and Type 2 Errors — 145
  - *United States v. Fatico* — 146
  - Notes and Problems — 154
  - *Certified Color Manufacturers Association v. Mathews* — 155
  - Notes and Problems — 162
- C. Chi-Square Testing — 163
  - *Chinese for Affirmative Action v. FCC* — 163
  - 1. Calculating the Chi-Square Value — 172
  - 2. Degrees of Freedom — 174
  - 3. Using the Chi-Square Table — 174
  - Notes and Problems — 178
  - Notes: Problems with Small Numbers — 180
    - 1. Combining Categories — 180
    - 2. The Yates Correction — 181
    - 3. Age-Discrimination Hypothetical — 182
    - 4. Causation — 182
  - *Chance v. Board of Examiners* — 182
  - Notes and Problems — 196
  - Problem: In re Kroger Co. — 197
- D. The Z Test — 198
  - *Gay v. Waiters' & Dairy Lunchmen's Union, Local 30* — 203
  - Notes and Problems — 228
  - Problem: Alabama Nursing Home Association v. Califano — 230

CHAPTER VI

INFERRING FROM A SAMPLE — 231

- A. Populations and Samples — 231
- B. Sample Means and Standard Errors — 232
- C. Confidence Intervals — 235
  - 1. The Standard Error of the Sample Mean — 236
  - 2. Choosing a Sample Size — 238
    - *Reserve Mining Co. v. EPA* — 240

	Notes and Problems	246
	Problem: Marathon Oil Co. v. EPA	248
	United States v. General Motors Corp.	249
	Notes and Problems	255
	Note: The Use of Student's *t* for Small Samples	257
	Problems: A.B.G. Instrument & Engineering, Inc. v. United States	261
	Presseisen v. Swarthmore College	262
	Notes and Problems	264
D.	Correlation Coefficients	264
	Boston Chapter, NAACP, Inc. v. Beecher	272
	United States v. City of Chicago	277
	Notes and Problems	282
	Problem: In re National Commission on Egg Nutrition	283
E.	Spearman Rank Correlation Coefficient	284
	Pennsylvania v. Local 542, International Union of Operating Engineers	284
	Notes and Problems	288
	Note: The Coefficient of Determination	291

CHAPTER VII

REGRESSION ANALYSIS — 293

A.	Introduction	293
B.	From Correlation Coefficients to Bivariate Regression	294
	1. Regression Coefficients: The *Coleman Motor* Case	294
	2. The Slope of the Regression Line	295
	3. The Intercept of the Regression Line	297
	4. The Error Term in the Regression Line	299
	5. Significance Level and Bivariate Regression Analysis	299
	a. Calculating the *t* Value	300
	b. The *t* Test	302
	c. Null Hypotheses Related to Legal Maxima and Minima	307

	South Dakota Public Utilities Commission v. Federal Energy Regulatory Commission	307
	Notes and Problems	314
C.	Multivariate Regression	315
	Vuyanich v. Republic National Bank of Dallas	316
	Presseisen v. Swarthmore College	331
	Notes and Problems	339
	1. Multicollinearity	339
	2. Categorical and Dummy Variables	341
	3. Interpreting Regression Results	343
	4. Chewning v. Seamans	345
	Agarwal v. Arthur G. McKee & Co.	345
	James v. Stockham Valves & Fitting Co. (I)	349
	James v. Stockham Valves & Fitting Co. (II)	351
	Notes and Problems	354
D.	Significance Testing: R^2 and F	355
	Northshore School District No. 417 v. Kinnear	357
	Notes and Problems	369
	Note: Corrected R^2	371
	Note: In re Quaker Oats Co.	371
E.	Autocorrelation and the Durbin-Watson Statistic	372
F.	Nonlinear Multiple Regression	376

CHAPTER VIII

HEALTHY SKEPTICISM   379

A.	Sources of Error in Legal Statistics	379
	1. Errors in Gathering Data	379
	2. Small Numbers Problems	382
B.	Specification Errors in Multiple Regression	384
	Vuyanich v. Republic National Bank of Dallas	385
	1. Choice of Variables and the Error Term	388
	2. Omitted Variables	389
	3. Heteroscedasticity	391
C.	Causation and Inference	392

APPENDIX A

STATISTICAL TABLES 397

Binomial Probabilities	398
Critical Values for χ^2	402
Z Values	404
Critical Values for Student's t	406
Critical Values for the Correlation Coefficient r	408
Critical Values for Rank Correlation Coefficient r_r	409
Critical Values for F	410

APPENDIX B

ANSWERS TO SELECTED PROBLEMS 415

Table of Cases	427
Substantive Law Index	429
Index	433

PREFACE

This book describes and explains the variety of ways in which statistical tools are used to prove factual issues relevant to legal disputes. It is completely self-contained; it gives examples of how statistics are used, shows how to calculate and interpret statistics, and provides numerous problems for the student to practice statistical skills. All the tables needed to test statistical conclusions are included in Appendix A as well as in the relevant textual location, and answers to problems requiring calculations are provided in Appendix B.

The emphasis of this text is on the basic understanding of the use of statistical tools rather than on their theoretical foundations. Because it is crucial that advocates know the weaknesses as well as the strengths of each statistical technique, the Notes following cases caution the student against inferring too much from a particular numerical result and suggest ways to sharpen the statistical questions in order to reach meaningful conclusions. The Notes and Problems are a crucial part of these materials. They not only provide an opportunity to test understanding of the concepts introduced in the cases but also provide new insights and modifications of the basic techniques. Nearly all basic statistical techniques used in courtrooms are described, though there are many statistical tests that are never mentioned. The goal is to provide enough background in the basic techniques that a student or practitioner will be able to learn more esoteric tests easily. The Substantive Law Index should help students identify the statistical considerations relevant to the areas in which statistics are most commonly applied.

In order to emphasize the statistical content of legal opinions included in this book, material that originally appeared as a footnote to a judicial opinion has occasionally been incorporated into the text of that opinion. Whenever this occurs, the footnote is indicated by boldface brackets and/or a cite to the original footnote number. In addition, string cites in cases are routinely omitted without notation.

ACKNOWLEDGMENTS

So many have shared in the preparation of this book that it hardly seems fair to put only the author's name on the cover. The number of hours my secretary, Sue Lederman, spent typing tables, numbers and formulas are incalculable, and while she did not always enthusiastically greet me as I brought another rewrite of a chapter to her desk, she invariably returned the work quickly. I think we reached a tacit appreciation for the occasional drudgery of her labors. I respect her diligence and appreciate her patient tolerance.

Michael Brown and Nancy Loeb, former research assistants and now lawyers, made valuable contributions to the selection of cases and preparation of problems. Nancy's editorial comments and cheerful encouragement also helped to shape this book.

Professor Roger Congelton of Clarkson University critically reviewed the discussions of regression analysis and provided useful guidance on that complex subject. Professor James Follain of the Maxwell School also examined those materials. I appreciate the willing assistance of both of these men. Any errors that remain in those sections are my own fault, of course, and are due to my own thickheadedness and my attempts to make the discussion as straightforward and understandable as possible. The stylistic and substantive comments made by reviewers contacted by Little, Brown and reaffirmed by my colleague David Baldus helped to crystalize the emphasis of the text, and I appreciate their wisdom.

Research assistants and other students who worked on various stages of manuscript preparation include, chronologically, Dale Thistlewaite, Beth Wolfson, Virginia Calvert, Sherri Sonin, Jill Kanoff, Richard Smith, and Kevin Theimann. I thank them and the people in my Interdisciplinary Legal Methods classes over the semesters who suffered through rough drafts and helped to mold the final product. I am impressed by the professionalism and dedication of Michael West, who was manuscript editor for this text. I look forward to working with him and the editorial staff at Little, Brown and Company in the future.

Acknowledgments

I wish to thank Dr. Paul Downing of Florida State University; I believe I acquired from him through osmosis a healthy attitude toward the research process.

I gratefully acknowledge the moral support of my colleagues at the Syracuse University College of Law and the Maxwell School. I came to Syracuse impressed by the apparent congeniality and companionableness of the faculty and have confirmed that initial impression. This book might never have appeared without the financial support of the Center for Interdisciplinary Legal Studies. The Center is dedicated to supporting the commingling of law and related disciplines in legal education and scholarship. This book is only the first of published works to emerge from the support grants of the Center.

I am grateful to the Literary Executor of the late Sir Ronald A. Fisher, F.R.S., and to Dr. Frank Yates, F.R.S., and to the Longman Group Ltd. of London for permission to reprint Tables II, III, IV, and VII from their book Statistical Tables for Biological, Agricultural and Medical Research (6th ed. 1974).

David W. Barnes

STATISTICS AS PROOF

CHAPTER I

INTRODUCTION

In 1897, Oliver Wendell Holmes predicted that the successful attorney of the future would be the master of statistics and economics. Modern use of statistical and social science evidence, directed at expanding judicial notice or required to present a prima facie case, illustrates Holmes' prescience. This text presents cases, problems, and examples of statistical analysis in cases involving labor relations, civil rights, carcinogenic additives to foods, antitrust, product safety, educational funding, drug smuggling, communications licensing, prison reform, environmental protection, and family relations, among other areas. Cases represent the broadest spectrum of private and public law practice in which the modern attorney is engaged.

Our focus is the evidentiary use of statistics in proving facts relevant to legal issues. The text is designed to help the practicing attorney present simple factual arguments in a mathematical form familiar to courts. At the very least, it will keep the practicing attorney from appearing like an illiterate when dealing with the statistician hired by either party as an expert witness. The reader will discover that statistical proof is indispensable to the judicial resolution of a variety of substantive legal issues. The best proof of the utility of these materials is the emphasis the courts quoted here place on statistics. But one might wonder why, if these methods are so important, he or she has never read such cases before in law school. The reason is that the legal profession, both practicing and teaching, is confused as to what "all those numbers" mean. Legal training has traditionally emphasized verbal skills, forcing attorneys who later found a need for quantitative tools to pick them up as they went along

Because the goal is literacy rather than expertise, anyone with a background in statistics or mathematics is likely to be frustrated by these materials. They include only enough background arithmetic to enable the blissfully ignorant to understand the ways statistical methods can be used to describe and draw conclusions from facts.

Some addition, subtraction, and division will be required. There may be a place or two where multiplication is required, but the numbers used in examples are chosen so as to be easily manipulated. When a client's case requires complex statistical calculations, get a calculator, a partner who can add, or an expert. While any of these aids would help you work out some of the problems presented here, you will not need them in order to appreciate the utility of statistics in the practice of law. You should be warned that some of the statistical procedures have many steps. Each step, however, is composed of relatively simple parts. An effort has been made to break each procedure into its component steps and to enumerate them in order to facilitate computation.

The book is designed for people who have, sometime in their dim, dark past, learned to perform basic mathematical tricks (e.g., multiplying your monthly rent by the number of months left in the school year, and then subtracting the total from available funds to determine whether you will be able to eat). It is designed for people who never took a statistics course or who, as bold undergraduates, ventured into the lion's den only to be torn to shreds. The reader with a background in statistics (such as a graduate level, e.g., MBA, course or two or three undergraduate courses) will benefit considerably less, because the tools are already familiar and our focus will be on tools rather than theory. However, even this pre-enlightened soul can learn from the integration of social science tools and law that the case method used here provides. The cases indicate the areas of substantive law in which statistical tools have been used and, in some instances, are indispensable.

A. HOW WE WILL PROCEED

The materials are organized along the lines of a traditional law casebook, with cases of increasing complexity as one progresses from one chapter to the next. After explanatory materials of the type to be found in a statistics book introduce a new set of statistical tools, a series of cases follows. The explanatory materials describe the statistical tools used in the cases in sufficient detail to enable the student to integrate the cases and materials. With any luck, leaving the integration task to the student will lead to an awareness on her or his part of the multiplicity of uses for a given statistical technique. At least it will require some thought. The cases will not make much sense unless the integration occurs. Questions and problems following each case assist the integration process.

A noted economist, Frank Knight, once said, "Only through varied iteration can alien concepts be forced on reluctant minds." Statistical concepts are unquestionably alien and many minds reluctant. Therefore the cases and materials are organized to cover the same concepts in a number of different ways. The notion of standard deviations, for instance, will be alien at first, but, by the time we get to multiple regression, it will seem like a long-lost friend.

Because no prior statistical training is assumed, we will proceed slowly. Statistical tools are explained in the simplest terms, and the student's keen analytical mind may be inspired to wonder "why?" with respect to some statistical formulas. The answer to that question does not usually appear in this text, since there are plenty of fairly straightforward statistics books that will take care of those problems. We will concentrate on the application, rather than the derivation, of statistical techniques.

In preparing each case, the reader's initial task will be a familiar one: spotting the issues. Rather than directing attention to the nuances of the substantive issues, the focus will be on methods of proof. In some cases, statistical evidence alone may be sufficient to establish a prima facie case. The goal is to recognize legal issues and determine how statistical proof can be used to resolve them.

B. TWO TYPES OF STATISTICS

There are two types of statistics, *descriptive* and *inferential*. Descriptive statistics are concerned with describing data by means of pictures and numbers. Inferential statistics aid in making decisions on the basis of imperfect information. Descriptive statistics provide ways of presenting information to triers of fact that are particularly succinct or appealing or persuasive. Inferential statistics allow finders of fact to draw conclusions about factual situations related to legal issues without having all the relevant and material information.

Descriptive statistics are all those graphs, charts, and numbers you skip over in newspaper articles, history books, and standardized tests. Chapter II explores ways in which the attorney can utilize descriptive statistics to display evidence in a manner that captures the attention of the finder of fact.

Inferential statistics allow conclusions to be drawn in cases where complete information is unavailable. Suppose, for example, an automobile-manufacturer client buys mirrors in large quantities from a supplier. When the manufacturer refuses to accept a particular delivery of 12 million mirrors, the supplier sues for breach of contract. As a defense, the manufacturer refers to a contractual term giving her the right to refuse to accept any shipment in which more than 3% of the mirrors are defective. The manufacturer does not check all of the mirrors before they are sent to the assembly line. It would take too long. Nor is it feasible simply to lay aside defective mirrors as they are encountered during the assembly of a car, because they can't be returned as easily after they are used, and it would be expensive to try to install a defective mirror, discover it was defective, then install another one. A quick and inexpensive method by which the manufacturer can determine the quality of the shipment is to take a random sample from the shipment of mirrors, and on the basis of the percentage of defective mirrors in the sample determine with some degree of confidence the percentage of defectives in the whole shipment. The rules for determining how many must be in the sample and how many must be defective in order to satisfy a court as to the existence of a contract breach are examples of the subject matter of inferential statistics. We have incomplete information (only a sample from the shipment), but the court still permits factual inferences to be drawn.[1]

Another sort of incomplete information appears in the employment-discrimination case. Often the missing data would provide testimony as to the discriminatory intent of the personnel department. While it may be impossible to prove that the personnel director hates women, the court allows plaintiffs to present statistical evidence from which the finder of fact may infer that illegal discrimination has occurred. The most straightforward cases are those in which no women are ever hired. But what do we infer from the fact that two or three women are hired for 16 available jobs, when women make up 35% of the qualified applicants? Courts have developed common law rules for such occasions. They also are the subject of these materials.

Because descriptive statistics are less mysterious and provide the basis for inferential statistics, we begin with figures, graphs, and simple numerical concepts.

---

1. This example was suggested in another context by H.T. Hayslett, Jr., Statistics Made Simple 7 (1968). The book is an excellent and straightforward "self-study" statistics text.

CHAPTER II

DESCRIPTIVE STATISTICS

A. INTRODUCTION

Alfred Conard is a noted advocate of the use of statistical techniques in legal analysis. His data are useful as a departure point. The data come from an article that describes the utility of mathematical techniques and that puts quantitative methods in perspective.

ALFRED F. CONARD, THE QUANTITATIVE
ANALYSIS OF JUSTICE
20 J. Legal Educ. 1 (1967)

A recent article by Paul A. Samuelson, the famous economist, says "economics, unlike the law, is *quantitative.*"

If Samuelson is right about the law, the law is unlike the physical sciences, unlike the life sciences, and unlike the social sciences, all of which have adopted quantitative analysis. Perhaps the law is a fine art, like sculpture, or a divine science, like theology. But if law is a human science, a behavioral science, or a policy science, it would be hard to believe that it could function at maximum effectiveness without quantification. This would be even harder to believe if we accepted Dean Pound's denomination of law as "social engineering," or Lon Fuller's as "the architecture of social design."

Most of us shy away from anything called "quantitative" because we believe that quantity is the enemy of quality. If the quantity of cars increases, we are sure the quality will decrease. If the number of students goes up, the quality is supposed to go down, and so on.

I will not tarry today to challenge this axiom, which you learned at your mother's knee. I will only ask you to consider the possibility that there is no such antipathy between quantitative *analysis* and qualitative *analysis*. On the contrary, each gives meaning to the other.

What Is Quantitative Analysis?

Let me illustrate by this little bottle, containing a clear fluid. *Qualitative* analysis has revealed that this bottle contains various elements, which include gold (a well-known medium of exchange), deuterium (a raw material for the hydrogen bomb), hydrogen and oxygen. This information suggests that the contents of the bottle may be quite valuable. However, *quantitative* analysis informs us that the oxygen in the bottle constitutes about 850,000 parts in a million, the hydrogen about 100,000 parts in a million, the deuterium about 20 parts in a million, and the gold about one part in a million. The fluid is commonly known as sea water.

In this instance, the quantitative analysis did not in any way detract from the qualitative analysis; on the contrary, it added to its significance. Quantitative analysis is never a substitute for qualitative; it is, rather, a further step in the same process.

What I would like to suggest today is that quantitative analysis has the same function in law which it has in chemistry — that it gives new and greater significance to the qualities of justice. But I don't propose to argue the point. Rather, I propose to present some quantitative observations, and ask you to judge whether or not they are significant to you. I will draw my instances from the area of automobile accidents, because that is a subject on which I happen to have some quantitative information.

Some Quantitative Observations on Injury Reparation

Qualitatively, you already know that a person who is injured by the negligence of another may have his claim determined by a jury and have that determination reviewed by an appellate court. Quantitatively, it might be interesting to know how many run the full course.

Figure 1 shows some answers for automobile accident personal injuries in Michigan in 1958. This figure summarizes the histories of 86,000 accident cases. The tall barber-pole on the left represents all these cases. The next barber-pole represents the 12,100 cases which were put in the hands of lawyers. In between these two barber-poles you see a couple of columns which are hanging in the air without visible means of support. The first one, which is in white, represents the 49,000 cases which were dropped without collecting anything, and the next one, which is solid black, represents 24-odd thousand cases which were paid off without a lawyer's intervention.

After 12,100 cases were put in the hands of lawyers, some of *them* were dropped and some settled, but 4,000 survived to become law-

FIGURE 1.
Persistence and Attrition of Potential Injury Claims

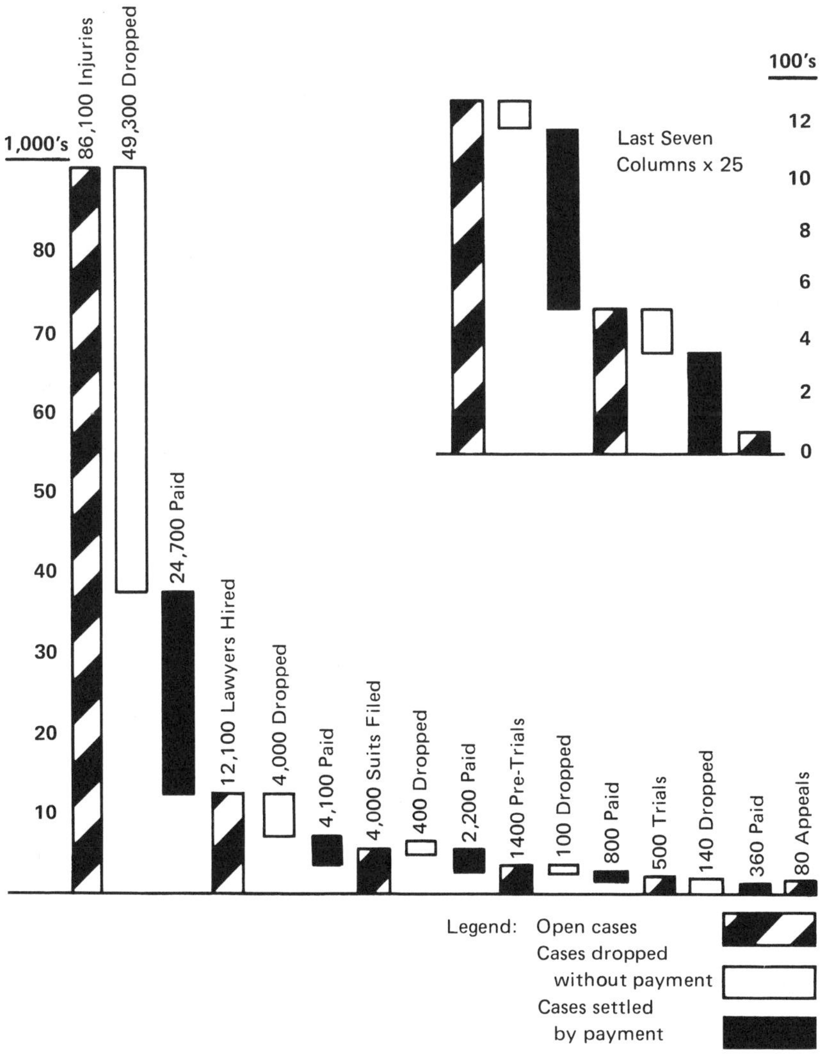

1,000's
86,100 Injuries
49,300 Dropped
24,700 Paid
12,100 Lawyers Hired
4,000 Dropped
4,100 Paid
4,000 Suits Filed
400 Dropped
2,200 Paid
1400 Pre-Trials
100 Dropped
800 Paid
500 Trials
140 Dropped
360 Paid
80 Appeals
80
70
60
50
40
30
20
10
100's
12
10
8
6
4
2
0
Last Seven
Columns x 25
Legend: Open cases
Cases dropped
without payment
Cases settled
by payment

suits. Some of these were dropped and some settled, so that only 1,400 got to a pre-trial conference, and only 500 to trial. By now, the columns have shrunk up so small that you can hardly measure them with your eye unless you look at the 25-fold enlargement in the upper right-hand corner. They keep getting smaller and smaller, until we get to that last little squiggle at the far right, which represents the 80 cases that were appealed. It is that one tenth of one percent from which is extracted the entire law of torts — at least as it was taught to me, and as I myself taught it many years ago.

Fortunately, some of the contemporary torts teachers are more catholic in their interests. Harry Kalven, for instance, has been investigating what goes on in the trials of cases. So he would be interested not only in that little squiggle of one tenth of one percent, but also in the preceding stump of a barber-pole, representing the 500 cases which reached the trial courtroom. This is, I presume, as far as the interests of an old-fashioned torts professor could possibly go.

However, my friend Maurice Rosenberg is a professor of procedure, and he is interested in what happens to cases between the time they are filed and the time they get to trial, an interest which is heightened by the fact that there are not nearly enough judges and courtrooms to try all of them. Thus, his interest extends back to the column indicating 4,000 cases, on which suit was filed. Hans Zeisel has also studied the attrition of cases between suit and trial.

Somewhere, I suppose, there is someone who is interested in the whole 12,100 cases which reach the hands of lawyers; probably Jerome Carlin would be interested in how the lawyers procure them, and in the tactics used by lawyers to secure settlements and fees.

Even if we include all these in our range of interest, we are still missing seven times as many cases as we are studying, because we are leaving out the 24,000 cases which are settled without lawyers at all, and the 49,000 cases which are dropped without a settlement. Both of these groups — the 24,000 and the 49,000 — would seem to be part of the workings of justice, in a broad sense, but have so far attracted very little attention in legal literature. . . .

Regardless of the stage at which injury cases are settled, we know that getting an adequate settlement is a long, hard process, filled with uncertainty. We sometimes wonder what the argument is about.

You have heard many views on this subject. Some of our distinguished colleagues have put the blame on the foggy idea of negligence. If we could get rid of it, everything would clear up. Others of our colleagues tell us that pain and suffering is the real problem. Another view is that causation is the root of the difficulty.

Who is right? A quantitative inquiry produced the data in Figure 4, which shows the weight which lawyers gave to different subjects of dispute. The first pair of bars represents the question of how the accident happened, including questions of negligence and contributory negligence. This question was *one* of the foci of dispute in about two-thirds of the cases, and was the primary focus in about half of the cases. But it was not the only focus. The valuation of pain and suffering divided the lawyers in almost an equal number of cases, although it was dominant in a smaller fraction. In over half the cases, medical

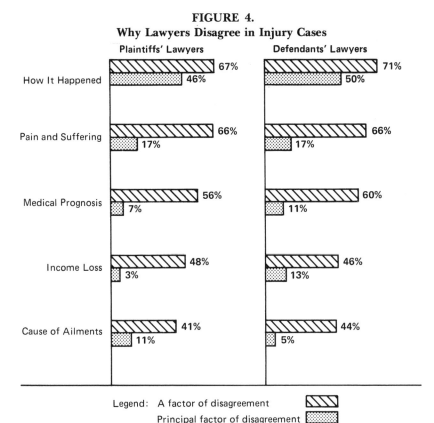

FIGURE 4.
Why Lawyers Disagree in Injury Cases

Legend: A factor of disagreement
         Principal factor of disagreement

prognosis was also disputed, and the amount of income loss was disputed in nearly half the cases. . . .

All loss shifting systems cost money, but few are as luxuriously expensive as the tort system. A rough comparison of systems is presented in Figure 7, which shows private loss insurance operating with about a 20% expense rate, public aid with about 4%, social security with about 2½%, tort liability with 56%, and workmen's compensation with about 33%. The last two are the only ones that are very much studied in law school, and we have good reason to maintain a friendly interest in them.

Looking at these other systems may prompt us to ask to what extent they also bear some of the burden of automobile injuries. Figure 8 presents a comparison of aggregate reparation with aggregate losses, in automobile injury cases. From the right-hand column, you can see that tort reparation nominally supplies about half of the total reparation. This was our estimate for 1958. This estimate probably understated the collateral sources even then, because some of the

FIGURE 7.
Aggregate Costs of Reparation for Death and Disability, 1960

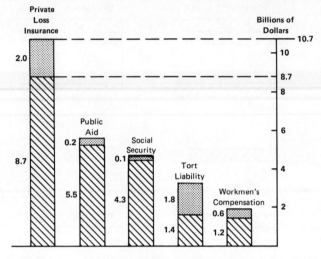

FIGURE 8.
Aggregate Economic Losses and Reparation in Personal Injury Accidents, 1958

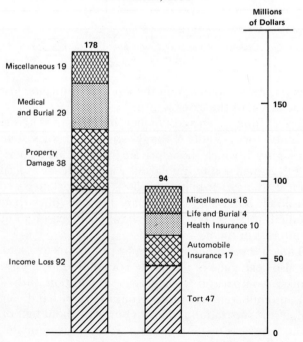

non-tort reparation comprised future payments which were discounted to present value. Today, the collateral sources are certainly much larger, because health insurance and social security have increased much more rapidly than tort liability in the intervening years.

The historical trends in this regard are revealed in Figure 9. This shows that automobile liability insurance is rising faster than either workmen's compensation or life insurance. But it is out-distanced in turn by social insurance and health insurance. This trend is certainly continuing. Social security got a further shot of adrenalin in 1966 from the Medicare program. So we can be certain that the tort liability share of reparation is today well under half of the total. For many injury victims, it is the tort payments which should be called "collateral benefits.". . .

FIGURE 9.
Relative Growth of Reparation Systems, 1940-1960

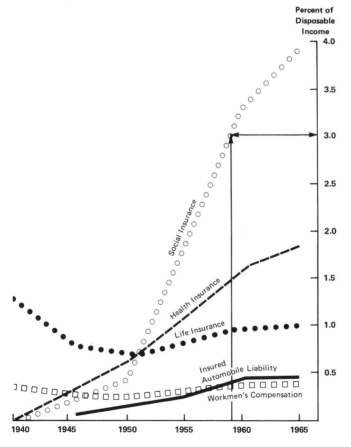

Must We All Become Statisticians?

If it is humiliating to think that the doctrinal questions over which we have fought are really not very vital, it would be even more distressing if we were to find that the research techniques in which we have invested the best years of our lives have become obsolete. Must we put behind us our skills of finding a case in point, distinguishing our opponent's cases, and brilliant dialectical demonstration, and learn instead how to draw a random sample, and calculate a standard deviation?

Nothing is more unlikely. One does not have to be expert in the science of probability in order to use statistics; it is probably better not to be one. In the Institute for Social Research with which I am familiar, the probability experts, who design population samples, are no more in charge of the main operation than are the anesthetists of a general hospital. They are called out to perform their mystical chores, and are then sent back to the numerarium. The "program directors," who plan and direct research, are political scientists, economists, or sociologists whose job is to know what are the questions that need to be answered. They could as well be lawyers, if lawyers had some quantitative questions to ask. A social research institute is like a hospital in which at every hour of the day quantitative analyses are being made of a number of body fluids which I need not name. But the attending physicians do not become chemists or serologists. They simply learn what the various quantities signify. Similarly, what we, as lawyers, need to do is to teach ourselves to think about some of the quantities which might be important. This is an illuminating and exciting exercise, even before we know the quantities.

There is hardly a field of study in which one does not gain new insights merely by thinking of the problem in quantitative terms. Consider, for instance, the raging contemporary conflict between civil liberties and the repression of crime. All of us recognize that we all have a legitimate interest in detecting and punishing crime, that we all have an equally legitimate interest in freedom of arrest for crimes of which we are not guilty, and that these interests conflict in the operations of the police. Those of us who have not yet lifted a lance on one side or the other of this perpetual tournament find ourselves on the horns of an insoluble dilemma between crime repression and civil rights. "Which is the higher value?" we ask ourselves. Whichever answer we give in one case seems to become mysteriously untenable in another which looks like it. Not only do we find no satisfactory answer, but we have difficulty in envisioning any means by which we could discover a satisfactory answer.

It has seemed to me, however, that quantitative analysis offers a useful way of thinking about this dilemma. Instead of asking myself whether freedom from arrest is more or less important than freedom from crime, I ask myself *how many* innocent citizens will suffer invasion of their liberties if police interrogation is permitted, and *how many* innocent citizens will suffer robbery or murder if police interrogation is suppressed. Then I slip different hypothetical figures into my analysis, and see how I feel about the result.

How I feel depends not only on the number of persons whom I think will be falsely arrested, but also on how I imagine that the innocent citizens will be treated when they are arrested. If I visualize their being courteously detained on the highway for twenty minutes while their cars are searched, I am willing to tolerate quite a few arrests of the innocent as an alternative to another robbery or murder. If I believe they will be kept for days or weeks in a stinking jail with depraved criminals, I will be prepared to accept very few arrests of the innocent as a price of higher crime detection.

Thinking about the problem in this way, I find that questions which seemed insoluble became quite manageable on certain assumptions of fact. I still don't know what the facts are, but I know what kinds of facts would help me solve the legal question which confronts the courts. I am also able to pinpoint the differences in presuppositions which divide some of my colleagues.

Those who generally favor (as a matter of law) reversing convictions for police errors seem also to believe that (as a matter of fact) the police habitually arrest and harass a large number of innocent persons for every guilty one they find. Those who generally favor affirming convictions commonly believe that very few people get arrested unless they have some connection with the crime, and that most of those who are arrested are pretty decently treated. It seems likely that the violent clashes between the legal opinions of these groups is attributable in substantial part to radical differences in their views of the quantitative facts. The avenue toward a greater agreement on principle may lie through a common understanding of the facts of life. This common understanding would be greatly promoted by quantitative observation and recording. . . .

B. HISTOGRAMS, TABLES, AND GRAPHS

Professor Conard's numbers can be presented in a variety of forms of tables and charts. For some reason, many of us find tables easier to read than charts. Figures 1, 7, and 8 in Conard's article are technically known as *histograms*. Histograms are characterized by their depiction of data in rectangular boxes that indicate the number of times a particular event occurs, i.e., the *frequency* of that event, measured in "cases" in Figure 1 and "dollars" in Figures 7 and 8. Conard's Figure 7 might also be displayed in a table like Exhibit II-1.

EXHIBIT II-1.
Aggregate Costs of Reparation for Death and Disability, 1960

Loss Shifting System Coverage Type	Cost (Billions of Dollars)	
	Expense of System	Compensation to Victim
Private Loss Insurance	2.0	8.7
Public Aid	0.2	5.5
Social Security	0.1	4.3
Tort Liability	1.8	1.4
Workmen's Compensation	0.6	1.2

For some reason, the table is easier to understand. But ask yourself whether it is equally effective in getting the message across. When comparing the size of the rectangles in Conard's Figure 7, it is striking that the Tort Liability box is so much shorter than the Private Loss Insurance box. The reader (or, when such evidence is presented to the court, the finder of fact) does not need to look at the specific numbers to be impressed by the importance of the insurance system. Greater visual impact on the jury just might win the case. The same is true of Conard's shading in the rectangles. Note that the top portion of each is shaded, while the bottom is striped. The shaded portion is the part of total cost of operating a compensation system that is administrative expenses and lawyers' fees. Again, it is striking that the expenses part of the torts rectangle (the shaded, top portion) is greater than the amount going to the victim. There is no need for the finder of fact to look at the numbers.

Look at the histogram in Conard's Figure 8 from the same perspective. The taller column represents total estimated losses from automobile accidents in 1958. Note how many losses go uncompensated. Aggregate reparations are in the much shorter rectangle. The colorful presentation of data is nothing more than creative lawyering.

Conard's Figure 9 is a different approach to pictorial representation of data. This type of picture we will refer to as a *graph*. The histogram is a type of graph, I suppose, but we will distinguish this graph from a histogram by noting that the data in a graph are compared by lines (or curves) rather than rectangles. In tabular form, the data in Figure 9 would look something like Exhibit II-2.

EXHIBIT II-2.
Relative Growth of Reparation Systems

		Reparation System (Percent of Disposable Income)			
Year	Workmen's Compensation	Insured Auto Liability	Life Insurance	Health Insurance	Social Insurance
1940	0.42	—	1.30	0.00	0.00
41	0.41	—	1.20	0.08	0.05
42	0.41	—	1.10	0.16	0.10
43	0.40	—	1.00	0.24	0.15
44	0.40	—	0.90	0.32	0.20
45	0.39	—	0.80	0.40	0.25
46	0.39	0.01	0.78	0.48	0.30
47	0.38	0.04	0.76	0.56	0.35
48	0.38	0.07	0.74	0.64	0.40
49	0.37	0.09	0.73	0.72	0.45
1950	0.37	0.12	0.72	0.80	0.50
51	0.36	0.15	0.71	0.88	0.80
52	0.36	0.18	0.70	0.96	1.10
53	0.37	0.21	0.73	1.04	1.30
54	0.37	0.24	0.76	1.12	1.60
55	0.38	0.27	0.79	1.20	1.90
56	0.38	0.30	0.82	1.28	2.20
57	0.39	0.34	0.86	1.32	2.50
58	0.39	0.37	0.90	1.40	2.80
59	0.39	0.40	0.95	1.50	3.00
1960	0.40	0.42	1.00	1.60	3.30
61	0.40	0.43	1.00	1.67	3.45
62	0.40	0.44	1.00	1.69	3.65
63	0.40	0.45	1.00	1.70	3.80

As you can see, the graph is a useful way to summarize a large quantity of data. All you need to be able to do is figure out how to read it. It may be that tables are easier to read because you still read from left to right; a graph, however, is a picture, and we are uncertain about where to begin. Consider what graphs and tables have in common.

For any table, histogram, or graph, first determine what the variables are. A *variable* is a characteristic or attribute that varies from item to item in a class of objects. Exhibit II-2 and Conard's Figure 9 have two classes of objects: reparation systems (the items of which are workmen's compensation, insured auto liability, life insurance, health insurance, and social insurance) and years (the items of which are 1940, 1941, and 1942, etc.). The variables describe what is being measured.

The other part of the graph, histogram, or table is the *measurement*. When each item (or category) for each variable is chosen (e.g., workmen's compensation for 1945), a measurement associated with that selection is given by the table, histogram, or graph. In order to understand the measurement, the table, histogram, or graph must indicate the *unit of measurement*. Figure 9 and Exhibit II-2 have percent of disposable income as the unit of measurement. The table in Exhibit II-1 and Figure 7 have billions of dollars as the unit of measurement.

Once the variables and unit of measurement are identified, the difference between a table and a histogram or graph can be explained. On tables, the measurement usually appears under the list of items or categories of one variable and to the right of the list of items or categories of the other variable. See Exhibits II-1 and II-2. On graphs or histograms, the measurement is usually shown on the vertical line on one side of the figure. On the histograms in Conard's Figures 1, 7, and 8, the measurements (thousands of cases, billions of dollars, and millions of dollars, respectively) are shown on the vertical line to one side of the colored rectangles. On the graph in Conard's Figure 9, the measurement (in percent of disposable income) is also on a vertical line to one side of the curves. Generally it makes no difference to which side of the curve or rectangle the measurements are shown, though traditionally the vertical axis rises from the left end of the horizontal axis, as in Conard's Figure 1. On each of these figures, items of one of the variables are indicated on the horizontal line beneath the rectangles or curves. The vertical line indicating the measurement is called the *ordinate axis*. The horizontal line indicating different items of the variable is called the *abscissa*.

With these terms understood, one can find any statistic by reference to a relevant well-designed table, histogram, or graph. First, determine the categories of the variables for which you want to find the measurement. On a table, trace a line down from the desired item or category of one variable (e.g., social insurance on Exhibit II-2), and another over from the desired item of the other variable (e.g., 1959 on Exhibit II-2). Where the lines meet is the measurement (e.g., 3.00% of disposable income).

On a graph, trace a line from the desired item of the variable measured on the abscissa (e.g., 1959 on Conard's Figure 9) up to the curve for the desired item of the other variable (e.g., social insurance on Figure 9) and another from the point where that line meets the curve over to the ordinate axis. Where the line hits the ordinate is the measurement. The graph has a separate curve for each item of the variable not indicated on the axes.

The histogram has a separate rectangle for each measurement. Consider Conard's Figure 7 and our Exhibit II-1. The table in Exhibit II-1 reveals two variables; the first, cost, has two categories, while the second, loss shifting system, has five categories. Under each of two cost categories you expect (and the table reveals) measurements for five system categories. Ten measurements are expected, and ten rectangles are shown in Figure 7. Though it is nowhere indicated specifically on the histogram, the variable depicted on the abscissa is loss shifting system (poor design by Professor Conard). Each of five tall rectangles (one for each system) is broken into two smaller rectangles (one for each cost category). Here, although again there is not a clue on the picture (there should be), the top part of each rectangle refers to expense while the bottom refers to compensation. The height of each rectangle is measured using the units (billions of dollars) provided on the ordinate axis. For the histogram, trace a line from the bottom of the rectangle for the desired category of each variable (e.g., the expense category of the private loss insurance) over to the ordinate and another over from the top of that rectangle. When the measurement of the line from the bottom (e.g., 8.7) is subtracted from the line from the top (e.g., 10.7), one has the measurement (e.g., 2.0 equals 10.7 minus 8.7) given in Exhibit II-1.

You may find it more difficult to read measurements from a histogram than from a graph, and easiest to read measurements from a table. On the other hand, the visual presentation of relative size makes graphs and histograms appeal more readily to the eye, to emphasize particular points that an advocate wishes to make.

This discussion and the excerpts from Conard's article have emphasized the utility of tables, histograms, and graphs in conveying information because descriptive statistics are like tools. Some descriptive methods are better suited to particular tasks. Tables, for instance, are useful for conveying precise information related to measurements. Histograms are useful in describing the relative magnitude of measurements of variables and categories within variables. Graphs quickly describe large quantities of data, allow for comparison among categories of variables, and highlight changes in magnitudes when small changes in a variable (e.g., year 1949 to year 1950 in Figure 9) occur. Remember that the same technique is used to interpret any of these methods of describing statistics. First determine what all the variables and categories are. Then determine the unit of measurement. With this information in hand, follow the directions given above for reading the measurement of a table, graph, or histogram. When designing a statistical visual aid, remember that all of this information must be indicated.

C. FLEXIBLE FIGURES

One of the amusing and useful features of graphs is that they are easily manipulated to highlight a particular point in your argument. Put yourself in the place of a plaintiff's attorney presenting the case on the damages issue of a torts claim to a jury. The state in which you are practicing permits inflation to be taken into account in computing loss of future income. Without getting into the details of discounting a future salary to its present value, consider graphic presentation of future and anticipated inflation based on the historical data. Assume that the inflation rate over the past ten years has been about 5%. Exhibit II-3, below, shows in a tabular format the consumer price

EXHIBIT II-3.
Table of Prices

Year	Percentage Change in Prices from Previous Year	Percentage Change from 1967	Price Index
1970	6%	116%	116
1971	4%	121%	121
1972	3%	125%	125
1973	6%	133%	133
1974	11%	148%	148
1975	9%	161%	161
1976	6%	171%	171
1977	6%	182%	182
1978	8%	195%	195
1979	11%	217%	217

index for each of the ten years that the plaintiff's attorney needs to show the jury. This index is established by picking an arbitrary year as the standard. So, if 1967 is the standard, the price index in 1967 is 100, i.e., 100% of the 1967 prices. If prices rise by 4% in 1968, the 1968 index is 104, i.e., 104% of 1967 prices.

Exhibit II-4 depicts the same data graphically; it shows a definite upward trend in prices that, you might argue, requires adjusting expected future income upwards. Now look at Exhibit II-5, which again

C. Flexible Figures 19

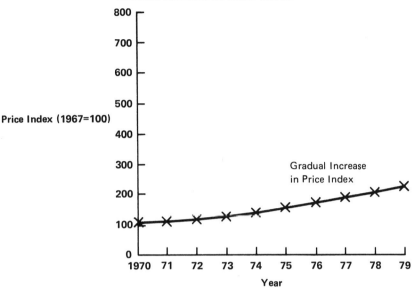

EXHIBIT II-4.
Gradual Increase in Price Index

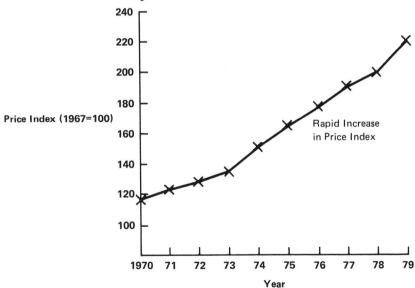

EXHIBIT II-5.
Rapid Increase in Price Index

20 Chapter II. Descriptive Statistics

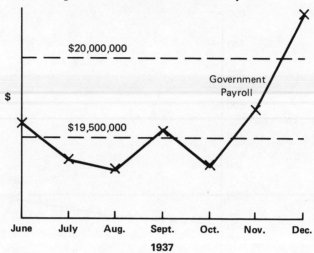

EXHIBIT II-6.
Rapid Increase in Government Payroll

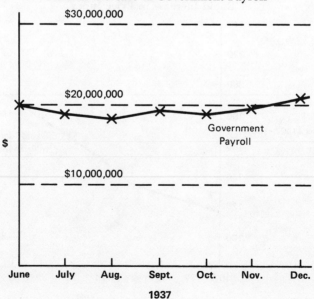

EXHIBIT II-7.
Gradual Increase in Government Payroll

presents the same data. It presents a much more impressive rate of inflation merely by changing the numbering of the ordinate axis and the scale of the abscissa. There is no untruth in Exhibit II-5, but it may be more persuasive.

Real-life examples of such use of graphs may be even more persuasive. Darrell Huff and Irving Geis reproduced a chart from an article in Dun's Review emphasizing the rapid expansion in the federal government.[1] The graph shows an increase in the government payroll during part of the 1937-1938 fiscal year from $19,500,000 to $20,200,000. As depicted in Exhibit II-6, it looks like a tremendous leap. This is undeniably a lot more impressive than the more objectively scaled version in Exhibit II-7.

There are no hard and fast rules about how to scale the axes. The graphs in Exhibits II-5 and II-6 tell the truth just as Exhibits II-4 and II-7 do, but their persuasive impacts may be much different.

Notes and Problems

1. The discussion above does not mention Conard's Figure 4, a "bar graph." To interpret such a graph, once again identify the variables and unit of measurement. As an exercise, try expressing the information in Figure 4 as a histogram and as a table. As an exercise, use the graph, histogram or table to answer the initial question, "What is the argument about?" i.e., "Why do so many cases go to trial?" Are most of the differences on the issue of liability or of damages?

2. As an exercise, try expressing Conard's Figures 1 and 8 in tabular form.

3. Pennsylvania v. O'Neill, 348 F. Supp. 1084 (E.D. Pa. 1972), was an action by the Commonwealth and several black citizens, representing a class of black police officers and unsuccessful applicants for positions in the city police department. They alleged that the department's hiring and promotion practices unconstitutionally discriminated against blacks. The plaintiffs won a preliminary injunction against the city, enjoining hiring or promotion of policemen except in a ratio approximating the ratio of black applicants for employment and the black population in Philadelphia. In this case, evidence presented by the plaintiff's expert statistician carried considerable weight with the court, which appended the following tables to the decision. Note that these measurements could have been presented with considerably greater impact in histogramic or graphic form. As an exer-

---

1. Huff, D. and Geis, I., How to Lie with Statistics, Chapter 5 (1954).

EXHIBIT II-8.
Probabilities of Rejection Given Factor on Background Exam

Factor	White	Black
Conviction	66.2%	90.5%
Arrests	72.0	93.0
Police Contacts	40.5	83.3
Traffic Offenses	32.4	67.5
Juvenile Delinquency	81.5	91.2
Juvenile Arrests	54.2	84.6
Juvenile Police Contacts	23.4	57.1
Court Martial Convictions	86.7	94.7
Summary Offenses in Military	50.0	83.5
Military Arrests	50.0	100.0
Military Discharge	75.3	94.4
No Valid Driver's License	57.8	89.3
Falsification of Application	63.1	87.4
Fired	74.5	94.2
Job Problems	70.3	89.4
Unemployed and/or on Welfare	38.2	75.0
Bad Credit	68.8	89.7
Education Academic Problems	52.3	85.8
Education Discipline Problems	59.9	91.9
Born out of Wedlock	36.3	66.7
Divorce	67.9	79.4
Illicit or Immoral Conduct	64.7	90.0
Alleged Threats or Violence	81.9	95.5
Improper Conduct of Friends or Relatives	41.8	71.5
Bad Appearance	45.3	77.2
Other	52.9	79.2

Source: Pennsylvania v. O'Neill, Appendix A. Table developed by plaintiff's expert, Dr. Siskin.

EXHIBIT II-9.
Probabilities of Rejection Given a Given Number of Factors

Number of Factors	White	Black
1	11.0%	29.5%
2	15.9	24.6
3	26.8	53.7
4	41.6	70.4
5	56.6	87.0
6	74.4	93.0
7	76.9	90.6
8	88.6	92.9
9 or more	92.2	94.1

Source: Pennsylvania v. O'Neill, Appendix B. Table developed by plaintiff's expert, Dr. Siskin.

C. Flexible Figures

EXHIBIT II-10.
Incidence of Factors by Race

Factor	White	Black	B%/W%
Convictions	6.3%	9.0%	1.4
Arrests	11.6	18.2	1.6
Police Contacts	1.7	1.7	1.0
Traffic Offenses	26.8	22.5	.8
Juvenile Delinquency	5.1	8.0	1.6
Juvenile Arrests	13.7	20.1	1.5
Juvenile Police Contacts	6.0	3.9	.7
Court Martial Convictions	.6	2.7	4.5
Summary Offenses in Military	15.5	21.5	1.4
Military Arrests	.4	1.5	3.8
Military Discharge	3.0	5.1	1.7
No Valid Driver's License	4.2	9.3	2.2
Falsification of Application	41.3	67.3	1.6
Fired	13.5	27.0	2.0
Job Problems	15.6	29.3	1.9
Unemployed and/or Welfare	22.3	23.7	1.1
Bad Credit	18.8	19.2	1.0
Education: Academic Problems	19.3	23.8	1.2
Education: Discipline Problems	13.8	19.0	1.4
Born out of Wedlock	4.5	3.4	.8
Divorce	3.2	4.8	1.5
Illicit or Immoral Conduct	9.7	29.4	3.0
Alleged Threats or Violence	3.0	6.2	2.1
Improper Conduct of Friends or Relatives	18.5	35.1	1.9
Bad Appearance	24.3	40.1	1.7
Other	56.3	78.7	1.4

Source: Pennsylvania v. O'Neill, Appendix C. Table developed by plaintiff's expert, Dr. Siskin.

cise, present the information regarding the first five factors in Exhibit II-8 in a histogram. Remember, the procedure is to describe what your variables are and what your unit of measurement is, then decide which variable to indicate along the abscissa.

4. Now try to express the information in Exhibit II-9 in graphic form. Use as the variables to be indicated on the abscissa and ordinate the number of factors and the percentage rejection, respectively. A different curve will be needed for each race. Be certain to label the curves and axes carefully.

5. Exhibit II-10 presents measurements of the incidence of the various factors by race. The large number of factors and numbers in this table makes it difficult to get an overall impression without care-

fully examining the numbers. The visual impact of a histogram or graph may provide a means for presenting this evidence for maximum effect on the trier of fact. Using this data, devise a visual exhibit for the court describing the incidence of several of these factors. One approach would be to use a bar graph like Figure 4 in Professor Conard's article. On a separate bar graph, present the data in column 3 (B%/W%) for several factors.

6. Consider the evidence you have summarized above in evaluating the propriety of the court's order that the City of Philadelphia be "preliminarily enjoined from hiring any policemen except in the ratio of at least one member of the black race for every two members of the caucasian race." Which of these data are of greatest practical significance in proving that the city discriminated against blacks? What additional statistical evidence would you like to have, if any?

7. Antitrust cases frequently include large quantities of data. The following data were taken from Coleman Motor Co. v. Chrysler Corp., 376 F. Supp. 546 (W.D. Pa. 1974), which used a statistical technique called *regression analysis* to figure damages — it will be examined in more detail in Chapter VII. In this case, a local automobile dealer alleged that Chrysler engaged in a restraint of trade in violation of §§1 and 2 of the Sherman Act by subsidizing some dealers in Allegheny County, Pa. Coleman was not subsidized and alleged injuries as a result. Exhibit II-11 reproduces a table from the case (376 F. Supp. at 557), showing the effect on Coleman's sales of the subsidy policy instituted in 1961. Design a histogram or a graph that displays this effect more dramatically.

PENNSYLVANIA v. LOCAL 542, INTERNATIONAL UNION OF OPERATING ENGINEERS
469 F. Supp. 329 (E.D. Pa. 1978)

[In Pennsylvania v. Local 542, International Union of Operating Engineers, an employment discrimination suit was instituted against a local union and contractors and trade associations by 12 blacks on behalf of a class of minority workers who alleged employment discrimination in the union's membership practices, the operation of its referral system, and hours and wages of minority workers. In arguing the merits of the Title VII claim, the plaintiffs' and defendants' experts engaged in a lengthy debate as to the relevance, accuracy, and utility of each other's statistics. We will get back to that debate later, once we have become familiar with more tools. For now, however, examine the following excerpt from the opinion and note the statistical tables ana-

EXHIBIT II-11.
Comparison—Percentages of Automobile and Truck Sales—Allegheny County

Dealer	1961	1962	1963	1964	1965	1966	1967	1968
Coleman	14.2%	7.4%	5.8%	6.0%	5.2%	5.6%	5.07%	3.8%
Financed Dealers	1.2%	39.0%	42.1%	45.1%	51.5%	52.5%	54.2%	51.9%

Source: Coleman Motor Co. v. Chrysler Corp., Table 6.

lyzed by the court. Consider alternative methods of drawing the court's attention to the underlying data by means of graphic presentation. This excerpt demonstrates one context in which the court considers statistical evidence.]

A. Leon HIGGINBOTHAM, Jr., Circuit Judge. . . .

G. STATISTICAL EVIDENCE

An important part of plaintiffs' case lies in its proof of discrimination in membership in Local 542 and its proof of discrimination in hours and wages of minority union members. Plaintiffs' proof on both issues is based in part on a statistical probability analysis. See International Brotherhood of Teamsters v. United States, 431 U.S. 324, 97 S. Ct. 1843, 53 L. Ed. 2d 396 (1977). For the sake of orderliness we will take the statistical proof of membership discrimination first.

Plaintiffs' expert, Dr. Bernard Siskin, is an Associate Professor at Temple University and Chairman of the Department of Statistics. He has written numerous articles and a text book in the field of statistics and has particular expertise in social statistics. At trial he presented a number of studies of 542's composition, ascertaining, inter alia, the number and proportion of minority union members, the pool of minority persons available for operating engineer work, the number and proportion of minority entries from 1966 to 1975, the numbers of minority persons entering the union via the various available means, and the comparative hours and wages of minority members. From such studies he was able to draw conclusions relating to the likelihood of discrimination in the 542 hiring hall system. This case, like most cases involving statistical proof, involves among the expert witnesses some evidentiary conflicts which require the fact finder in turn to make findings of credibility. *Upon balance I find Dr. Siskin's testimony to be credible, persuasive and accurate on these subjects and on the other subjects of his testimony. The testimony of the other experts who differ does not cause me to repudiate or modify my findings as to Dr. Siskin's credibility.*

1. MEMBERSHIP DISPARITIES

Based on the computer tapes of a "Master List of Active Members" provided to Dr. Siskin by the union (with key punch errors being corrected through the annual pension files), Dr. Siskin determined the union membership to be 6,051 as of December 31, 1971. This figure includes the parent body, branches A through D, and the Registered Apprenticeship enrollees. (Holders of registrant books are not counted in this tabulation and are not considered by the union as members.) There were two hundred and thirty-five black members. Thirty-three more were members of another minority. Thus the composition among active members of Local 542 at the end of 1971 was 3.9% black. When all minorities are considered, the minority composition of the union was 4.4%. These 1971 percentages are very nearly identical when pension welfare, and honorary members (inactive) are included in the membership definition. The following table reflects the composition of 542, including pension, welfare and honorary members, for the years 1966-71:

Year	Membership	Number of Minorities	Percent Minority
1966	5092	174	3.42
1967	5385	182	3.38
1968	5703	216	3.79
1969	5995	261	4.35
1970	6192	255	4.12
1971	6453	289	4.48
1972	6631	307	4.6
1973	6942	344	5.0
1974	7066	356	5.0
1975	6974	336	4.8

Based on the 1970 census data for the area covered by Local 542's jurisdiction, the total population is 7,729,115 of which 888,370 or 11.5% are black and 33,073 or .4% are members of another minority group. Limiting the population to males between the ages of 18 to 65, in order to define more precisely the pool of potential applicants, the figures become 11.0% black and 11.5% minority. The 11.0% and 11.5% figures are a conservative statement of the available black or minority labor pool. This was Dr. Siskin's conclusion and I agree. Siskin's best estimate, however, was 12.7 to 13.4% black and a total of 13.2 to 13.9% minority, taking into account by his calculations labor force participation rates, the census undercount, education and occupation and disregarding particularly the suggested factors of automobile and telephone ownership.

Using the conservative 11.0% and 11.5% result, the membership of Local 542 as of 1971 was grossly disproportionate to the 3.9% black and 4.4% minority percentage in the labor pool. The likelihood that such a disparity would occur by chance is less than 1 in 100 trillion, less than 10^{-23}. Of course, if Siskin's greater percentage figures, which I find on a preponderance of the evidence to be accurate, were used, the disparity would increase still further. Notably the disparity remained extremely significant as against the conservative labor pool figures as of 1975 when the minority percentage (four years after the initiation of this suit) had risen to only 4.8%.

2. DISCRIMINATION IN ENTRY

In addition to demonstrating this gross disparity, plaintiffs' expert, Dr. Siskin, compiled data principally from union exhibits indicating the ratios of minority entry into the union (parent, A, B, C, and D branches and the RA program). During the period 1966-1971, 2601 new members entered 542. Minorities entered at the following rates:

[Year]	Blacks	Minorities
1966	5.1%	5.5%
1967	2.6%	2.8%
1968	5.8%	5.8%
1969	8.2%	9.8%
1970	3.8%	4.8%
1971	6.8%	8.7%

The total black entries for the union as a whole were 143 or 5.5% over the course of these years and the total minority number was 166 (6.4%). The direction of the entry rate (increasing or decreasing) fluctuated during this period, so it is not possible to deduce with absolute certainty any definite trend, although in the last three years before suit (1969-1971) the average entry rate was 6.4% black and 7.9% minority. The difference between this average and the conservative labor pool percentage is statistically significant at less than 1 in 1,000,000 (i.e., the probability that the difference can be explained by chance is less than one out of one million). On the basis of random entry one might reasonably expect almost 300 minority entries between 1966 and 1971. The difference between this expected number and the lesser actual number is 133 — 44%. By the end of 1974, the minority membership of Local 542 was 4.3% black and 4.8% minority out of 6725 members. As will be developed below, there are no other factors which would fairly require a finding that this gross disparity is not the result of discrimination. There is no sufficient evidence to explain that this disparity occurred because of any valid job-related qualifications not possessed either by members of the minority labor pool or by actual applicants for union membership.

C. Flexible Figures

Even after suit, from 1972-1974, the minority entry ratio into the union is at 7.4%, not far from the 6.4% pre-suit figure applicable between 1966 and 1971 and less than the 7.9% rate between 1969 and 1971. In 1972, 5.9% of entrants were minority; in 1973, 8.1%; in 1974, 8.0%; and in 1975, based on union data after excluding reinstatement and withdrawals, only 1.5% (adjusted from 2.8%).

CHAPTER III

INTRODUCTION TO INFERENTIAL STATISTICS

A. INFERENTIAL STATISTICS AS PROOF

In Pennsylvania v. Local 542, International Union of Operating Engineers, above, the court's next step after presenting the data related to membership was to determine whether the disproportionately small number of minority members was significant enough to constitute illegal employment discrimination. This sounds like it must be a judgment call on the part of the court. The issue is: how outrageous must a disparity be before it is judged illegal?

Inferential statistics do not prove what caused the particular result that is of interest to the factfinder. Nor do they establish that a particular event occurred by chance. Inferential statistics do, however, indicate the *likelihood* that a particular history of events could occur by chance. A strong relationship between two events, such as race and the probability of being hired or the whereabouts of the defendant on a particular day and the facts surrounding the commission of a crime, tempts the logical mind to infer a causal connection between the events. By eliminating chance as an alternative causal explanation and by showing that there is a weak relationship between other events and the outcome of interest, statistics aid the inference of an inculpatory explanation of the observed occurrence. The opposing attorney's task is then to show that there is little connection between the events, that there are innocent explanations for the relationships observed, or that the circumstantial evidence offered as proof by the other party is unpersuasive.

1. The Role of the Statistician

The actual methodology of statistics is frequently misunderstood. The gathering of numerical data, like the gathering of evidence more traditionally accepted in a courtroom, is only the first stage of the process. Both Sherlock Holmes and Perry Mason would return to their offices after foraging around the countryside to examine pieces of evidence collected, tie them together, and draw inferences from them. These famous detectives would test their conclusions by confronting suspected parties and provoking them into violence, or in Perry Mason's case, into leaping up in court and confessing to the vile deed. The statistical method is to provoke the numbers into leaping from the page and pointing an accusatory finger at the culpable party.

The statistician's task is to provide the lawyer with different perspectives on the data. A statistician unfamiliar with the law may need considerable guidance from the lawyer in order to collect relevant information and present it in a convincing and acceptable format for the courtroom. The statistician who is more experienced with legal procedure and the issues involved in a particular area of substantive law may have developed the same feeling for the niceties of proof that the litigator has. In either case, the lawyer's role is to guide the strategy of confronting the opposing party and the finder of fact with the inferences drawn from the numerical data.

Inferential statistics provide a more objective basis for drawing inferences than the hunches of the great foraging detectives. When the numerical evidence does not conform precisely to a hypothesis or legal theory or when the connection between the data and the causal theory is unclear, inferential methods measure the degree of uncertainty implicit in drawing a particular inference, thereby quantifying the risks of error involved in reaching certain conclusions. Wayne Curtis and Lowell Wilson, in an article exploring the use of statisticians in litigation,[1] describe the statistician's role as the expert who determines the size of a sample large enough to reach a result with a given level of accuracy, draws that sample from the population as a whole in a scientifically acceptable way, extracts useful information from the raw data obtained from a sample, specifies confidence intervals around conclusions arrived at by the sampling process in order to establish the uncertainty involved in those estimates, and evaluates the significance of differences between numerical results obtained through sampling processes. This work, however, cannot take place

---

1. W. Curtis and L. Wilson, The Use of Statistics and Statisticians in the Litigation Process, 20 Jurimetrics J. 109-120 (1979).

without the lawyer guiding the statistician to ensure the legal relevance of the results.

2. *Examples of Inferential Statistical Proof*

Courts have grown accustomed to having inferential statistical proof offered in a tremendous variety of cases. The tax court in News Publishing Co. v. United States, 81-1 U.S.T.C. 9435 (Ct. Cl. 1981), sampled salaries of executives in similar publishing companies to estimate appropriate compensation for the taxpayer's chief executive officer and to determine the deductibility from the corporate tax of a bonus paid to that executive. In South Dakota Public Utilities Commission v. Federal Energy Regulatory Commission, 668 F.2d 333 (8th Cir. 1981), estimates of future natural gas reserves were examined to determine the propriety of accelerated depreciation of a pipeline. Damages suffered by plaintiffs in antitrust cases such as Coleman Motor Co. v. Chrysler Corp., 376 F. Supp. 546 (W.D. Pa. 1974), can be determined by sampling the profit experience of previous years, and in torts actions such as Orchard View Farms v. Martin Marietta Aluminim Corp., 500 F. Supp. 984 (D. Ore. 1980), by examination of the effects of similar particulate trespass on other fruit trees. These are examples of inference because the finder of fact is urged to draw conclusions based on partial information, a sampling of trees, profits, gas discoveries, or salaries.

Quantification made necessary by increasing federal regulation designed to protect safety and health has resulted in frequent analysis of inferential statistics in environmental protection cases. In Marathon Oil Co. v. EPA, 564 F.2d 1253 (9th Cir. 1977), the oil-drilling company challenged estimates made by the EPA of the quantities of oil, mud, grease, and soaps that would wash off the most modern offshore oil-drilling platform as effluent. In Reserve Mining Co. v. EPA, 514 F.2d 492 (8th Cir. 1975), samples of concentrations of airborne asbestiform fibers at different locations were examined to determine the air polluting effects of the company's taconite ore processing plant. Standards for new source performance, examined in National Lime Association v. EPA, 627 F.2d 416 (D.C. Cir. 1980), and in Ethyl Corp. v. EPA, 541 F.2d 1 (D.C. Cir. 1976), were established by statistical tests, the propriety of which was debated by lawyers in those cases.

In regulating commerce, the Federal Trade Commission frequently confronts statistical evidence in evaluating manufacturers' claims and the propriety of their business practices. The FTC examined statistical support of product superiority claims in In re Coca Cola, Inc., 83 F.T.C. 746 (1975) (vitamin C content of Hi-C fruit drink), in In re ITT Continential Baking Co., 83 F.T.C. 865 (1973) (healthfulness of Wonder Bread), and in In re Pfizer, Inc., 81 F.T.C. 23 (1973) (efficacy of the product, Un-Burn, compared to a placebo). It analyzed statistical claims in antitrust actions ranging from price discrimination, In re United Fruit Co., 82 F.T.C. 53 (1973), to mergers, In re Sterling Drug, Inc., 80 F.T.C. 477 (1972). Courts have also examined the results of statistical inference in trade cases involving the Food and Drug Administration, Certified Color Manufacturers' Association v. Mathews, 543 F.2d 284 (D.C. Cir. 1976) (safety of Red Dye No. 2), and the National Highway Traffic Safety Commission, United States v. General Motors Corp., 377 F. Supp. 242 (D.D.C. 1974) (safety of disc wheels).

In contracts cases, statistical inference is used to examine whether the contract has been adequately performed, A.B.G. Instruments & Engineering, Inc. v. United States, 593 F.2d 394 (Ct. Cl. 1979) (number of defective parts in shipment estimated by sampling), or whether the buyer has been mislead as to specifications, In re Forte-Fairbairn, Inc., 62 F.T.C. 1146 (1963) (identity of fibers shipped to buyer confirmed by sampling). As in other areas, potential use of statistics in this area is limited only by the advocate's imagination.

The most notorious use of statistics, however, is in civil rights cases. In such cases a court is asked to infer from a statistical underrepresentation of a particular race, age, sex, or national group in the context of juries, employment, housing, schooling, or voting. Often the issue is whether the number of persons in a particular category observed in a job or school district or on a jury or voting list is significantly smaller than expected. Statistical inference answers this question by indicating the probability of observing so few people of a given category if the selection process is not discriminatory. Similar methods are used in the employment context to determine the fairness of hiring procedures, salary determinations, job assignments, and promotions. Because judges writing opinions in civil rights cases are particularly explicit about the statistical tests and inferential methods employed, many such cases are included in this text.

3. Two Illustrations of Statistical Inference

The inferential technique can be illustrated by the settlement of

A. Inferential Statistics as Proof   35

an insurance claim involving a trainload of kegs of beer that became derailed during a snowstorm. The insurance policy of the shipper covered the wholesale value of all kegs that were rendered unusable by the accident and, if more than 10% of the kegs were damaged, the complete cost of clean-up and salvage. The train was 37 cars long and contained 2,500 kegs of beer. In order to determine exactly how many of the kegs were damaged, the claimant could have examined each of the 2,500 kegs to see if the seals were broken, indicating a rupture of the keg. The claimant's lawyer suggested instead that they examine the evidence from just 50 kegs, save time and money on counting, and multiply the number of damaged kegs by 50 to get the total. Twelve kegs or 24% of the 50 kegs examined were found to be damaged. The inference involved was that, since these 50 cases were typical of all kegs on the train, then 24% or 600 of all of the kegs aboard were damaged.

The problem with this estimate of 600 damaged kegs is that the insurance company had no idea how reliable the estimate was. But inferential statistics calculated by the claimant's lawyer enabled him to say that if these kegs were indeed representative of all kegs on the train, then they could be 95% certain that at least 447 kegs had been damaged. This estimate and its associated confidence level served as a basis for the award of damages. Inferential statistics indicated that they could be 95% certain that more than 13% of the total shipment was damaged, thereby supporting the claim for clean-up and salvage costs.

A second example of the use of inferential techniques emerged in a torts claim. The plaintiff alleged that the defendant's libel had damaged his business, resulting in a decrease in profits of $857,000. The defendant responded that, since average profits for the plaintiff's business were equal to $2,755,000, a small fluctuation of $857,000 could occur just by chance. By calculating a statistical measure of the typical fluctuation around this average, the plaintiff demonstrated to the court that profits typically varied by about $221,000 a year over the 50 years he had been in business. The inference involved was that the $857,000 variation could not have occurred by chance. If no other explanation were forthcoming, the court was asked to conclude that the defendant's libel was the cause of the loss. Statistical inference enabled the plaintiff to support his argument by calculating that the probability of a decline of this magnitude in profits due solely to random business fluctuations was less than one chance in a thousand. Estimates like this provide an objective measure of the strength of the inference to be made by the factfinder when exact causality cannot be proved.

B. "SIMPLE" STATISTICAL EVIDENCE

In some cases, sophisticated statistical tests are developed to aid the interpretation of facts relevant to the resolution of legal questions. Often, however, the resolution of legal issues depends on the interpretation of pre-existing statistics. While reading Craig v. Boren, below, note the different interpretations given to the same statistical evidence by Justice Brennan and Justice Rehnquist. What exactly is the statistical issue before the Court? Is all of the evidence presented relevant to those issues? Is the roadside survey a fair sampling of drivers? What do these statistics prove? What inferences are the Justices drawing from the data?

CRAIG v. BOREN
429 U.S. 190 (1976)

MR. JUSTICE BRENNAN delivered the opinion of the Court.

The interaction of two sections of an Oklahoma statute, Okla. Stat., Tit. 37, §§241 and 245 (1958 and Supp. 1976), prohibits the sale of "nonintoxicating" 3.2% beer to males under the age of 21 and to females under the age of 18. The question to be decided is whether such a gender-based differential constitutes a denial to males 18-20 years of age of the equal protection of the laws in violation of the Fourteenth Amendment. . . .

Analysis may appropriately begin with the reminder that Reed [v. Reed] emphasized that statutory classifications that distinguish between males and females are "subject to scrutiny under the Equal Protection Clause." 404 U.S., at 75. To withstand constitutional challenge, previous cases establish that classifications by gender must serve important governmental objectives and must be substantially related to achievement of those objectives. . . .

The District Court recognized that Reed v. Reed was controlling. In applying the teachings of that case, the court found the requisite important governmental objective in the traffic-safety goal proffered by the Oklahoma Attorney General. It then concluded that the statistics introduced by the appellees established that the gender-based distinction was substantially related to achievement of that goal.

We accept for purposes of discussion the District Court's identification of the objective underlying §§241 and 245 as the enhancement

B. "Simple" Statistical Evidence   37

of traffic safety. Clearly, the protection of public health and safety represents an important function of state and local governments. However, appellees' statistics in our view cannot support the conclusion that the gender-based distinction closely serves to achieve that objective and therefore the distinction cannot under *Reed* withstand equal protection challenge.

The appellees introduced a variety of statistical surveys. First, an analysis of arrest statistics for 1973 demonstrated that 18-20-year-old male arrests for "driving under the influence" and "drunkenness" substantially exceeded female arrests for that same age period.[8] Similarly, youths aged 17-21 were found to be overrepresented among those killed or injured in traffic accidents, with males again numerically exceeding females in this regard.[9] Third, a random roadside survey in Oklahoma City revealed that young males were more inclined to drive and drink beer than were their female counterparts. Fourth, Federal Bureau of Investigation nationwide statistics exhibited a notable increase in arrests for "driving under the influence."[11] Finally, statistical evidence gathered in other jurisdictions, particularly Minnesota and Michigan, was offered to corroborate Oklahoma's experience by indicating the pervasiveness of youthful participation in motor vehicle accidents following the imbibing of alcohol. Conceding that "the case is not free from doubt," 399 F. Supp., at 1314, the District Court nonetheless concluded that this statistical showing substantiated "a rational basis for the legislative judgment underlying the challenged classification." Id., at 1307.

Even were this statistical evidence accepted as accurate, it nevertheless offers only a weak answer to the equal protection question presented here. The most focused and relevant of the statistical surveys, arrests of 18-20-year-olds for alcohol-related driving offenses, exemplifies the ultimate unpersuasiveness of this evidentiary record. Viewed in terms of the correlation between sex and the actual activity that Oklahoma seeks to regulate — driving while under the influence

---

8. The disparities in 18-20-year-old male-female arrests were substantial for both categories of offenses: 427 versus 24 for driving under the influence of alcohol, and 966 versus 102 for drunkenness. Even if we assume that a legislature may rely on such arrest data in some situations, these figures do not offer support for a differential age line, for the disproportionate arrests of males persisted at older ages; indeed, in the case of arrests for drunkenness, the figures for all ages indicated "even more male involvement in such arrests at later ages." 399 F. Supp., at 1309. See also n. 14, infra.

9. This survey drew no correlation between the accident figures for any age group and levels of intoxication found in those killed or injured.

11. The FBI made no attempt to relate these arrest figures either to beer drinking or to an 18-21 age differential, but rather found that male arrests for all ages exceeded 90% of the total.

of alcohol — the statistics broadly establish that .18% of females and 2% of males in that age group were arrested for that offense. While such a disparity is not trivial in a statistical sense, it hardly can form the basis for employment of a gender line as a classifying device. Certainly if maleness is to serve as a proxy for drinking and driving, a correlation of 2% must be considered an unduly tenuous "fit."[12] Indeed, prior cases have consistently rejected the use of sex as a decision-making factor even though the statutes in question certainly rested on far more predictive empirical relationships than this.[13]

Moreover, the statistics exhibit a variety of other shortcomings that seriously impugn their value to equal protection analysis. Setting aside the obvious methodological problems,[14] the surveys do not adequately justify the salient features of Oklahoma's gender-based traffic-safety law. None purports to measure the use and dangerousness of 3.2% beer as opposed to alcohol generally, a detail that is of particular importance since, in light of its low alcohol level, Oklahoma apparently considers the 3.2% beverage to be "nonintoxicating." Okla. Stat., Tit. 37, §163.1 (1958). Moreover, many of the studies, while graphically documenting the unfortunate increase in driving while under the influence of alcohol, make no effort to relate their findings to age-sex differentials as involved here. Indeed, the only survey that explicitly centered its attention upon young drivers

---

12. Obviously, arrest statistics do not embrace all individuals who drink and drive. But for purposes of analysis, this "underinclusiveness" must be discounted somewhat by the shortcomings inherent in this statistical sample, see n. 14, infra. In any event, we decide this case in light of the evidence offered by Oklahoma and know of no way of extrapolating these arrest statistics to take into account the driving and drinking population at large, including those who avoided arrest.

13. For example, we can conjecture that in *Reed,* Idaho's apparent premise that women lacked experience in formal business matters (particularly compared to men) would have proved to be accurate in substantially more than 2% of all cases. And in both Frontiero [v. Richardson, 411 U.S. 677 (1973)] and [Weinberger v.] Wiesenfeld [, 420 U.S. 636 (1975)], we expressly found appellees' empirical defense of mandatory dependency tests for men but not women to be unsatisfactory, even though we recognized that husbands are still far less likely to be dependent on their wives than vice versa. See, e.g., 411 U.S., at 688-690.

14. The very social stereotypes that find reflection in age-differential laws, are likely substantially to distort the accuracy of these comparative statistics. Hence "reckless" young men who drink and drive are transformed into arrest statistics, whereas their female counterparts are chivalrously escorted home. Moreover, the Oklahoma surveys, gathered under a regime where the age-differential law in question has been in effect, are lacking in controls necessary for appraisal of the actual effectiveness of the male 3.2% beer prohibition. In this regard, the disproportionately high arrest statistics for young males — and, indeed, the growing alcohol-related arrest figures for all ages and sexes — simply may be taken to document the relative futility of controlling driving behavior by the 3.2% beer statute and like legislation, although we obviously have no means of estimating how many individuals, if any, actually were prevented from drinking by these laws.

and their use of beer — albeit apparently not of the diluted 3.2% variety — reached results that hardly can be viewed as impressive in justifying either a gender or age classification.[16]

There is no reason to belabor this line of analysis. It is unrealistic to expect either members of the judiciary or state officials to be well versed in the rigors of experimental or statistical technique. But this merely illustrates that proving broad sociological propositions by statistics is a dubious business, and one that inevitably is in tension with the normative philosophy that underlies the Equal Protection Clause. Suffice to say that the showing offered by the appellees does not satisfy us that sex represents a legitimate, accurate proxy for the regulation of drinking and driving. In fact, when it is further recognized that Oklahoma's statute prohibits only the selling of 3.2% beer to young males and not their drinking the beverage once acquired (even after purchase by their 18-20-year-old female companions), the relationship between gender and traffic safety becomes far too tenuous to satisfy *Reed's* requirement that the gender-based difference be substantially related to achievement of the statutory objective.

We hold, therefore, that under *Reed,* Oklahoma's 3.2% beer statute invidiously discriminates against males 18-20 years of age. . . .

We conclude that the gender-based differential contained in Okla. Stat., Tit. 37, §245 (1976 Supp.) constitutes a denial of the equal protection of the laws to males aged 18-20 and reverse the judgment of the District Court.

It is so ordered.

MR. JUSTICE REHNQUIST, dissenting.

The Court's disposition of this case is objectionable on two grounds. First is its conclusion that *men* challenging a gender-based

---

16. The random roadside survey of drivers conducted in Oklahoma City during August 1972 found that 78% of drivers under 20 were male. Turning to an evaluaton of their drinking habits and factoring out nondrinkers, 84% of the males versus 77% of the females expressed a preference for beer. Further 16.5% of the men and 11.4% of the women had consumed some alcoholic beverage within two hours of the interview. Finally, a blood alcohol concentration greater than .01% was discovered in 14.6% of the males compared to 11.5% of the females. "The 1973 figures, although they contain some variations, reflect essentially the same pattern." 399 F. Supp., at 1309. Plainly these statistical disparities between the sexes are not substantial. Moreover, when the 18-20 age boundaries are lifted and all drivers analyzed, the 1972 roadside survey indicates that male drinking rose slightly whereas female exposure to alcohol remained relatively constant. Again, in 1973, the survey established that "compared to all drivers interviewed, . . . the under-20 age group generally showed a lower involvement with alcohol in terms of having drunk within the past two hours or having a significant BAC (blood alcohol content)." Ibid. In sum, this survey provides little support for a gender line among teenagers and actually runs counter to the imposition of drinking restrictions based upon age.

statute which treats them less favorably than women may invoke a more stringent standard of judicial review than pertains to most other types of classifications. Second is the Court's enunciation of this standard, without citation to any source, as being that "classifications by gender must serve *important* governmental objectives and must be *substantially* related to achievement of those objectives." The only redeeming feature of the Court's opinion, to my mind, is that it apparently signals a retreat by those who joined the plurality opinion in Frontiero v. Richardson, 411 U.S. 677 (1973), from their view that sex is a "suspect" classification for purposes of equal protection analysis. I think the Oklahoma statute challenged here need pass only the "rational basis" equal protection analysis expounded in cases such as McGowan v. Maryland, 366 U.S. 420 (1961), and Williamson v. Lee Optical Co., 348 U.S. 483 (1955), and I believe that it is constitutional under that analysis. . . .

The applicable rational-basis test is one which

> permits the States a wide scope of discretion in enacting laws which affect some groups of citizens differently than others. The constitutional safeguard is offended only if the classification rests on grounds wholly irrelevant to the achievement of the State's objective. State legislatures are presumed to have acted within their constitutional power despite the fact that, in practice, their laws result in some inequality. A statutory discrimination will not be set aside if any state of facts reasonably may be conceived to justify it.

McGowan v. Maryland, 366 U.S., at 425-426 (citations omitted). . . .

The Court "accept[s] for purposes of discussion" the District Court's finding that the purpose of the provisions in question was traffic safety, and proceeds to examine the statistical evidence in the record in order to decide if "the gender-based distinction *closely* serves to achieve that objective." (Whether there is a difference between laws which "closely serv[e]" objectives and those which are only "substantially related" to their achievement we are not told.) I believe that a more traditional type of scrutiny is appropriate in this case, and I think that the Court would have done well here to heed its own warning that "[i]t is unrealistic to expect . . . members of the judiciary . . . to be well versed in the rigors of experimental or statistical technique." One need not immerse oneself in the fine points of statistical analysis, however, in order to see the weaknesses in the Court's attempted denigration of the evidence at hand.

One survey of arrest statistics assembled in 1973 indicated that males in the 18-20 age group were arrested for "driving under the influence" almost 18 times as often as their female counterparts, and for "drunkenness" in a ratio of almost 10 to 1. Accepting, as the Court does, appellants' comparison of the total figures with 1973 Oklahoma census data, this survey indicates a 2% arrest rate among males in the age group, as compared to a .18% rate among females.

Other surveys indicated (1) that over the five-year period from 1967 to 1972, nationwide arrests among those under 18 for drunken driving increased 138%, and that 93% of all persons arrested for drunken driving were male; (2) that youths in the 17-21 age group were overrepresented among those killed or injured in Oklahoma traffic accidents, that male casualties substantially exceeded female, and that deaths in this age group continued to rise while overall traffic deaths declined; (3) that over three-fourths of the drivers under 20 in the Oklahoma City area are males, and that each of them, on average, drives half again as many miles per year as their female counterparts; (4) that four-fifths of male drivers under 20 in the Oklahoma City area state a drink preference for beer, while about three-fifths of female drivers of that age state the same preference; and (5) that the percentage of male drivers under 20 admitting to drinking within two hours of driving was half again larger than the percentage for females, and that the percentage of male drivers of that age group with a blood alcohol content greater than .01% was almost half again larger than for female drivers.

The Court's criticism of the statistics relied on by the District Court conveys the impression that a legislature in enacting a new law is to be subjected to the judicial equivalent of a doctoral examination in statistics. Legislatures are not held to any rules of evidence such as those which may govern courts or other administrative bodies, and are entitled to draw factual conclusions on the basis of the determination of probable cause which an arrest by a police officer normally represents. In this situation, they could reasonably infer that the incidence of drunk driving is a good deal higher than the incidence of arrest.

And while, as the Court observes, relying on a report to a Presidential Commission which it cites in a footnote, such statistics may be distorted as a result of stereotyping, the legislature is not required to prove before a court that its statistics are perfect. In any event, if stereotypes are as pervasive as the Court suggests, they may in turn influence the conduct of the men and women in question, and cause the young men to conform to the wild and reckless image which is their stereotype.

The Court also complains of insufficient integration of the various surveys on several counts — that the injury and death figures are in no way directly correlated with intoxication; that the national figures for drunk driving contain no breakdown for the 18-21-year-old group; and that the arrest records for intoxication are not tied to the consumption of 3.2% beer. But the State of Oklahoma — and certainly this Court for purposes of equal protection review — can surely take notice of the fact that drunkenness is a significant cause of traffic casualties, and that youthful offenders have participated in the increase of the drunk-driving problem. On this latter point, the survey data indicating increased driving casualties among 18-21-year-olds, while overall casualties dropped, are not irrelevant.

Nor is it unreasonable to conclude from the expressed preference for beer by four-fifths of the age-group males that that beverage was a predominant source of their intoxication-related arrests. Taking that as the predicate, the State could reasonably bar those males from any purchases of alcoholic beer, including that of the 3.2% variety. This Court lacks the expertise or the data to evaluate the intoxicating properties of that beverage, and in that posture our only appropriate course is to defer to the reasonable inference supporting the statute — that taken in sufficient quantity this beer has the same effect as any alcoholic beverage.

Quite apart from these alleged methodological deficiencies in the statistical evidence, the Court appears to hold that that evidence, on its face, fails to support the distinction drawn in the statute. The Court notes that only 2% of males (as against .18% of females) in the age group were arrested for drunk driving, and that this very low figure establishes "an unduly tenuous 'fit'" between maleness and drunk driving in the 18-20-year-old group. On this point the Court misconceives the nature of the equal protection inquiry.

The rationality of a statutory classification for equal protection purposes does not depend upon the statistical "fit" between the class and the trait sought to be singled out. It turns on whether there may be a sufficiently higher incidence of the trait within the included class than in the excluded class to justify different treatment. Therefore the present equal protection challenge to this gender-based discrimination poses only the question whether the incidence of drunk driving among young men is sufficiently greater than among young women to justify differential treatment. Notwithstanding the Court's critique of the statistical evidence, that evidence suggests clear differences between the drinking and driving habits of young men and women. Those differences are grounds enough for the State reasonably to conclude that young males pose by far the greater drunk-driving hazard, both in terms of sheer numbers and in terms of hazard on a per-driver basis. The gender-based difference in treatment in this case is therefore not irrational.

The Court's argument that a 2% correlation between maleness and drunk driving is constitutionally insufficient therefore does not pose an equal protection issue concerning discrimination between males and females. The clearest demonstration of this is the fact that the precise argument made by the Court would be equally applicable to a flat bar on such purchases by *anyone,* male or female, in the 18-20 age group; in fact it would apply *a fortiori* in that case given the even more "tenuous 'fit'" between drunk-driving arrests and femaleness. The statistics indicate that about 1% of the age group population as a whole is arrested. What the Court's argument is relevant to is not equal protection, but due process — whether there are enough persons in the category who drive while drunk to justify a bar against purchases by all members of the group.

Cast in those terms, the argument carries little weight, in light of our decisions indicating that such questions call for a balance of the State's interest against the harm resulting from any overinclusiveness or underinclusiveness. The personal interest harmed here is very minor — the present legislation implicates only the right to purchase 3.2% beer, certainly a far cry from the important personal interests which have on occasion supported this Court's invalidation of statutes on similar reasoning. And the state interest involved is significant — the prevention of injury and death on the highways. . . .

The Oklahoma Legislature could have believed that 18-20-year-old males drive substantially more, and tend more often to be intoxicated than their female counterparts; that they prefer beer and admit to drinking and driving at a higher rate than females; and that they suffer traffic injuries out of proportion to the part they make up of the population. Under the appropriate rational-basis test for equal protection, it is neither irrational nor arbitrary to bar them from making purchases of 3.2% beer, which purchases might in many cases be made by a young man who immediately returns to his vehicle with the beverage in his possession. The record does not give any good indication of the true proportion of males in the age group who drink and drive (except that it is no doubt greater than the 2% who are arrested), but whatever it may be I cannot see that the mere purchase right involved could conceivably raise a due process question. There being no violation of either equal protection or due process, the statute should accordingly be upheld.

Notes and Problems

1. Evaluation of the statistical arguments in Craig v. Boren is simplified by listing every statistical point raised and, beside each point, describing how the opposing parties evaluated that evidence.

Consider the point that 2% of young male drivers and only .18% of young female drivers were arrested for drunk driving. The difference in interpretation between Justice Brennan and Rehnquist is based on varying legal as well as mathematical viewpoints. Brennan was evaluating an equal protection claim when he questioned the appropriateness of categorizing young males as more dangerous drivers when only 2% of them had been arrested for such an offense. Rehnquist cast his argument in a due process light; he focused on the 10-to-1 ratio of males to females arrested for drunkenness to determine whether separate treatment is "rationally based." Interpretation of a particular statistic is often a matter of emphasis. The strategy of appropriate emphasis to persuade the finder of fact of a particular viewpoint is a familiar one to the practicing attorney.

2. Exhibits III-1 to III-4 present the data on which Justice Brennan relies in drawing his statistical conclusions. They are taken

EXHIBIT III-1.
Persons Arrested by Age and Sex for the Months September, October, November, and December, 1973 in the State of Oklahoma for Alcohol Related Offenses*

		18 yrs.	19 yrs.	20 yrs.	Total Persons Arrested 18-65 and over
Driving under the influence	Male	152	107	168	5,400
	Female	14	2	8	499
Drunkenness	Male	340	321	305	14,713
	Female	39	33	30	1,278

* The information contained in this report is summarized crime statistical data compiled from sixty-four (64) Sheriff's Departments and one hundred thirty (130) Police Departments in the State of Oklahoma. 84% population coverage.
Source: Walker v. Hall, excerpted from defendants' Exhibit 1.

EXHIBIT III-2.
Oklahoma City Police Department Arrest Statistics for the Year 1973*

Classification of Offenses	Sex	Age 18 yrs.	19 yrs.	20 yrs.	Total for All Ages
Driving under the influence	Male	47	54	72	3,206
	Female	10	1	5	279
Drunkenness	Male	102	104	96	9,413
	Female	18	22	19	823

* Includes those released without having been formally charged.
Source: Walker v. Hall, excerpted from defendants' Exhibit 2.

EXHIBIT III-3.
Partial Data Summary, 1972 and 1973 Roadside Survey, Oklahoma City, Oklahoma, OMEC, Inc.

Respondent:	Males Less Than 20 Years		All Males		Females Less Than 20 Years		All Females	
Year:	1972	1973	1972	1973	1972	1973	1972	1973
Drink Alcoholic Beverages — Yes (Percent)	70.4	75.6	75.5	75.2	68.6	70.6	59.9	60.2
Drank Alcoholic Beverages in Last 2 Hours — Yes (Percent)	16.5	15.6	21.2	21.6	11.4	11.7	14.1	10.9
BAC Greater Than .01 (Percent)	14.6	8.4	17.5	16.7	11.5	7.4	10.8	7.8
BAC Greater Than or Equal To .05 Given BAC Greater Than .01 (Percent)	29.7	50.0	51.2	52.3	14.3	0.0	36.8	40.7
Average Miles Driven	15,670	16,794	19,360	20,042	10,471	10,456	10,803	11,399
Average Days/Week Driving Vehicle	6.8	6.7	6.7	6.7	6.7	6.5	6.5	6.4
Number of Participants	243	238	1246	1161	70	68	354	349

Source: Walker v. Hall, Table 1 from defendants' Exhibit 3.

EXHIBIT III-4.
Oklahoma Statewide Summary of Motor Vehicle Traffic Collisions*

1972

	Total				Driver			
	Killed		Injured		Killed		Injured	
	Male	Fem.	Male	Fem.	Male	Fem.	Male	Fem.
Municipal	34	8	1640	1277	16	4	932	637
Other	82	26	1171	639	49	10	681	261
Statewide	116	34	2811	1916	65	14	1613	898

1973

	Killed		Injured		Killed		Injured	
Municipal	34	13	1796	1373	16	6	995	695
Other	91	27	1277	676	53	9	745	292
Statewide	125	40	3073	2049	69	15	1740	987

* Number of persons killed and injured, aged 17-21 years old.
Source: Walker v. Hall, excerpted from defendants' Exhibits 4 and 5.

from the evidentiary exhibits offered by the defendants in the district court where the case was heard under the name Walker v. Hall, 399 F. Supp. 1304, at 1309 et seq. (W.D. Okla. 1975). How compelling are they? How relevant are they? What weakness do you see in the defendants' roadside survey (Exhibit III-3)? Does any of the evidence considered in this case prove that having a higher male drinking age for 3.2% beer results in greater traffic safety? Is it proved that placing the same restrictions on females would not also result in greater traffic safety? Is the evidence strong enough to justify the different treatment of men and women?

C. USING STATISTICAL STUDIES

In Ballew v. Georgia, below, evidence of statistical analyses made by professional statisticians aided the Court in establishing guidelines for jury size. Note how the statistical studies affect the Court's perception of Sixth Amendment rights. The willingness of the Court to use statistical analysis in such a policymaking context opens new sources of evidence for the advocate. Not that reference to simple statistics by the Supreme Court is a recent phenomenon; Brown v. Board of Education, 347 U.S. 483 (1954), is often alluded to as a classic case in this area. But Ballew v. Georgia represented a new wave in the ability of advocates to use relatively complex statistical analysis, developed not for the purpose of the instant case but as a scholarly exercise, to support a particular policy option. First, consider Hans Zeisel's statistical analysis of the impact of varying jury size. Then note how Justice Blackmun incorporates such research into his opinion and Justice Powell's reply.

HANS ZEISEL, . . . AND THEN THERE WERE NONE: THE DIMINUTION OF THE FEDERAL JURY
38 U. Chi. L. Rev. 710 (1971)[2]

Goneril: Hear me, my lord. What need you five-and-twenty? ten? or five?
Regan: What need one?
King Lear, Act II, Scene IV

In a dramatic move sponsored by the Chief Justice of the United States Supreme Court, seventeen of the federal district courts will reduce the size of their civil juries from twelve members to six. Immediately following the Chief Justice's announcement, Representative William Lloyd Scott of Virginia introduced a bill in Congress to provide for six-member juries in all federal trials, both civil and criminal, except in cases involving capital offenses. On the state level, the New Jersey Supreme Court called for an amendment to the state constitution that would allow the legislature to reduce the size of all juries and to end jury trials in civil cases. Moreover, at least one of the federal district courts has already been experimenting with six-member juries in criminal trials, albeit by encouraged agreement between prosecution and defense.

Juries with less than twelve members, of course, are not foreign to our experience. Some state courts try small civil claims and minor criminal cases before six-member juries; four states even try non-capital felony cases before juries that have fewer than twelve members. Despite this background, experimentation with the jury was previously confined to the states. Until the present time, the federal jury appeared to be immutable.

The reasons presently given for reduction of the size of federal civil juries are to expedite jury trials and to lessen their cost. With respect to the latter, the Chief Justice has estimated that contracting

---

2. Copyright © 1971 by the University of Chicago Law Review. Reprinted with permission.

the size of federal civil juries to six would result in an annual savings of four million dollars. While this may seem to be a substantial sum, it is only 2.4 per cent of the total federal judicial budget, and little more than a thousandth part of one per cent of the total federal budget.

With respect to minimizing delay, the smaller jury would merely decrease the time required for impaneling. Although there are no data available on the time consumed in impaneling juries, we do have accurate data indicating that federal district court judges spend eight per cent of their total working time in trying civil jury cases. Estimating that impaneling the jurors takes, on the average, about ten per cent of the trial time, one discovers that only eight-tenths of one per cent of the federal district judges' total working time is presently consumed by impaneling civil juries. On first impression, it might seem that reducing the twelve jurors to six would save half of that impaneling time. But in many federal courts the jurors are examined primarily by the judge, who directs most of his questions to all jurors at the same time. In this situation the savings would be minimal, since it takes no more time to ask a question of twelve jurors than to ask it of six. In any event, we are discussing an amount which is less than half of the impaneling time — at best four-tenths, more likely three-tenths, of one per cent of the judge's working time.

Thus, neither the amount of money nor the amount of time that would be saved can adequately explain and justify the reform recommended by the Chief Justice. However, when viewed in the broader context of the other proposals to limit the functioning of the jury, the decision of the federal district courts to adopt the six-member jury appears as a significant step toward a drastic reduction of the American jury system in general. Under these circumstances, this initial reform deserves careful scrutiny on its own merits.

I. Williams v. Florida

The last pieces of the legal foundation for the six-member jury were laid in the Supreme Court's decision in Williams v. Florida. Williams, accused and subsequently convicted of robbery, had made a pre-trial motion to impanel a twelve-member jury instead of the six-member jury prescribed by Florida law for non-capital cases. The motion had been denied. In affirming the conviction, the Supreme Court ruled that the sixth amendment's guarantee of trial by jury does not require that jury membership be fixed at twelve. In sweeping language, the Court removed the constitutional obstacles to decreasing the size of federal or state juries in both civil and criminal cases. First, the Court summarized its historical discussion by stating that the

C. Using Statistical Studies   49

twelve-member jury appears to have been a "historical accident, unrelated to the great purposes which gave rise to the jury in the first place."

History, however, might have embodied more wisdom than the Court would allow. It might be more than an accident that after centuries of trial and error the size of the jury at common law came to be fixed at twelve. A primary function of the jury was to represent the community as broadly as possible; yet at the same time, it had to remain a group of manageable size. Twelve might have been, and might still be, the upper limit beyond which the difficulty of self-management becomes insuperable under the burdensome condition of a trial. On this view, twelve would be the number that optimizes the jury's two conflicting goals — to represent the community and to remain manageable.

Having disposed of the rationality of the number twelve, the Court proceeded:

> Nothing in this history suggests, then, that we do violence to the letter of the Constitution by turning to other than purely historical considerations to determine which features of the jury system, as it existed at common law, were preserved in the Constitution. *The relevant inquiry,* as we see it, *must be the function that the particular feature performs and its relation to the purposes of the jury trial.*

After a casual reference to empirical data, to which we will devote our attention presently, the Court concluded that while the jury should comprise a cross-section of the community, a six-member jury does not perceptibly differ in this respect from a twelve-member jury:

> [W]hile in theory the number of viewpoints represented on a randomly selected jury ought to increase as the size of the jury increases, in practice the difference between the 12-man and the six-man jury in terms of the cross-section of the community represented seems likely to be negligible. Even the 12-man jury cannot insure representation of every distinct voice in the community, particularly given the use of the peremptory challenge. As long as arbitrary exclusions of a particular class from the jury rolls are forbidden . . . the concern that the cross-section will be significantly diminished if the jury is decreased in size from 12 to six seems an unrealistic one.

Here, then, is the Court's reference to empirical data:

> What few experiments have occurred — usually in the civil area — indicate that there is no discernable difference between the results reached by the two different-sized juries.

The Court cites, impressively enough, six items: (1) Judge Wiehl's article on "The Six Man Jury" in the Gonzaga Law Review; (2) Judge Tamm's "The Five-Man Civil Jury: A Proposed Constitutional Amendment" in the Georgetown Law Journal; (3) Cronin's piece on "Six-Member Juries in District Courts" in the Boston Bar Journal; (4) "Six-Member Juries Tried in Massachusetts District Court," in the Journal of the American Judicature Society; (5) "New Jersey Experiments with Six-Man Jury," in the Bulletin of the Section of Judicial Administration of the American Bar Association; and (6) Judge Phillips' article on "A Jury of Six in All Cases" in the Connecticut Bar Journal.

It is worthwhile to disinter the substance buried in these citations:

(1) Judge Wiehl approvingly cites Charles Joiner's Civil Justice and the Jury, in which Joiner somewhat disingenuously states that "it could easily be argued that a six-man jury would deliberate equally as well as one of twelve." Since Joiner had no evidence for his conclusion, Judge Wiehl also does not have any.

(2) Judge Tamm had presided over condemnation trials in the District of Columbia in which five-man juries are used and found them satisfactory.

(3) Cronin relates that the Massachusetts legislature had authorized, on an experimental basis, the use of six-member juries for civil cases in the District Court of Worcester, a civil court of limited jurisdiction. Forty-three such trials were conducted, and the highest verdict was for a sum of $2,500. The clerk of the court is said to have reported that "the six-member jury verdicts are about the same as those returned by regular twelve-member juries." Three lawyers also testified that they could not detect any differences in verdicts, one because "the panel is drawn from the regular Superior Court panel of jurors," another because "[t]here seems to be no particular reason why the size of a finding would be affected by a six-man jury." All those trials, it seems, were given preferential scheduling to endear them to counsel.

(4) The Court's fourth cited authority consists of an abbreviated summary of the Massachusetts experiment and concludes that "the lawyers who use the District Court, as well as the clerk, report that the verdicts are not different than those returned by twelve-member juries."

(5) The ABA Bulletin contains the statement that "the Monmouth [New Jersey] County Court has experimented with the use of a six-man jury in a [sic] civil negligence case."

(6) Judge Phillips summarizes the economic advantages derived from the Connecticut law that permits litigants to opt for a six-member jury in civil cases. He advocates a mandatory reduction in

jury size, but never even mentions the problem of possible differences in verdicts in comparison to the twelve-member jury.

This is scant evidence by any standards. The several thousand verdicts by criminal juries each year in Florida, Louisiana, and Utah would have provided better evidence. Of course, no such evidence was produced at the trial court, but the Court could conceivably have asked sua sponte for such a study. Even without specific data, however, it is possible to demonstrate that the six-member jury must be expected to perform quite differently than the twelve-member jury in several important respects.

II. SIX-MEMBER CIVIL JURIES IN A STRATIFIED COMMUNITY

The jury system is predicated on the insight that people see and evaluate things differently. It is one function of the jury to bring these divergent perceptions and evaluations to the trial process. If all people weighed trial evidence in the same manner, a jury of one would be as good as a jury of twelve because there would never be any disagreement among them. In fact, we know the opposite to be true, if not from observation of our community then from the performance of our juries. Two-thirds of all juries find their vote split in the first ballot in a criminal case. We have no comparable data on the liability vote of the civil jury, but we do know that the evaluation of damages usually covers a broad range.

There is, therefore, good reason to believe that the jury, to some extent, brings into the courtroom the differences in perception which exist in the community. To see how the six-member jury performs this function in comparison with a twelve-member jury, it will be useful to begin with a simple model of a stratified society. We shall assume that 90% of the community share identical viewpoints and that the remaining 10% have a different viewpoint. Even a jury of twelve, of course, is too small to represent all community views, but it can be shown that the smaller the size of the jury, the less frequently it even approaches community representation.

Juries, especially federal juries, are chosen by lottery from the pool of available jurors. Suppose, now, we were to draw 100 twelve-member juries and 100 six-member juries from a population that had a 10% minority. Of the 100 twelve-member juries, approximately 72 would have at least one representative of that minority; while of the 100 six-member juries, only 47 would have one. It is clear, then, that however limited a twelve-member jury is in representing the full spectrum of the community, the six-member jury is even more limited, and not by a "negligible" margin.

Whether this difference in degree of community representation results in a difference in civil verdicts is, of course, even more important. To explore this question we will slightly complicate our stratification model of the community. We know from experience and from many careful studies that the values different people place on the harm done in a personal injury case are likely to diverge considerably. Table 1 assumes that with respect to the evaluation of a particular claim the community is divided into six groups of equal size. We shall make a further assumption, very close to reality, that whatever the composition of the jury, the damages it awards will lie around the average of the evaluations of all individual jurors. Again, we shall simulate 100 random selections of the two types of juries — the twelve-member and the six-member jury. This time, however, we shall be interested in the average of the individual evaluations of all members of any given jury. Thus, we shall record a twelve-member jury consisting of 6 persons who would evaluate an injury at $1,000, and 6 who would evaluate the same injury at $2,000, as $1,500 — the average of 6 × $1,000 + 6 × $2,000. Table 2 indicates the relative

TABLE 1.
Community Evaluation of Personal Injury Claim

	Fraction of Community	Dollar Evaluation
	$1/6$	1,000
	$1/6$	2,000
	$1/6$	3,000
	$1/6$	4,000
	$1/6$	5,000
	$1/6$	6,000
Average	$1/1$	3,500

TABLE 2.
Per Cent Distribution of the Average Evaluation by 100 Randomly Selected Juries

Interval	Twelve-Member	Six-Member
$1,000-1,499	——	0.1
$1,500-1,999	0.1	1.4
$2,000-2,499	2.0	6.4
$2,500-2,999	13.7	16.4
$3,000-3,499	34.2	25.7
$3,500-3,999	34.2	25.7
$4,000-4,499	13.7	16.4
$4,500-4,999	2.0	6.4
$5,000-5,499	0.1	1.4
$5,500-6,000	——	0.1
Totals	100.0	100.0

spread of these averages around the middle value for all juries of $3,500.

It is easy to see that the six-member juries show a considerably wider variation of "verdicts" than the twelve-member juries. For instance, 68.4% of the twelve-member jury evaluations fall between $3,000 and $4,000, while only 51.4% of the six-member jury evaluations fall in this range. Almost 16% of the six-member juries will reach verdicts that will fall into the extreme levels of more than $4,500 or less than $2,500, as against only a little over 4% of the twelve-member juries. The appropriate statistical measure of this variation is the so-called standard deviation.[37] The actual distribution pattern will always depend on the kind of stratification that is relevant in a particular case, but whatever the circumstances, the six-member jury will always have a standard deviation that is greater by about 42%. This is the result of a more general principle that is by now well known to readers of such statistics as public opinion polls — namely, that the size of any sample is inversely related to its margin of error.[38]

Lest it be thought that standard deviation is merely part of a statistician's game that has no counterpart in reality, it will be useful to provide the appropriate translation of the term into lawyer's language. It is the measure of the gamble the lawyer takes when he goes to trial. The "gambling" notion is seldom made explicit because normally each case is tried only once, but lawyers are quite conscious of the degree to which jury verdicts may vary. To obtain a measure of this variability, trial lawyers were asked to think about their next civil case and to estimate how they would expect ten different juries to decide.[39] These estimates show, as a rule, a considerable amount of vari-

37. The standard deviation is the square root of all squared deviations from the group average.
38. A reduction of the size of a sample by $1/2$ increases the margin of error by the square root of $2/1$, or simply of 2 — that is, by a factor of 1.42, or by 42%.
39. Data collected by the University of Chicago Jury Project, on file at the University of Chicago Law School. The following is an excerpt from the questionnaire and a sample answer:
Which, in your estimate, will be the most likely award in this case after trial?

$25,000

Of course, you cannot be certain that this will be the verdict. If you had to try this case ten different times before ten different juries, you would expect some variation in the verdicts. What do you think these verdicts would look like?

$ 0	$ 20,000	$25,000	$ 25,000
$ 35,000	$ 35,000	$50,000	$100,000
$100,000	$100,000		

Another way of providing an estimate of the variability of jury verdicts would be to allow properly selected extra-juries — for instance by observation of closed-circuit television screens — to try the same case after simulated deliberations.

ation, which should not come as a surprise since a case goes to the jury on the very ground that reasonable men may differ in its resolution. Whatever the extent of the "gamble" incurred through the twelve-member jury, we must expect that it will be significantly greater with a six-member jury.

This increase in the "gamble" might well have an interesting side effect; it could increase the incidence of jury waiver and thereby reduce the frequency of jury trials. The trial lawyer survey also suggests that lawyers expect the variation of verdicts returned by twelve-member juries to be of about the same magnitude as the variation expected if the same case were tried in bench trials before different judges. If the jury size is reduced from twelve to six, this perception of the approximate balance between jury and bench trial will be disturbed. Henceforth, the "gamble" with a jury will be significantly greater than the "gamble" with a judge and, as a result, more lawyers might waive their right to a jury, perhaps a consequence not unexpected by those who initiated the reform.

In addition to the tendency to be less representative and to produce more varied damage verdicts, the six-member jury is likely to yield fewer examples of that treasured, paradoxical phenomenon — the hung jury. Hung juries almost always arise from situations in which there were originally several dissenters. Even if only one holds out, his having once been the member of a group is essential in sustaining him against the majority's efforts to make the verdict unanimous. Fewer hung juries can be expected in six-member juries for two reasons: first, as discussed earlier, there will be fewer holders of minority positions on the jury; second, if a dissenter appears, he is more likely to be the only one on the jury. Lacking any associate to support his position, he is more likely to abandon it.

In *Williams* the Court cites the studies conducted in connection with The American Jury to support its proposition that "jurors in the minority on the first ballot are likely to be influenced by the proportional size of the majority aligned against them." It is only fair to point out that the findings were quite different:

> Nevertheless, for one or two jurors to hold out to the end, it would appear necessary that they had companionship at the beginning of the deliberations. The juror psychology recalls a famous series of experiments by the psychologist Asch and others which showed that in an ambiguous situation a member of a group will doubt and finally disbelieve his own correct observation if all other members of the group claim that he must have been mistaken. To maintain his original position, not only before others but even before himself, it is necessary for him to have at least one ally.

C. Using Statistical Studies 55

The distinction is crucial in this respect: If it is only the *proportion* that matters, then one versus five is the equivalent of two versus ten; but if the original companionship of an ally is essential, then one versus five is far less likely to produce a hung jury than two versus ten. As stated previously, the probability of having at least one member of a minority (which comprises 10% of the population) on the twelve-member jury is 72 out of every 100, as against 47 on six-member juries. The discrepancy is aggravated by the fact that the expectation of having *more than one* minority member on the twelve-member jury is 34 out of every 100, as against only 11 on six-member juries.

This was one hypothesis, among those developed in this article, that could be put to an immediate test. A survey was initiated to establish the frequency of hung juries in criminal jury trials that had gone to verdict since January 1, 1969 in the Miami Circuit Court, the largest Florida court. The results were 7 hung juries in 290 trials before six-member juries. The comparison with the national average provides startling and gratifying confirmations of this prediction:

TABLE 3.
Hung Juries in Criminal Jury Trials

Twelve-member	5.0%
Six-member	2.4%

On grounds of economy, one might welcome any reduction in the number of hung juries. One should understand, however, that such reduction is but the combined result of less representative, more homogeneous juries and of a reduced ability to resist the pressure for unanimity.

From the foregoing discussion, it would appear that the Court's holding in *Williams* rests on a poor foundation. In several important respects, the six-member jury performs differently than the twelve-member jury. The Court probably suspected that some differences in composition and performance would exist between the types of juries but thought them negligibly small. It would seem that the Court has underestimated their magnitude.

III. CRIMINAL SIX-MEMBER JURIES AND MAJORITY VERDICTS

Neither the reduction of the size of the criminal jury nor the adoption of majority verdicts are presently being considered by the federal courts. Yet it is not premature to explore the consequences that would accompany these two changes. Despite the Court's emphasis of the importance of the unanimity requirement in *Williams*, there are indications, as Justice Harlan has recognized, that this requirement could fall, and the reduction of the criminal jury may similarly be within purview.

The above analysis of the six-member civil jury applies with minor variations to the criminal jury, where the stratification does not pertain to the dollar-evaluation of a claim, but to the perception of the gravity of the charged crime and, more importantly, to the differing standards of "reasonable doubt." To obtain a conviction under the unanimity rule, the prosecutor must persuade the juror with the highest standard. Considering the class of jurors who are most difficult to convince as a "minority" in terms of our model, it is evident that fewer six-member juries will contain representatives of that minority. Consequently, a six-member jury provides a lesser safeguard for the defendant than a twelve-member jury. Careful study of the operation of juries with less than twelve members in such states as Florida, Louisiana, and Utah should confirm this hypothesis by revealing that these juries yield fewer hung juries, more findings of guilt, and among them relatively fewer convictions for the lesser included offense than are rendered in comparable cases by twelve-member juries.

With respect to the abandonment of the unanimity rule, there is again considerable experience in the state courts. Many states allow majority verdicts in civil trials before both twelve- and six-member juries, some states permit such verdicts in minor criminal trials, and two states allow majority verdicts even in felony trials for non-capital offenses. The important element to observe is that the abandonment of the unanimity rule is but another way of reducing the size of the jury. But it is reduction with a vengeance, for a majority verdict requirement is far more effective in nullifying the potency of minority viewpoints than is the outright reduction of a jury to a size equivalent to the majority that is allowed to agree on a verdict. Minority viewpoints fare better on a jury of ten that must be unanimous than on a jury of twelve where ten members must agree on a verdict.

An example will elucidate this proposition. Suppose we again assume a minority position held by 10% of the population, and that two sets of juries are drawn: 100 twelve-member juries and 100 ten-member juries. In Table 4 are the frequencies with which the minority view can be expected to be represented.

Looking first at the 100 twelve-member juries, we expect to find 38 juries with one minority representative, 23 with two, and 11 with three or more. If these twelve-member juries must be unanimous to reach a verdict, the majority will have to reckon with at least one minority member in $38 + 23 + 11 = 72$ out of the 100 cases. If these

TABLE 4.
Expected Number of Minority Jurors*
on 100 Randomly Selected Juries

Number	Twelve-member	Ten-member
None	28	35
One	38	39
Two	23	19
Three or more	11	7
Totals	100	100

* Representing 10% of the population.

juries are permitted to reach a verdict by agreement of ten jurors, then the majority will be able simply to disregard the minority position in 38 + 23 = 61 of the 72 cases. Only in the 11 cases in which we must expect three or more minority jurors will they be able to influence the verdict.

Let us now turn to the ten-member juries. Here only 39 + 19 + 7 = 65 of the 100 juries are expected to have at least one minority member. But if we assume that the ten-member jury must be unanimous, then the ten-member jury will give the 10% minority a chance to influence the verdict almost six times as frequently (65:11) as in the case of the twelve-member jury with majority rule.

No present proposals envisage combining size reduction and majority rule for federal juries. Yet the two jury-enfeebling measures do exist jointly in some of our state civil courts of limited jurisdiction, and there no longer appears to be any constitutional guarantee against such an extension even to federal criminal juries. This powerful combination has been brought to dramatic public attention by the special military court martial jury which tried Lieutenant Calley after the My Lai affair in Vietnam. Calley was tried on a criminal charge — a capital one at that — before a six-member jury authorized to reach a verdict by the agreement of only four jurors, a form which allows the majority to disregard a minority position as long as it does not have at least three representatives on the jury. In our hypothetical community with a 10% minority, fewer than 2 out of every 100 Calley-type juries will have more than two minority representatives if the juries are randomly selected from the population. Even if we assume a minority position that is held by 30% of the eligible jurors, only about every fourth Calley-type jury will effectively represent that minority. One might wonder why the men who drafted the rules for this type of court martial jury went to the extreme. Might one of their motives have been that such a jury, more than any other, could be expected to circumvent or conceal a disturbing minority position?

A significant characteristic of the Calley-type jury is its ability to hide the fact that the jury's findings may not have been unanimous; whether the verdict of the Calley jury was unanimous is still unknown. The formula for announcing the verdict reads merely; "Upon secret written ballot, two-thirds of the members present at the time the vote was taken. . . ." Only one voting constellation out of seven possible ones (from 6:0 to 0:6) results in a hung jury—the 3:3 constellation. All the others result in a verdict. The reason for the extraordinary length of the Calley jury's deliberation may have been its desire to achieve unanimity in a trial for a capital offense, even if that aspect of the verdict remained unpublished.

IV. CONCLUSION

We have shown that the change in verdicts that might be expected from the reduction of the twelve-member jury to six members is by no means negligible. We have also considered another potential modification of the federal jury, the majority verdict, and the possible combined application of both. However, the thrust of this article must not be misread. Its purpose is not to advocate any of the possible forms of federal juries. It pleads neither for the twelve-member jury nor for the smaller one, neither for the unanimity requirement nor for the majority verdict. All these solutions are possible, as is shown by the variety of rules adopted by our states. The legal systems of most countries do not have any jury trials; to be sure, their mode of selecting judges differs radically from ours.

The purpose of this article, rather, is simply to make clear that all these modifications make for differences in adjudication that appear to be negligible only to superficial scrutiny. Both in the short and in the long run our judicial system has many options, but every solution has its own balance sheet of advantages and costs. What is necessary is that we, and with us the United States Supreme Court, see both with equal clarity.

BALLEW v. GEORGIA
435 U.S. 223 (1977)

[By a five-person jury and judgment of the Criminal Court, Fulton County, Georgia, petitioner, Claude Davis Ballew, was convicted of distributing and exhibiting obscene materials, a film entitled "Behind the Green Door."— ED.)]

MR. JUSTICE BLACKMUN delivered the opinon of the Court. . . .

This case presents the issue whether a state criminal trial to a jury of only five persons deprives the accused of the right to trial by jury guaranteed to him by the Sixth and Fourteenth Amendments. Our resolution of the issue requires an application of principles enunciated in Williams v. Florida, 399 U.S. 78 (1970), where the use of a six-person jury in a state criminal trial was upheld against similar constitutional attack. . . .

II

The Fourteenth Amendment guarantees the right of trial by jury in all state nonpetty criminal cases. Duncan v. Louisiana, 391 U.S. 145, 159-162 (1968). The Court in *Duncan* applied this Sixth Amendment right to the States because "trial by jury in criminal cases is fundamental to the American scheme of justice." Id., at 149. The right attaches in the present case because the maximum penalty for violating §26-2101, as it existed at the time of the alleged offenses, exceeded six months' imprisonment.

In Williams v. Florida, 399 U.S., at 100, the Court reaffirmed that the "purpose of the jury trial, as we noted in *Duncan,* is to prevent oppression by the Government. 'Providing an accused with the right to be tried by a jury of his peers gave him an inestimable safeguard against the corrupt or overzealous prosecutor and against the compliant, biased, or eccentric judge.' Duncan v. Louisiana , [391 U.S.] at 156." This purpose is attained by the participation of the community in determinations of guilt and by the application of the common sense of laymen who, as jurors, consider the case. Williams v. Florida, 399 U.S., at 100.

Williams held that these functions and this purpose could be fulfilled by a jury of six members. As the Court's opinion in that case explained at some length, id., at 86-90, common-law juries included 12 members by historical accident, "unrelated to the great purposes which gave rise to the jury in the first place." Id., at 89-90. The Court's earlier cases that had *assumed* the number 12 to be constitutionally compelled were set to one side because they had not considered history and the function of the jury. Id., at 90-92. Rather than requiring 12 members, then, the Sixth Amendment mandated a jury only of sufficient size to promote group deliberation, to insulate members from outside intimidation, and to provide a representative cross-section of the community. Id., at 100. Although recognizing that by 1970 little empirical research had evaluated jury performance, the Court found no evidence that the reliability of jury verdicts diminished with six-member panels. Nor did the Court anticipate significant differences in result, including the frequency of "hung" juries. Id., at 101-102, and nn. 47 and 48. Because the reduction in size did not threaten exclusion of any particular class from jury roles, concern that the representative or cross-section character of the jury would suffer with a decrease to six members seemed "an unrealistic one." Id., at 102. As a consequence, the six-person jury was held not to violate the Sixth and Fourteenth Amendments.

III

When the Court in *Williams* permitted the reduction in jury size — or, to put it another way, when it held that a jury of six was not unconstitutional — it expressly reserved ruling on the issue whether a number smaller than six passed constitutional scrutiny. Id., at 91 n. 28. The Court refused to speculate when this so-called "slippery slope" would become too steep. We face now, however, the two-fold question whether a further reduction in the size of the state criminal trial jury does make the grade too dangerous, that is, whether it inhibits the functioning of the jury as an institution to a significant degree, and, if so, whether any state interest counterbalances and justifies the disruption so as to preserve its constitutionality.

Williams v. Florida and Colgrove v. Battin, 413 U.S. 149 (1973) (where the Court held that a jury of six members did not violate the Seventh Amendment right to a jury trial in a civil case), generated a quantity of scholarly work on jury size.[10] These writings do not draw or identify a bright line below which the number of jurors would not be able to function as required by the standards enunciated in *Williams*. On the other hand, they raise significant questions about the wisdom and constitutionality of a reduction below six. We examine these concerns:

First, recent empirical data suggest that progressively smaller juries are less likely to foster effective group deliberation. At some point, this decline leads to inaccurate factfinding and incorrect appli-

---

10. . . . Zeisel, . . . And Then There Were None: The Diminution of the Federal Jury, 38 U. Chi. L. Rev. 710 (1971). . . .

Some of these studies have been pressed upon us by the parties.

We have considered them carefully because they provide the only basis, besides judicial hunch, for a decision about whether smaller and smaller juries will be able to fulfill the purpose and functions of the Sixth Amendment. Without an examination about how juries and small groups actually work, we would not understand the basis for the conclusion of Mr. Justice Powell that "a line has to be drawn somewhere." We also note that the Chief Justice did not shrink from the use of empirical data in Williams v. Florida, 399 U.S. 78, 100-102, 105 (1970), when the data were used to support the constitutionality of the six-person criminal jury, or in Colgrove v. Battin, 413 U.S. 149, 158-160 (1973), a decision also joined by Mr. Justice Rehnquist.

cation of the common sense of the community to the facts. Generally, a positive correlation exists between group size and the quality of both group performance and group productivity. A variety of explanations have been offered for this conclusion. Several are particularly applicable in the jury setting. The smaller the group, the less likely are members to make critical contributions necessary for the solution of a given problem. Because most juries are not permitted to take notes, memory is important for accurate jury deliberations. As juries decrease in size, then, they are less likely to have members who remember each of the important pieces of evidence or argument. Furthermore, the smaller the group, the less likely it is to overcome the biases of its members to obtain an accurate result. When individual and group decisionmaking were compared, it was seen that groups performed better because prejudices of individuals were frequently counterbalanced, and objectivity resulted. Groups also exhibited increased motivation and self-criticism. All these advantages, except, perhaps, self-motivation, tend to diminish as the size of the group diminishes. Because juries frequently face complex problems laden with value choices, the benefits are important and should be retained. In particular, the counterbalancing of various biases is critical to the accurate application of the common sense of the community to the facts of any given case.

Second, the data now raise doubts about the accuracy of the results achieved by smaller and smaller panels. Statistical studies suggest that the risk of convicting an innocent person (Type I error) rises as the size of the jury diminishes. Because the risk of not convicting a guilty person (Type II error) increases with the size of the panel, an optimal jury size can be selected as a function of the interaction between the two risks. Nagel and Neef concluded that the optimal size, for the purpose of minimizing errors, should vary with the importance attached to the two types of mistakes. After weighting Type I error as 10 times more significant than Type II, perhaps not an unreasonable assumption, they concluded that the optimal jury size was between six and eight. As the size diminished to five and below, the weighted sum of errors increased because of the enlarging risk of the conviction of innocent defendants.

Another doubt about progressively smaller juries arises from the increasing inconsistency that results from the decreases. Saks argued that the "more a jury type fosters consistency, the greater will be the proportion of juries which select the correct (i.e., the same) verdict and the fewer 'errors' will be made." From his mock trials held before undergraduates and former jurors, he computed the percentage of "correct" decisions rendered by 12-person and 6-person panels. In the student experiment, 12-person groups reached correct verdicts 83% of the time; 6-person panels reached correct verdicts 69% of the time. The results for the former-juror study were 71% for the 12-person groups and 57% for the 6-person groups. Working with statistics described in H. Kalven & H. Zeisel, The American Jury 460 (1966), Nagel and Neef tested the average conviction propensity of juries, that is, the likelihood that any given jury of a set would convict the defendant. They found that half of all 12-person juries would have average conviction propensities that varied by no more than 20 points. Half of all six-person juries, on the other hand, had average conviction propensities varying by 30 points, a difference they found significant in both real and percentage terms. Lempert reached similar results when he considered the likelihood of juries to compromise over the various views of their members, an important phenomenon for the fulfillment of the commonsense function. In civil trials averaging occurs with respect to damages amounts. In criminal trials it relates to numbers of counts and lesser included offenses. And he predicted that compromises would be more consistent when larger juries were employed. For example, 12-person juries could be expected to reach extreme compromises in 4% of the cases, while 6-person panels would reach extreme results in 16%. All three of these post-*Williams* studies, therefore, raise significant doubts about the consistency and reliability of the decisions of smaller juries.

Third, the data suggest that the verdicts of jury deliberation in criminal cases will vary as juries become smaller, and that the variance amounts to an imbalance to the detriment of one side, the defense. Both Lempert and Zeisel found that the number of hung juries would diminish as the panels decreased in size. Zeisel said that the number would be cut in half — from 5% to 2.4% with a decrease from 12 to 6 members. Both studies emphasized that juries in criminal cases generally hang with only one, or more likely two, jurors remaining unconvinced of guilt. Also, group theory suggests that a person in the minority will adhere to his position more frequently when he has at least one other person supporting his argument. In the jury setting the significance of this tendency is demonstrated by the following figures: If a minority viewpoint is shared by 10% of the community, 28.2% of 12-member juries may be expected to have no minority representation, but 53.1% of 6-member juries would have none. Thirty-four percent of 12-member panels could be expected to have two minority members, while only 11% of 6-member panels would have two. As the numbers diminish below six, even fewer panels would have one member with the minority viewpoint and still fewer would have two. The chance for hung juries would decline accordingly.

Fourth, what has just been said about the presence of minority viewpoint as juries decrease in size foretells problems not only for jury decisionmaking, but also for the representation of minority groups in the community. The Court repeatedly has held that meaningful community participation cannot be attained with the exclusion of minorities or other identifiable groups from jury service. "It is part of the established tradition in the use of juries as instruments of public justice that the jury be a body truly representative of the community." Smith v. Texas, 311 U.S. 128, 130 (1940). The exclusion of elements of the community from participation "contravenes the very idea of a jury . . . composed of 'the peers or equals of the person whose rights it is selected or summoned to determine.'" Carter v. Jury Comm'n, 396 U.S. 320, 330 (1970), quoting Strauder v. West Virginia, 100 U.S. 303, 308 (1880). Although the Court in *Williams* concluded that the six-person jury did not fail to represent adequately a cross-section of the community, the opportunity for meaningful and appropriate representation does decrease with the size of the panels. Thus, if a minority group constitutes 10% of the community, 53.1% of randomly selected six-member juries could be expected to have no minority representative among their members, and 89% not to have two. Further reduction in size will erect additional barriers to representation.

Fifth, several authors have identified in jury research methodological problems tending to mask differences in the operation of smaller and larger juries. For example, because the judicial system handles so many clear cases, decisionmakers will reach similar results through similar analyses most of the time. One study concluded that smaller and larger juries could disagree in their verdicts in no more than 14% of the cases. Disparities, therefore, appear in only small percentages. Nationwide, however, these small percentages will represent a large number of cases. And it is with respect to those cases that the jury trial right has its greatest value. When the case is close, and the guilt or innocence of the defendant is not readily apparent, a properly functioning jury system will insure evaluation by the sense of the community and will also tend to insure accurate factfinding.

Studies that aggregate data also risk masking case-by-case differences in jury deliberations. The authors, H. Kalven and H. Zeisel, of The American Jury (1966), examined the judge-jury disagreement. They found that judges held for plaintiffs 57% of the time and that juries held for plaintiffs 59%, an insignificant difference. Yet case-by-case comparison revealed judge-jury disagreement in 22% of the cases. Id., at 63, cited in Lempert. This casts doubt on the conclusion of another study that compared the aggregate results of civil cases tried before 6-member juries with those of 12-member jury trials. The investigator in that study had claimed support for his hypothesis that damages awards did not vary with the reduction in jury size. Although some might say that figures in the aggregate may have supported this conclusion, a closer view of the cases reveals greater variation in the results of the smaller panels, i.e., a standard deviation of $58,335 for the 6-member juries, and of $24,834 for the 12-member juries. Again, the averages masked significant case-by-case differences that must be considered when evaluating jury function and performance.

IV

While we adhere to, and reaffirm our holding in Williams v. Florida, these studies, most of which have been made since *Williams* was decided in 1970, lead us to conclude that the purpose and functioning of the jury in a criminal trial is seriously impaired, and to a constitutional degree, by a reduction in size to below six members. We readily admit that we do not pretend to discern a clear line between six members and five. But the assembled data raise substantial doubt about the reliability and appropriate representation of panels smaller than six. Because of the fundamental importance of the jury trial to the American system of criminal justice, any further reduction that promotes inaccurate and possibly biased decisionmaking, that causes untoward differences in verdicts, and that prevents juries from truly representing their communities, attains constitutional significance. . . .

[T]he empirical data cited by Georgia do not relieve our doubts. The State relies on the Saks study for the proposition that a decline in the number of jurors will not affect the aggregate number of convictions or hung juries. This conclusion, however, is only one of several in the Saks study; that study eventually concludes:

> Larger juries (size twelve) are preferable to smaller juries (six). They produce longer deliberations, more communication, far better community representation, and, possibly, greater verdict reliability (consistency).

Saks 107.

Far from relieving our concerns, then, the Saks study supports the conclusion that further reduction in jury size threatens Sixth and Fourteenth Amendment interests.

Methodological problems prevent reliance on the three studies that do purport to bolster Georgia's position. The reliability of the two Michigan studies cited by the State has been criticized elsewhere. The critical problem with the Michigan laboratory experiment, which used a mock civil trial, was the apparent clarity of the case. Not one of the juries found for the plaintiff in the tort suit; this masked any potential difference in the decisionmaking of larger and smaller panels. The results also have been doubted because in the experiment only students composed the juries, only 16 juries were tested, and only a video tape of the mock trial was presented. The statistical review of the results of actual jury trials in Michigan erroneously aggregated outcomes. It is also said that it failed to take account of important changes of court procedure initiated at the time of the reduction in size from 12 to 6 members. The Davis study, which employed a mock criminal trial for rape, also presented an extreme set of facts so that none of the panels rendered a guilty verdict. None of these three reports, therefore, convinces us that a reduction in the number of jurors below six will not affect to a constitutional degree the functioning of juries in criminal trials. . . .

The judgment of the Court of Appeals is reversed, and the case is remanded for further proceedings not inconsistent with this opinion.

It is so ordered.

JUSTICE POWELL, concurring. . . . I have reservations as to the wisdom — as well as the necessity — of Mr. Justice Blackmun's heavy reliance on numerology derived from statistical studies. Moreover, neither the validity nor the methodology employed by the studies cited was subjected to the traditional testing mechanisms of the adversary process. The studies relied on merely represent unexamined findings of persons interested in the jury system.

D. BINOMIAL PROBABILITIES

In the preceding article, Hans Zeisel presents a number of statistical conclusions without explaining how they were calculated. For instance, he states that of the 100 12-member juries, approximately 72 would have at least one representative of the minority; while of the 100 six-member juries, only 47 would have one. How does he know that? Well, if he computed it by hand, which is unlikely, he would have used a complicated formula called the *binomial formula*, which can be found in many of the more theoretical statistics books. It is more likely that he used a table such as that which appears in Exhibit III-5, a table of binomial probabilities. This awesome-looking table

EXHIBIT III-5.
Binomial Probabilities

n	x	.05	.10	.15	.20	.25	.30	.35	.40	.45	.50
1	0	.9500	.9000	.8500	.8000	.7500	.7000	.6500	.6000	.5500	.5000
	1	.0500	.1000	.1500	.2000	.2500	.3000	.3500	.4000	.4500	.5000
2	0	.9025	.8100	.7225	.6400	.5625	.4900	.4225	.3600	.3025	.2500
	1	.0950	.1800	.2550	.3200	.3750	.4200	.4550	.4800	.4950	.5000
	2	.0025	.0100	.0225	.0400	.0625	.0900	.1225	.1600	.2025	.2500
3	0	.8574	.7290	.6141	.5120	.4219	.3430	.2746	.2160	.1664	.1250
	1	.1354	.2430	.3251	.3840	.4219	.4410	.4436	.4320	.4084	.3750
	2	.0071	.0270	.0574	.0960	.1406	.1890	.2389	.2880	.3341	.3750
	3	.0001	.0010	.0034	.0080	.0156	.0270	.0429	.0640	.0911	.1250
4	0	.8145	.6561	.5220	.4096	.3164	.2401	.1785	.1296	.0915	.0625
	1	.1715	.2916	.3685	.4096	.4219	.4116	.3845	.3456	.2995	.2500
	2	.0135	.0486	.0975	.1536	.2109	.2646	.3105	.3456	.3675	.3750
	3	.0005	.0036	.0115	.0256	.0469	.0756	.1115	.1536	.2005	.2500
	4	.0000	.0001	.0005	.0016	.0039	.0081	.0150	.0256	.0410	.0625
5	0	.7738	.5905	.4437	.3277	.2373	.1681	.1160	.0778	.0503	.0312
	1	.2036	.3280	.3915	.4096	.3955	.3602	.3124	.2592	.2059	.1562
	2	.0214	.0729	.1382	.2048	.2637	.3087	.3364	.3456	.3369	.3125
	3	.0011	.0081	.0244	.0512	.0879	.1323	.1811	.2304	.2757	.3125
	4	.0000	.0004	.0022	.0064	.0146	.0284	.0488	.0768	.1128	.1562
	5	.0000	.0000	.0001	.0003	.0010	.0024	.0053	.0102	.0185	.0312

n	x										
6	0	.7351	.5314	.3771	.2621	.1780	.1176	.0754	.0467	.0277	.0156
	1	.2321	.3543	.3993	.3932	.3560	.3025	.2437	.1866	.1359	.0938
	2	.0305	.0984	.1762	.2458	.2966	.3241	.3280	.3110	.2780	.2344
	3	.0021	.0146	.0415	.0819	.1318	.1852	.2355	.2765	.3032	.3125
	4	.0001	.0012	.0055	.0154	.0330	.0595	.0951	.1382	.1861	.2344
	5	.0000	.0001	.0004	.0015	.0044	.0102	.0205	.0369	.0609	.0938
	6	.0000	.0000	.0000	.0001	.0002	.0007	.0018	.0041	.0083	.0156
7	0	.6983	.4783	.3206	.2097	.1335	.0824	.0490	.0280	.0152	.0078
	1	.2573	.3720	.3960	.3670	.3115	.2471	.1848	.1306	.0872	.0547
	2	.0406	.1240	.2097	.2753	.3115	.3177	.2985	.2613	.2140	.1641
	3	.0036	.0230	.0617	.1147	.1730	.2269	.2679	.2903	.2918	.2734
	4	.0002	.0026	.0109	.0287	.0577	.0972	.1442	.1935	.2388	.2734
	5	.0000	.0002	.0012	.0043	.0115	.0250	.0466	.0774	.1172	.1641
	6	.0000	.0000	.0001	.0004	.0013	.0036	.0084	.0172	.0320	.0547
	7	.0000	.0000	.0000	.0000	.0001	.0002	.0006	.0016	.0037	.0078
8	0	.6634	.4305	.2725	.1678	.1001	.0576	.0319	.0168	.0084	.0039
	1	.2793	.3826	.3847	.3355	.2670	.1977	.1373	.0896	.0548	.0312
	2	.0515	.1488	.2376	.2936	.3115	.2965	.2587	.2090	.1569	.1094
	3	.0054	.0331	.0839	.1468	.2076	.2541	.2786	.2787	.2568	.2188
	4	.0004	.0046	.0185	.0459	.0865	.1361	.1875	.2322	.2627	.2734
	5	.0000	.0004	.0026	.0092	.0231	.0467	.0808	.1239	.1719	.2188
	6	.0000	.0000	.0002	.0011	.0038	.0100	.0217	.0413	.0703	.1094
	7	.0000	.0000	.0000	.0001	.0004	.0012	.0033	.0079	.0164	.0312
	8	.0000	.0000	.0000	.0000	.0000	.0001	.0002	.0007	.0017	.0039

(continued)

EXHIBIT III-5.
Binomial Probabilities (continued)

n	x	.05	.10	.15	.20	.25	P .30	.35	.40	.45	.50
9	0	.6302	.3874	.2316	.1342	.0751	.0404	.0207	.0101	.0046	.0020
	1	.2985	.3874	.3679	.3020	.2253	.1556	.1004	.0605	.0339	.0176
	2	.0629	.1722	.2597	.3020	.3003	.2668	.2162	.1612	.1110	.0703
	3	.0077	.0446	.1069	.1762	.2336	.2668	.2716	.2508	.2119	.1641
	4	.0006	.0074	.0283	.0661	.1168	.1715	.2194	.2508	.2600	.2461
	5	.0000	.0008	.0050	.0165	.0389	.0735	.1181	.1672	.2128	.2461
	6	.0000	.0001	.0006	.0028	.0087	.0210	.0424	.0743	.1160	.1641
	7	.0000	.0000	.0000	.0003	.0012	.0039	.0098	.0212	.0407	.0703
	8	.0000	.0000	.0000	.0000	.0001	.0004	.0013	.0035	.0083	.0176
	9	.0000	.0000	.0000	.0000	.0000	.0000	.0001	.0003	.0008	.0020
10	0	.5987	.3487	.1969	.1074	.0563	.0282	.0135	.0060	.0025	.0010
	1	.3151	.3874	.3474	.2684	.1877	.1211	.0725	.0403	.0207	.0098
	2	.0746	.1937	.2759	.3020	.2816	.2335	.1757	.1209	.0763	.0439
	3	.0105	.0574	.1298	.2013	.2503	.2668	.2522	.2150	.1665	.1172
	4	.0010	.0112	.0401	.0881	.1460	.2001	.2377	.2508	.2384	.2051
	5	.0001	.0015	.0085	.0264	.0584	.1029	.1536	.2007	.2340	.2461
	6	.0000	.0001	.0012	.0055	.0162	.0368	.0689	.1115	.1596	.2051
	7	.0000	.0000	.0001	.0008	.0031	.0090	.0212	.0425	.0746	.1172
	8	.0000	.0000	.0000	.0001	.0004	.0014	.0043	.0106	.0229	.0439
	9	.0000	.0000	.0000	.0000	.0000	.0001	.0005	.0016	.0042	.0098
	10	.0000	.0000	.0000	.0000	.0000	.0000	.0000	.0001	.0003	.0010

n	k										
11	0	.5688	.3138	.1673	.0859	.0422	.0198	.0088	.0036	.0014	.0004
	1	.3293	.3835	.3248	.2362	.1549	.0932	.0518	.0266	.0125	.0055
	2	.0867	.2131	.2866	.2953	.2581	.1998	.1395	.0887	.0513	.0269
	3	.0137	.0710	.1517	.2215	.2581	.2568	.2254	.1774	.1259	.0806
	4	.0014	.0158	.0536	.1107	.1721	.2201	.2428	.2365	.2060	.1611
	5	.0001	.0025	.0132	.0388	.0803	.1321	.1830	.2207	.2360	.2256
	6	.0000	.0003	.0023	.0097	.0268	.0566	.0985	.1471	.1931	.2256
	7	.0000	.0000	.0003	.0017	.0064	.0173	.0379	.0701	.1128	.1611
	8	.0000	.0000	.0000	.0002	.0011	.0037	.0102	.0234	.0462	.0806
	9	.0000	.0000	.0000	.0000	.0001	.0005	.0018	.0052	.0126	.0269
	10	.0000	.0000	.0000	.0000	.0000	.0000	.0002	.0007	.0021	.0054
	11	.0000	.0000	.0000	.0000	.0000	.0000	.0000	.0000	.0002	.0005
12	0	.5404	.2824	.1422	.0687	.0317	.0138	.0057	.0022	.0008	.0002
	1	.3413	.3766	.3012	.2062	.1267	.0712	.0368	.0174	.0075	.0029
	2	.0988	.2301	.2924	.2835	.2323	.1678	.1088	.0639	.0339	.0161
	3	.0173	.0852	.1720	.2362	.2581	.2397	.1954	.1419	.0923	.0537
	4	.0021	.0213	.0683	.1329	.1936	.2311	.2367	.2128	.1700	.1208
	5	.0002	.0038	.0193	.0532	.1032	.1585	.2039	.2270	.2225	.1934
	6	.0000	.0005	.0040	.0155	.0401	.0792	.1281	.1766	.2124	.2256
	7	.0000	.0000	.0006	.0033	.0115	.0291	.0591	.1009	.1489	.1934
	8	.0000	.0000	.0001	.0005	.0024	.0078	.0199	.0420	.0762	.1208
	9	.0000	.0000	.0000	.0001	.0004	.0015	.0048	.0125	.0277	.0537
	10	.0000	.0000	.0000	.0000	.0000	.0002	.0008	.0025	.0068	.0161
	11	.0000	.0000	.0000	.0000	.0000	.0000	.0001	.0003	.0010	.0029
	12	.0000	.0000	.0000	.0000	.0000	.0000	.0000	.0000	.0001	.0002

Source: CRC Handbook of Tables for Probability and Statistics 183-184 (W. Beyer ed., 2d ed. 1968). Copyright © 1968 by The Chemical Rubber Co., CRC Press, Inc. Reprinted with permission.

can be used to establish the probability of an event's occuring whenever there are only two possible outcomes, e.g., either a minority person is selected for jury duty or a majority member is selected.[3]

Here is how it works. The top row on each side of the table is labeled P. The symbol P stands for the percentage of your community that comprises minority members (or, in general, the percentage of the time a particular event occurs). The percentages given in this table are 5%, 10%, 15%, 20%, 25%, 30%, 35%, 40%, 45%, and 50%. The symbol n, on the left-hand column, stands for the number of people in the jury (or, in general, the number of times an experiment, such as choosing a jury member, is repeated, e.g., six times for a six-person jury). Binomial tables go to much higher numbers than 12, but that is the maximum jury size considered here. The symbol x, also on the left-hand column, measures the number of minority group members you want to have in the jury (in general, the number of occurrences of one of the two possible events). The figures inside the table indicate, for any combination of percentage minority (P), jury size (n), and number of minority members (x), the probability of getting that many members.

Zeisel assumes a 10% minority population, so look at the ".10" column. Go down that column to jury size "6" and look at the probability of having no (i.e., zero) minorities on a jury. The table shows .5314, which indicates a 53% chance of having no minorities. That means that there is a 47% chance (100% − 53% = 47%) of having at least one minority member on the jury. If you add them up, you will find that this probability is the same as the total of the probabilities of having either one, two, three, four, five, or six minority members. You can follow the same procedure to find the chance of having at least one minority member on a 12-person jury, i.e.:

$$.3766 + .2301 + .0852 + .0213 + .0038 + .0005 =$$
$$1.00 - .2824 = .7176$$

that is, approximately a 72% chance of having at least one minority member. If you choose a large number of 12-person juries, you can expect 72% of them to have at least one minority member and 28% to have no minority members. These statistics show that the smaller the jury size, the lower the likelihood of having at least one minority member.

---

3. Because there are more than two values for the observations in Zeisel's Table 2 (six values actually: one for each dollar evaluation), the probabilities in Table 2 of Zeisel's article cannot be calculated by using the binomial probabilities table. The calculations involved there are not relevant here, though the conclusions drawn from that table are of practical importance.

If having a diverse jury is our goal, we can determine how certain we want to be that every minority is represented and, by reference to such a table, select a jury big enough that its chances of not having the minorities represented is acceptably small. A 20-person jury has an 88% chance of having at least one member of a 10% minority.

Even if you don't care about juries, a binomial table can be useful. Imagine that for every A you get in a law school course, your expected future yearly income rises by $10,000. If you take seven courses in your first year ($n = 7$), and if 15% of the class gets A's ($P = .15$), and if your chances are as good as anyone else's (due to the arbitrariness of professorial grading), then you can figure your expected future yearly income by reference to this table. The likelihood of your making $10,000 a year more than the poor slob with no A's is 39.6%, $20,000 a year more is 21%, $30,000 a year more is 6%, $40,000 more is 1%, and $50,000 more is .12%. The chances of your making $70,000 a year more are negligible, so if this monetary difference isn't worth the extra effort, don't bother.

Notes and Problems

1. If you interview with the public defender's office in a given city, your chance of getting a job is 30%. What is the chance of your getting *at least one* job offer if you fly to six cities? The probability of a job in each is 30%, so $P = .30$. The repetitions of the experiment, i.e., the number of cities visited, is six, so $n = 6$. The number of desired outcomes is either one, two, three, four, five, or six, because we want to know the probability of getting *at least one* job offer. What is it? An alternative method is to see what is the probability of getting no job offers, $x = 0$. Subtracting this probability from 100% gives the same answer.

2. As an exercise, consider the following problems. How many of the cities in Note 1 must you visit so that your chances of getting at least one job offer are 50%? How many to get a 50% chance of a choice of jobs?

3. Professor Zeisel's imaginary two-group community may be less complicated than a real city, which has many minority groups. If we desire a representative jury, we must ask two questions before deciding what size jury to have. They are: (1) how small a group is important enough to worry about? and (2) what percentage of juries must have at least one representative of this group? Consider a town with a population that is 35% Republicans, 39% Democrats, 20% Anarchists, 5% Satanists, and 1% apolitical, irreligious statisticians. If we

decide that, for Sixth Amendment purposes, a minority group smaller than 5% need not be represented at all, and that a minority of 5% or greater should be represented one-fourth of the time, then we can use the binomial table to indicate the smallest acceptable jury. You might say at this point, "Wait a minute. I thought the binomial table could be used only when there were just two categories of people." Quite right. But we can arrange this problem so it can be handled with only two categories. After all, we can forget the apolitical, irreligious statisticians because they are only a 1% group. And we need not worry about the Anarchists, because if juries are large enough to represent a much smaller minority (the Satanists), then the Anarchists will be adequately represented. So we need look only to the smallest group with which we are concerned, the 5% Satanists. This gives us two groups, Satanists and others.

To use the binomial table, proceed as follows:

Step 1. *Decide what it is we want to know.* Here the question is, what is the smallest jury size that will have at least one representative Satanist 25% of the time? This means that for every jury size we look at, we must look at the probability of having one or more than one Satanist juror.

Step 2. *Decide what P is.* Here, P, the probability of our event occurring, a Satanist being randomly chosen from the jury pool, is 5%, because Satanists are 5% of the population.

Step 3. *Decide what x is.* Here, x is the number of times our measured event, a Satanist being selected, occurs, i.e., $x = 1$, then 2, then 3, . . . , up to the total number of jurors in each jury size considered.

Step 4. *Decide what n is.* Here, n, the number of times an experiment (the drawing of a juror's name) is repeated, stands for the jury size. And that is what we need to calculate; n is the unknown.

Step 5. *Find the unknown.* Go down the $P = .05$ column until you find a jury size, n, where the sum of the probabilities of $x = 1$, then 2, then 3, . . . , up to n, is equal to or greater than 25%. You will see that a jury size of six is the smallest acceptable jury. For when $P = .05$ and $n = 6$, the probability of one Satanist juror, $x = 1$, is .2321, of 2 is .0305, of 3 is .0021, of 4 is .0001, of 5 is .0000, and of 6 is .0000. Adding these gives .2749 or 27.49%. A jury of five is too small because the minority representation occurs only on 22.62% (.2036 + .0214 + .0011 + .0000 + .0000 = .2262) of the juries. A jury of seven is unnecessarily large for our purposes.

We know from previous discussions that the probability of having one or more of a minority is the same as 100% minus the probability of having zero. So an easier way to do this same calculation is to go down the $P = .05$ column until you find an $x = 0$ where the probability is less than .75. For a jury of size $n = 1$ and $p = .05$, the probability of getting no Satanists, $x = 0$, is .9500 or 95%. For juries of sizes $n = 2, 3, 4, 5$, and 6 and $p = .05$, the probabilities of $x = 0$ are .9025, .8574, .8145, .7738 and .7381, respectively. A jury of size six is therefore the smallest acceptable jury. Juries without Satanists will occur 73.51% of the time, so Satanists will be represented 27.49% (100% − 73.51%) of the time. A jury of five is too small because this minority representation occurs only 22.62% (100% − 77.38%) of the time. Again, a jury of seven is unnecessarily large.

4. As an exercise, determine the smallest jury size needed to represent the Anarchists 90% of the time.

5. For 70 years, your client has been selling petunias to inner-city stores at twice the price he sells to suburban stores. These petunias are identical, and there is no cost justification for the price differential. The Federal Trade Commission has finally caught on to this practice and decided to prosecute, unless your client agrees to plant petunias free of charge in all the city parks next year. Conviction would mean great humiliation to your client, and he will settle if the chances of conviction are greater than 10%. What do you advise?

While the hypothetical is a little strange, the situation is not at all unusual. A trial lawyer needs to know how a jury is likely to decide before advising a client on such a matter. Your survey of jurors reveals a community of 35% vegephiles, who believe that buying and selling plants is immoral and are certain to convict your client. The other 65% are geriophiles, who would never convict anyone over 60. If nine-person juries are the rule in your jurisdiction and they decide by majority rule, how do you advise your client? Remember first to decide what it is you want to know. Here, what is the chance of getting five or more vegephiles on the jury? Then decide what p, n, and x are.

6. If you practice in a jurisdiction with a majority-rule, nine-person jury, made up 10% of people who favor the death penalty and 90% of those who do not, how do you advise your client, who is sure to be found guilty of first-degree murder if the case goes to trial but who will plead guilty to a lesser charge if the chance of getting the death penalty is greater than 5%? Assume that the jury is empowered both to render the verdict and determine the punishment.

7. A fellow attorney seeks your assistance on five cases she currently has in various stages of litigation. She offers to pay you $1,000.00 per case that she wins or a lump sum of $2,500.00 for all five cases regardless of her success. If the probability of winning any given case is 40%, which fee arrangement would you select?

8. If the probability of passing the bar exam in any given state is 50%, in how many states will you have to take the exam to be at least 90% certain of admittance to at least one bar?

CHAPTER IV

STANDARD DEVIATIONS

In the previous chapter, we encountered the notion of standard deviation, probably the single most important statistical concept used in legal cases. Professor Zeisel used this concept to show how much the outcomes of alternative jury compositions vary from the typical value for all juries. He defined the standard deviation as the square root of all squared deviations from the group average. (See page 53 above). In order to understand the utility of this mathematical expression for proving legal claims, we must first be more precise about the group average.

A. GROUP AVERAGES

There are two commonly accepted ways of describing the group average. The first measure of group average is the median. The *median* of a group of numbers is the middle number, when all the numbers are arranged in order of size. Consider two lists of numbers:

(1) 4.0, 3.5, 3.2, 2.7, 1.0, and
(2) 3.7, 3.1, 2.4, 2.3.

These numbers are arranged in order of size, and for list (1), 3.2 is obviously the middle, i.e., median, number. There are always an equal number of larger and smaller numbers above and below the median. For list (2) or for any list with an even number of entries, there are two middle numbers, 3.1 and 2.4. The median of such a group is the *mean* or simple arithmetic average of the two middle numbers; it is calculated in this case by adding the two numbers together and dividing by two, i.e.:

$$\frac{(3.1 + 2.4)}{2} = \frac{5.5}{2} = 2.75$$

This simple arithmetic average of the two middle numbers is the median for list (2).

The *mean* of a group of numbers, then, is the simple arithmetic average of all numbers in the group. It is calculated by adding all the individual numbers in the group and dividing by the number of individual numbers. For list (1) the mean is:

$$\frac{(4.0 + 3.5 + 3.2 + 2.7 + 1.0)}{5} = \frac{14.4}{5} = 2.88$$

For list (2) the mean is:

$$\frac{(3.7 + 3.1 + 2.4 + 2.3)}{4} = \frac{11.5}{4} = 2.875$$

The mean may be either larger or smaller than the median. In some cases, the mean may even equal the median, as in Professor Zeisel's Table 1 (see page 52 above). Because the median and mean reflect slightly different characteristics of the group of numbers, the coincidence of equality is of no particular significance.

B. STATISTICAL NOTATION

In order to make statistical calculations easier, it is helpful first to learn some shorthand language. Three symbols are relevant in calculating standard deviations: M, X_i, and Σ. M stands for the mean, which has already been explained. Now, when we subtract the mean from something, we can write "something $- M$," a shorthand that saves time and ink. X_i (read X sub i) stands for each of the numbers in a list. Consider another list of numbers:

(3) 9, 6, 7, 2, 2, 6, 3

Let us call this list the X list and give each entry a name as follows:

Give 9 the name X_1 (read X sub one),

Give 6 (the first 6) the name X_2,
Give 7 the name X_3,
Give 2 (the first 2) the name X_4,
Give 2 (the second 2) the name X_5,
Give 6 (the second 6) the name X_6, and
Give 3 the name X_7.

The "sub 1," "sub 2," etc., refer to the location on the list. We can refer to the number 7 by referring to its place on the list, i.e., X_3, the third number on the list. X_i stands for the ith number on the list, so when $i = 4$, X_i stands for the number 2.

We use the X_i shorthand in combination with Σ (called *summation* or *sigma*). Having previously denoted list (3) as the X list, Σ (9 + 6 + 7 + 2 + 2 + 6 + 3) means "add all the numbers in the X list." Since X_i stands for the various numbers on the X list, we can, in shorthand, write ΣX_i, instead of $\Sigma(9 + 6 + 7 + 2 + 2 + 6 + 3)$. Since X_i refers to this particular list, we know that ΣX_i equals 35.

If we refer to list (2) as the Y list, what is ΣY_i? The answer is 11.5.

C. VARIABILITY

Table 2 in Professor Zeisel's article demonstrated that the six-member juries show a considerably wider variation of verdicts than the 12-member juries, i.e., more six-member juries reached verdicts far away from the mean of $3500 than did 12-member juries. This variation in results can be measured in a number of ways. One is to compute the *range,* the distance between the largest and smallest numbers. For list (3), above, the range is from 2 to 9, a spread of 7. Consider an additional list:

(4) 9, 9, 9, 9, 9, 9, 2

For list (4) the range is still from 2 to 9, although one can see that there are more different numbers in list (3), namely, 9, 6, 7, 2, 2, 6, 3.

A way to take into account all the different numbers in the list is to calculate the *variance,* a statistical measure denoting not only the range but also the variability within the range. The variance uses the mean as a standard of comparison to see how far each number deviates from the mean. For mathematical reasons into which we needn't delve,[1] the deviations from the mean for each number are squared[2] as the first step in calculating the variance.

A deviation from the mean for any item from the X list is $X_i - M$ (subtract the mean from each number). The squared deviation from the mean for any number is $(X_i - M)^2$. This means: subtract the mean from each number and then square the result. The statistic called the *variance* summarizes the squared deviations from the mean by averaging them, i.e., finding the mean of the squared deviations. We know the mean of anything is:

$$\frac{\Sigma \text{ (the items in ``anything'')}}{\text{the number of items}}$$

If we refer to the number of items (more formally, the number of observations) on our list as N, then the variance of the X list is:

$$\frac{\Sigma (X_i - M)^2}{N}$$

We square the deviation of each value from the mean, then add the squared deviations, then divide the result by the number of values in the X list.

For example, we can compute the variance for list (3) as follows:

Step 1. Find the mean, M of the items, i.e.:

$$\frac{\Sigma X_i}{N} = \frac{(9 + 6 + 7 + 2 + 2 + 6 + 3)}{7} = \frac{35}{7} = 5$$

Step 2. Find the deviation from the mean, X_i minus M, for each item, i.e.:

for
$X_1, X_1 - M = 9 - 5 = 4$
$X_2, X_2 - M = 6 - 5 = 1$
$X_3, X_3 - M = 7 - 5 = 2$
$X_4, X_4 - M = 2 - 5 = -3$
$X_5, X_5 - M = 2 - 5 = -3$
$X_6, X_6 - M = 6 - 5 = 1$
$X_7, X_7 - M = 3 - 5 = -2$

---

1. Two reasons are (1) to emphasize large variations from the mean and (2) to eliminate the minus signs.

2. Recall that the square of a number is the result obtained from multiplying that number by itself, e.g., 9 squared (denoted 9^2) is 81. The superscript 2 denotes the square of a number, $9^2 = 9 \times 9$. Similarly, $(-3)^2 = (-3) \times (-3) = 9$, and $9^4 = 9 \times 9 \times 9 \times 9$.

Step 3. Square each deviation, i.e.:

for X_1, $4^2 = 16$
X_2, $1^2 = 1$
X_3, $2^2 = 4$
X_4, $(-3)^2 = 9$
X_5, $(-3)^2 = 9$
X_6, $1^2 = 1$
X_7, $(-2)^2 = 4$

Step 4. Sum the squared deviations and divide by N to find the mean of the squared deviations, i.e.:

$$\frac{\Sigma (X_i - M)^2}{N} = \frac{\Sigma (16 + 1 + 4 + 9 + 9 + 1 + 4)}{7} = \frac{44}{7} = 6^2/_7$$

The variance of the items of list (3) is $6^2/_7$. The variance is denoted by the symbol SD^2.

D. CALCULATING A STANDARD DEVIATION

The concept of variance is an important one in statistics. Once you have calculated the variance for a list of numbers (generally, for a number of observations of the population you are examining), there are a number of other useful statistics you can calculate. A problem with the variance is that the number you calculate does not seem to have any real connection with the original list of numbers, particularly because you have squared all the deviations from the mean. Squaring seems somewhat arbitrary and so inflates many values that the variance may be larger than many of the numbers in the list. For instance, $6^2/_7$ (the variance of list (3)) does not seem to bear any obvious relationship to the numbers in list (3), which were 9, 6, 7, 2, 2, 6, and 3. A more impressive example is the variance of the values in list (5):

(5)   412, 52, 59, 70, 80

If you follow the four steps for calculating the variance of this list, you will find SD^2 equals 19,328.64.

The *standard deviation* is a more understandable number and statistical concept. It is also the concept relied on most heavily by the largest variety of courts and an indispensable element of proof in many cases. The term *standard deviation* is somewhat self-explanatory, although it could also be called "typical," "usual," or "average" deviation. One measure of how much variation there is in a sample (a list of numbers measuring some characteristic of the population in question) is how much the typical number on the list varies (deviates) from the mean. This typical deviation sets a standard by which other deviations are judged. Using this concept, we can see how deviant a particular number (observation) is from the "expected" value, i.e., from the mean. This concept may have had its first uses related to law in jury-discrimination cases. Our observations in such cases focus on the number of minority members on each jury. We compare the number on a given jury to the expected number (from the percentage of minority members in the community). If the actual number *deviates widely* from the expected number, we might suspect some discrimination. The legal problem is to determine what legally constitutes a *wide* deviation. The courts' answers have been to compare the observed deviation to the standard deviation.

In order to find this single number, which is representative of the deviations from the mean found on a particular list (i.e., in a given "population" of numbers), simply take the square root of the variance. This makes intuitive sense. The variance was a number that seemed somehow to have been blown out of proportion by the squaring process. Taking the square root brings the number calculated as the variance back to scale. The standard deviation is the square root of the variance and is denoted SD. The variance is:

$$SD^2 = \frac{\Sigma (X_i - M)^2}{N}$$

So the standard deviation is:

$$SD = \sqrt{SD^2} = \sqrt{\frac{\Sigma (X_i - M)^2}{N}}$$

Once you calculate the variance, just take its square root. Most calculators, even inexpensive ones, find square roots at the touch of a button. The variance for list (3) is $6^2/7$ (or 6.28, the decimal equivalent), and the square root is $\sqrt{6.28}$ or 2.51. This is the standard deviation.

E. A SPECIAL CASE: THE BINOMIAL DISTRIBUTION

Lists (1) through (5), used above, are meant to be examples of measurements one might make in observing a population such as a group of people. List (1), 4.0, 3.5, 3.2, 2.7, 1.0, for instance, might be the grade-point averages of the five people sitting in front of you. List (3), 9, 6, 7, 2, 2, 6, 3 might measure the number of times each of your friends has skipped an afternoon class to go drink beer. List (5), 412, 52, 59, 70, 80, might measure the number of times each day last week that you wished you had gotten a job instead of going to school.

Note that the items on these lists can take a wide range of values. On other lists, however, the possible values of the items may be limited. For instance, a list of the sex of each class member can only have one of two values, M or F. To give a numerical value to the sex variable, we might assign females the number 1 and males the number 0. Then, for a small class, the list might look like this:

(6) 1, 0, 0, 1, 1, 0, 1, 1, 1, 1, 0

where the values of the items are restricted to the values 1 and 0.

Whenever the items observed can only have two values, the items have what is referred to as a *binomial distribution*. You need not worry about what the term signifies, only about its practical implications. One is that we can use the binomial probability table discussed in the previous chapter to calculate the likelihood that certain observations would occur. Another implication is that we can use a simplified method for calculating the standard deviation.

The standard deviation represents the typical deviation from the mean or expected value for a population (or list of numbers). Consider list (6), which had only two possible values for the observation. The list may measure the occurrence in the population of a male or female, a minority or majority person, a professor or student, or a lawyer or nonlawyer. If females, majority members, students, or non-lawyers are assigned the value 1, while the males, minorities, professors, or lawyers are assigned the value 0, then the expected value, the mean, of such a list is equal to the percentage of females, majority members, students, or lawyers in the population.

For instance, if a community is 10% minority members and if majority members are assigned the value 1, then the mean of a list of values for the people in the community would be 0.9. This indicates, for instance, that for a community of 12,000 people, 1200 (10% of

12,000) is the expected number of minority people, while 10,800 (90% of 12,000) is the expected number of majority people.

These mean values are particularly easy to calculate for a binomial distribution. The standard deviation is equally straightforward, though the reason for the method of calculation is not intuitively obvious. Don't worry about it. For a binomial distribution, i.e., a population which can have only two values, 1 and 0, the standard deviation is equal to the square root of three figures multiplied together. The figures are (1) the total size of the population with which we are concerned, (2) the percentage of people in the first group, e.g., majority members or females or nonlawyers, and (3) the percentage of people not in the first group. If we let N stand for the total number of people in the group we are examining, and P_1 stand for the percentage of people in the first group, and P_2 stand for the percentage not in the first group, we get the following formula for the standard deviation of a population with a binomial distribution:

$$SD = \sqrt{N \times P_1 \times P_2}$$

Castaneda v. Partida is a leading case using standard deviations.

F. STANDARD DEVIATIONS AS STATISTICAL PROOF

CASTANEDA v. PARTIDA
430 U.S. 483 (1976)

Mr. Justice Blackmun delivered the opinion of the Court.

The sole issue presented in this case is whether the State of Texas, in the person of petitioner, the Sheriff of Hidalgo County, successfully rebutted respondent prisoner's prima facie showing of discrimination against Mexican-Americans in the state grand jury selection process.[1] In his brief, petitioner, in claiming effective rebuttal, asserts:

---

1. Under Texas' "key man" system for selecting grand juries, jury commissioners are appointed by a state district judge to select prospective jurors from different portions of the county, after which the district judge proceeds to test their qualifications. A grand juror, in addition to being a citizen of the State and of the county in which he is to serve and a qualified voter in the county, must be "of sound mind and good moral character," be literate, have no prior felony conviction, and be under no pending indictment or other accusation.

F. Standard Deviations as Statistical Proof

This list [of the grand jurors that indicated respondent] indicates that 50 percent of the names appearing thereon were Spanish. The record indicates that 3 of the 5 jury commissioners, 5 of the grand jurors who returned the indictment, 7 of the petit jurors, the judge presiding at the trial, and the Sheriff who served notice on the grand jurors to appear had Spanish surnames. . . .

Respondent, Rodrigo Partida, was indicted in March 1972 by the grand jury of the 92d District Court of Hidalgo County for the crime of burglary of a private residence at night with intent to rape. Hidalgo is one of the border counties of southern Texas. After a trial before a petit jury, respondent was convicted and sentenced to eight years in the custody of the Texas Department of Corrections. He first raised his claim of discrimination in the grand jury selection process on a motion for new trial in the State District Court. In support of his motion, respondent testified about the general existence of discrimination against Mexican-Americans in that area of Texas and introduced statistics from the 1970 census and the Hidalgo County grand jury records. The census figures show that in 1970, the population of Hidalgo County was 181,535. United States Bureau of the Census, 1970 Census of Population, Characteristics of the Population, vol. 1, pt. 45, §1, Table 119, p. 914. Persons of Spanish language or Spanish surname totaled 143,611; Ibid., and id., Table 129, p. 1092. On the assumption that all the persons of Spanish language or Spanish surname were Mexican-Americans, these figures show that 79.1% of the county's population was Mexican-American.

Respondent's data compiled from the Hidalgo County grand jury records from 1962 to 1972 showed that over that period, the average percentage of Spanish-surnamed grand jurors was 39%. [The statistics for grand jury composition can be organized as follows:

Year	No. Persons on Grand Jury List	Av. No. Spanish Surnamed per List	Percentage Spanish Surnamed
1962	16	6	37.5%
1963	16	5.75	35.9%
1964	16	4.75	29.7%
1965	16.2	5	30.9%
1966	20	7.5	37.5%
1967	20.25	7.25	35.8%
1968	20	6.6	33%
1969	20	10	50%
1970	20	8	40%
1971	20	9.4	47%
1972	20	10.5	52.5%

Of the 870 persons who were summoned to serve as grand jurors over the 11-year period, 339, or 39%, were Spanish surnamed. (fn. 7)] In the 2½-year period during which the District Judge who impaneled the jury that indicted respondent was in charge, the average percentage was 45.5%. On the list from which the grand jury that indicted respondent was selected, 50% were Spanish surnamed. The last set of data that respondent introduced, again from the 1970 census, illustrated a number of ways in which Mexican-Americans tend to be underprivileged, including poverty-level incomes, less desirable jobs, substandard housing, and lower levels of education. The State offered no evidence at all either attacking respondent's allegations of discrimination or demonstrating that his statistics were unreliable in any way. . . .

. . . [I]n order to show that an equal protection violation has occurred in the context of grand jury selection, the defendant must show that the procedure employed resulted in substantial underrepresentation of his race or of the identifiable group to which he belongs. The first step is to establish that the group is one that is a recognizable, distinct class, singled out for different treatment under the laws, as written or as applied. Hernandez v. Texas, 347 U.S., at 478-479. Next, the degree of underrepresentation must be proved, by comparing the proportion of the group in the total population to the proportion called to serve as grand jurors, over a significant period of time. Id., at 480. This method of proof, sometimes called the "rule of exclusion," has been held to be available as a method of proving discrimination in jury selection against a delineated class. Hernandez v. Texas, 347 U.S., at 480. Finally, as noted above, a selection procedure that is susceptible of abuse or is not racially neutral supports the presumption of discrimination raised by the statistical showing. Washington v. Davis, 426 U.S., at 241; Alexander v. Louisiana, 405 U.S., at 630. Once the defendant has shown substantial underrepresentation of his group, he has made out a prima facie case of discriminatory purpose, and the burden then shifts to the State to rebut that case.

B. In this case, it is no longer open to dispute that Mexican-Americans are a clearly identifiable class. . . .

The disparity proved by the 1970 census statistics showed that the population of the county was 79.1% Mexican-American, but that, over an 11-year period, only 39% of the persons summoned for grand jury service were Mexican-American. This difference of 40% is greater than that found significant in Turner v. Fouche, 396 U.S. 346 (1970) (60% Negroes in the general population, 37% on the grand jury lists). Since the State presented no evidence showing why the 11-year period was not reliable, we take it as the relevant base for comparison. The mathematical disparities that have been accepted by this Court as adequate for a prima facie case have all been within the range presented here. For example, in Whitus v. Georgia, 385 U.S. 545 (1967), the number of Negroes listed on the tax digest amounted to 27.1% of the taxpayers, but only 9.1% of those on the grand jury venire. The disparity was held to be sufficient to make out a prima facie case of discrimination. See Sims v. Georgia, 389 U.S. 404 (1967) (24.4% of tax lists, 4.7% of grand jury lists); Jones v. Georgia, 389 U.S. 24 (1967) (19.7% of tax lists, 5% of jury list). We agree with the District Court and the Court of Appeals that the proof in this case was enough to establish a prima facie case of discrimination against the Mexican-Americans in the Hidalgo County grand jury selection.

[If the jurors were drawn randomly from the general population, then the number of Mexican-Americans in the sample could be modeled by a binomial distribution. See Finkelstein, The Application of Statistical Decision Theory to the Jury Discrimination Cases, 80 Harv. L. Rev. 338, 353-356 (1966). See generally P. Hoel, Introduction to Mathematical Statistics 58-61, 79-86 (4th ed. 1971); F. Mosteller, R. Rourke, & G. Thomas, Probability with Statistical Applications 130-146, 270-291 (2d ed. 1970). Given that 79.1% of the population is Mexican-American, the expected number of Mexican-Americans among the 870 persons summoned to serve as grand jurors over the 11-year period is approximately 688. The observed number is 339. Of course, in any given drawing some fluctuation from the expected number is predicted. The important point, however, is that the statistical model shows that the results of a random drawing are likely to fall in the vicinity of the expected value. See F. Mosteller, R. Rourke, & G. Thomas, supra, at 270-290. The measure of the predicted fluctuations from the expected value is the standard deviation, defined for the binomial distribution as the square root of the product of the total number in the sample (here 870) times the probability of selecting a Mexican-American (0.791) times the probability of selecting a non-Mexican-American (0.209). Id., at 213. Thus, in this case the standard deviation is approximately 12. As a general rule for such large samples, if the difference between the expected value and the observed number is greater than two or three standard deviations, then the hypothesis that the jury drawing was random would be suspect to a social scientist. The 11-year data here reflect a difference between the expected and observed number of Mexican-Americans of approximately 29 standard deviations. A detailed calculation reveals that the likelihood that such a substantial departure from the expected value would occur by chance is less than 1 in 10^{140}.

The data for the 2½-year period during which the State District

Judge supervised the selection process similarly support the inference that the exclusion of Mexican-Americans did not occur by chance. Of 220 persons called to serve as grand jurors, only 100 were Mexican-Americans. The expected Mexican-American representation is approximately 174 and the standard deviation, as calculated from the binomial model, is approximately six. The discrepancy between the expected and observed values is more than 12 standard deviations. Again, a detailed calculation shows that the likelihood of drawing not more than 100 Mexican-Americans by chance is negligible, being less than 1 in 10^{25}. (fn. 17)]

Supporting this conclusion is the fact that the Texas system of selecting grand jurors is highly subjective. The facial constitutionality of the key-man system, of course, has been accepted by this Court. Nevertheless, the Court has noted that the system is susceptible of abuse as applied. See Hernandez v. Texas, 347 U.S., at 479. Additionally, as noted, persons with Spanish surnames are readily identifiable.

The showing made by respondent therefore shifted the burden of proof to the State to dispel the inference of intentional discrimination. Inexplicably, the State introduced practically no evidence. The testimony of the State District Judge dealt principally with the selection of the jury commissioners and the instructions given to them. The commissioners themselves were not called to testify. A case such as Swain v. Alabama, 380 U.S., at 207 n.4, 209, illustrates the potential usefulness of such testimony, when it sets out in detail the procedures followed by the commissioners. The opinion of the Texas Court of Criminal Appeals is particularly revealing as to the lack of rebuttal evidence in the record:

> How many of those listed in the census figures with Mexican-American names were not citizens of the state, but were so-called "wet-backs" from the south side of the Rio Grande; how many were migrant workers and not residents of Hidalgo County; how many were illiterate and could not read and write; how many were not of sound mind and good moral character; how many had been convicted of a felony or were under indictment or legal accusation for theft or a felony; *none of these facts appear in the record.*

506 S.W.2d, at 211 (emphasis added). In fact, the census figures showed that only a small part of the population reported for Hidalgo County was not native born. Without some testimony from the grand jury commissioners about the method by which they determined the other qualifications for grand jurors prior to the statutory time for testing qualifications, it is impossible to draw any inference about literacy, sound mind and moral character, and criminal record from the

statistics about the population as a whole. These are questions of disputed fact that present problems not amenable to resolution by an appellate court. We emphasize, however, that we are not saying that the statistical disparities proved here could never be explained in another case; we are simply saying that the State did not do so in this case.

C. In light of our holding that respondent proved a prima facie case of discrimination that was not rebutted by any of the evidence presently in the record, we have only to consider whether the District Court's "governing majority" theory filled the evidentiary gap. In our view, it did not dispel the presumption of purposeful discrimination in the circumstances of this case. Because of the many facets of human motivation, it would be unwise to presume as a matter of law that human beings of one definable group will not discriminate against other members of their group. . . .

Rather than relying on an approach to the jury discrimination question that is as faintly defined as the "governing majority" theory is on this record, we prefer to look at all the facts that bear on the issue, such as the statistical disparities, the method of selection, and any other relevant testimony as to the manner in which the selection process was implemented. Under this standard, the proof offered by respondent was sufficient to demonstrate a prima facie case of discrimination in grand jury selection. Since the State failed to rebut the presumption of purposeful discrimination by competent testimony, despite two opportunities to do so, we affirm the Court of Appeals' holding of a denial of equal protection of the law in the grand jury selection process in respondent's case.

It is so ordered.

MR. CHIEF JUSTICE BURGER, with whom MR. JUSTICE POWELL and MR. JUSTICE REHNQUIST join, dissenting.

. . . What the majority characterizes as a prima facie case of discrimination simply will not "wash." The decisions of this Court suggest, and common sense demands, that *eligible* population statistics, not gross population figures, provide the relevant starting point. In Alexander v. Louisiana, 405 U.S. 625, 630 (1972), for example, the Court in an opinion by Mr. Justice White looked to the "proportion of blacks in the *eligible* population. . . ." (Emphasis supplied.)

The failure to produce evidence relating to the eligible population in Hidalgo County undermines respondent's claim that any statistical "disparity" existed in the first instance. Particularly where, as here, substantial numbers of members of the identifiable class actually served on grand jury panels, the burden rightly rests upon the challenger to show a meaningful statistical disparity. After all, the presumption of constitutionality attaching to all state procedures has even greater force under the circumstances presented here, where exactly one-half the members of the grand jury list now challenged by respondent were members of the allegedly excluded class of Mexican-Americans.

The Court has not previously been called upon to deal at length with the sort of statistics required of persons challenging a grand jury selection system. The reason is that in our prior cases there was little doubt that members of identifiable minority groups had been excluded in large numbers. In Alexander v. Louisiana, supra, the challenger's venire included only one member of the identifiable class and the grand jury that indicted him had none. In Turner v. Fouche, 396 U.S. 346 (1970); Jones v. Georgia, 389 U.S. 24 (1967); Sims v. Georgia, 389 U.S. 404 (1967); and Whitus v. Georgia, 385 U.S. 545 (1967), there was at best only token inclusion of Negroes on grand jury lists. The case before us, in contrast, involves neither tokenism nor absolute exclusion; rather, the State has used a selection system resulting in the inclusion of large numbers of Spanish-surnamed citizens on grand jury lists. In this situation, it is particularly incumbent on respondent to adduce precise statistics demonstrating a significant disparity. To do that, respondent was obligated to demonstrate that disproportionately large numbers of eligible individuals were excluded systematically from grand jury service.

Respondent offered no evidence whatever in this respect. He therefore could not have established any meaningful case of discrimination, prima facie or otherwise. In contrast to respondent's approach, which the Court's opinion accepts without analysis, the Census Bureau's statistics for 1970 demonstrate that of the *adults* in Hidalgo County, 72%, not 79.1% as respondent implies, are Spanish surnamed. At the outset, therefore, respondent's gross population figures are manifestly overinclusive.

But that is only the beginning. Respondent offered no evidence whatever with respect to other basic qualifications for grand jury service.[1] The statistics relied on in the Court's opinion suggest that 22.9% of Spanish-surnamed persons over age 25 in Hidalgo County have had no schooling at all. Since one requirement of grand jurors in Texas is literacy in the English language, approximately 20% of

---

1. The burden of establishing a prima facie case obviously rested on respondent. It will not do to produce patently overinclusive figures and thereby seek to shift the burden to the State. Rather, a prima facie case is established only when the challenger shows a disparity between the percentage of minority persons in the eligible population and the percentage of minority individuals on the grand jury.

adult-age Mexican-Americans are very likely disqualified on that ground alone.

The Court's reliance on respondent's overbroad statistics is not the sole defect. As previously noted, one-half of the members of respondent's grand jury list bore Mexican-American surnames. Other grand jury lists at about the same time as respondent's indictment in March 1972 were *predominantly Mexican-American.* Thus, with respect to the September 1971 grand jury list, 70% of the prospective grand jurors were Mexican-American. In the January 1972 Term, 55% were Mexican-American. Since respondent was indicted in 1972, by what appears to have been a truly representative grand jury, the mechanical use of Hidalgo County's practices some 10 years earlier seems to me entirely indefensible. We do not know, and on this record we cannot know, whether respondent's 1970 gross population figures, which served as the basis for establishing the "disparity" complained of in this case, had any applicability at all to the period prior to 1970. Accordingly, for all we know, the 1970 figures may be totally inaccurate as to prior years; if so, the apparent disparity alleged by respondent would be increased improperly.

Therefore, I disagree both with the Court's assumption that respondent established a prima facie case and with the Court's implicit approval of respondent's method for showing an allegedly disproportionate impact of Hidalgo County's selection system upon Mexican-Americans.

Notes and Problems

1. Footnote 8, from the majority opinion in Castaneda v. Partida, contains Justice Blackmun's response to the dissent:

> At oral argument, counsel for petitioner suggested that the data regarding educational background explained the discrepancy between the percentage of Mexican-Americans in the total population and the percentage on the grand jury lists. For a variety of reasons, we cannot accept that suggestion. First, under the Texas method of selecting grand jurors, qualifications are not tested until the persons on the list appear in the District Court. Prior to that time, assuming an unbiased selection procedure, persons of all educational characteristics should appear on the list. If the jury commissioners actually exercised some means of winnowing those who lacked the ability to read and write, it was incumbent on the State to call the commissioners and to have them explain how this was done. In the absence of any evidence in the record to this effect, we shall not assume that the only people excluded from grand jury service were the illiterate.

Second, it is difficult to draw valid inferences from the raw census data, since the data are incomplete in some places and the definition of "literacy" would undoubtedly be the subject of some dispute in any event. The State's failure to discuss the literacy problem at any point prior to oral argument compounds the difficulties. One gap in the data occurs with respect to the younger persons in the jury pool. The census reports for educational background cover only those who are 25 years of age and above. Yet the only age limitation on eligibility for grand jury service is qualification to vote. During the period to which the census figures apply, a person became qualified to vote at age 21. . . . It is not improbable that the educational characteristics of persons in the younger age group would prove to be favorable to Mexican-Americans.

Finally, even assuming that the statistics for persons age 25 and over are sufficiently representative to be useful, a significant discrepancy still exists between the number of Spanish-surnamed people and the level of representation on grand jury lists. Table 83 of the 1970 census shows that of a total of 80,049 persons in that age group, 13,205 have no schooling. Table 97 shows that of the 55,949 Spanish-surnamed persons in the group, 12,817 have no schooling. This means that of the 24,100 persons of all other races and ethnic groups, 388 have no schooling. Translated into percentages, 22.9% of the Spanish-surnamed persons have no schooling, and 1.6% of the others have no schooling. This means that 43,132 of the Spanish-surnamed persons have some schooling and 23,712 of the others have some schooling. The Spanish-surnamed persons thus represent 65% of the 66,844 with some schooling, and the others 35%. The 65% figure still creates a significant disparity when compared to the 39% representation on grand juries shown over the 11-year period involved here.

The suggestion is made in the dissenting opinion of The Chief Justice that reliance on eligible population figures and allowance for literacy would defeat respondent's prima facie showing of discrimination. But the 65% to 39% disparity between Mexican-Americans *over* the age of 25 who have some schooling and Mexican-Americans represented on the grand jury venires takes both of The Chief Justice's concerns into account. Statistical analysis indicates that the discrepancy is significant. If one assumes that Mexican-Americans constitute only 65% of the jury pool, then a detailed calculation reveals that the likelihood that so substantial a discrepancy would occur by chance is less than 1 in 10^{50}.

We prefer not to rely on the 65% to 39% disparity, however, since there are so many implicit assumptions in this analysis, and we consider it inappropriate for us, as an appellate tribunal, to undertake this kind of inquiry without a record below in which those assumptions were tested. We rest, instead, on the fact that the record does not show any way by which the educational characteristics are taken into account in the compilation of the grand jury lists, since the procedure established by the State provides that literacy is tested only after the group of 20 are summoned.

To calculate the number of standard deviations resulting from the 40.1% absolute disparity described in the main text of its opinion, the majority had to determine N, the number of people in the jury pool, P_1, the percent of Spanish-surnamed people in Hidalgo County, and P_2, the percent of non-Spanish-surnamed people in Hidalgo County. Footnote 8, quoted above, as well as the dissenting opinion of Chief Justice Burger reveal considerable dispute as to the proper numbers to be used in the calculation. For instance, if one uses as N an estimate of the number selected for the grand jury pool in the year of Rodrigo Partida's indictment, and as P_1 and P_2 the percentages of "eligible" Spanish- and non-Spanish-surnamed people in the county, respectively, and as the observed value the reported percentage of "minority" members in the indictment year, the results are strikingly different. These alternative figures suggest an observed number of Spanish-surnamed potential jurors that is only 2.01 standard deviations from the expected number. If you compare this to the 29 standard deviations used in the majority's calculation, you can see that there is considerable room for advocacy in establishing the proper numbers to be used in these critical calculations.

2. As an exercise, dig out from the opinions those numbers that present the strongest and weakest case for and against each party. Note that your task as an advocate is to convince the court that your numbers are the more plausible.

3. In addition to taking the advocate's role, you might also consider the Justices' positions. Which figures are more persuasive? Is there just one set of correct numbers? Once you determine which set is the most convincing, calculate the number of standard deviations that the actual figures are from the expected values.

You may find the following nine-step procedure helpful in applying the *Castaneda* test:

Step 1. Find N. N is the size of the group which is the *result* of discrimination. In *Castaneda*, N is the number of people in the jury pool for the 11-year period, 870. The jury pool is the "group of suspect composition" because we suspect there was discrimination in selecting the members of this group.

Step 2. Find P_1. P_1 is the underlying percentage of the allegedly discriminated-against group in the population. In *Castaneda*, 79.1% of the population was Spanish-surnamed, so $P_1 = .791$.

Step 3. Find P_2. P_2 is the percentage of the underlying population not in P_1. In *Castaneda*, P_2 is the percentage of non-Spanish-surnamed people, so P_2 equals 1 minus .791, or .209.

Step 4. Find the number of the allegedly discriminated-against group you would expect to find in the group of suspect composition if there were no discrimination. The expected number is always P_1 times N. In *Castaneda* we would expect .791 times 870 Spanish-surnamed people in the jury pool if the jury pool reflected the underlying population, so the expected number is 688.

Step 5. Find the number of the allegedly discriminated-against group you actually observed in the group of suspect composition. You would find this as part of the discovery process. The *Castaneda* opinion reports that 339 Spanish-surnamed people were actually chosen for this group, so the observed number is 339.

Step 6. Calculate the disparity by subtracting the expected number from the observed number. In *Castaneda,* 339 minus 688 is minus 349, so the disparity is -349 people. The minus sign indicates an underrepresentation.

Step 7. Calculate one standard deviation by the formula $SD = \sqrt{N \times P_1 \times P_2}$. In *Castaneda* this figure amounts to $\sqrt{870 \times .791 \times .209} = 12$, so one standard deviation is 12 people.

Step 8. Calculate the number of standard deviations the observed is from the expected by dividing the disparity by one standard deviation. For *Castaneda,* divide the -349 person disparity by 12 people in one standard deviation: $-349/12$ equals -29. The observed number is 29 standard deviations below what we expected.

Step 9. Compare the calculated number of standard deviations to the two-or-three-standard-deviations guideline. For *Castaneda,* 29 is greater than two or three, so we suspect that the jury pool members were selected discriminatorily.

Use this procedure to apply the Court's test to whatever set of correct numbers you find most convincing. Should this case have come out differently, based on your interpretation of the statistical evidence?

4. Assume that all law school applicants have equal qualifications. Six hundred students were admitted to Private U. College of Law over a three-year period. Fifteen percent of the applicants during those years were minority students. Twenty-five minority students were admitted. Is there any evidence of discrimination against minorities in admissions at Private U., using the *Castaneda* test?

5. Last year at Nearby Hospital, 250 nurses were hired. Twenty-five percent of the applicants were male. The rest were

female. The males and females had equal qualifications. Ninety males were hired. Is there any evidence of discrimination against males in hiring at Nearby?

6. The standard-deviation calculations do not indicate what caused the underrepresentation. This test only indicates the probability that the disparity between expected and observed values occurred by chance. The court reports that the randomness of an observation more than two or three standard deviations from the expected value would be suspect to a social scientist, suggesting that such a disparity would also be suspect to a court. If this test does not prove what caused the disparity, is the Court in effect ruling that causation is irrelevant in a case like this? Or is causation only irrelevant when there is such a tremendous disparity as that calculated by the majority? Under your recalculation in Note 3 above, is the disparity small enough that the "governing majority theory" propounded by petitioner, that is, that a majority racial group in power would not discriminate against members of its own race (see page 87 above), should be given more weight? Or should that theory continue to be disregarded for other reasons?

UNITED STATES v. GOFF
509 F.2d 825 (5th Cir. 1975)

THORNBERRY, Circuit Judge:

In these appeals two criminal defendants claim that we must reverse their mail fraud convictions[1] since the jury selection process in their cases violated the standards of the Jury Selection Act of 1968. 28 U.S.C. §1861 et seq. In the district court, appellants properly filed a timely motion to dismiss the indictments for failure to comply with the Jury Selection Act. 28 U.S.C. §1867(a). They based their challenge to the Eastern District of Louisiana's jury pool on inadequate representation of blacks, and poor people. The district court denied the motion, finding that there was no substantial underrepresentation of blacks, and that poor people did not constitute a cognizable class under 28 U.S.C. §1862. . . .

UNDERREPRESENTATION OF BLACKS

Appellants' statistics showed that in 1970 (the date of grand jury selection) blacks comprised 26.33% of the voting age population and

---

1. The government charged that the appellants fraudulently received welfare benefits, a violation of 18 U.S.C. §1341.

21.06% of the registered voters in the Eastern District of Louisiana. Thus there was a 5.27% absolute differential between blacks in the voting age population and registered black voters, and a 20.02 percentage underrepresentation of blacks on the jury list.[3] The court below found that this underrepresentation was not substantial within the meaning of the Jury Selection Act after assessing the impact of this underrepresentation on a grand jury of twenty-three persons. A grand jury that statistically mirrored the voter registration list would contain 4.6 black persons, while one statistically mirroring the voting age population would contain 6.0 black persons. The court concluded that this amount of underrepresentation was not sufficiently substantial to require the Eastern District of Louisiana to supplement its voter registration list. 28 U.S.C. §1863. We hold that the district court properly disposed of that issue.

UNDERREPRESENTATION OF FOOD STAMP RECIPIENTS

The district court also examined the underrepresentation of the class of food stamp recipients in the Eastern District of Louisiana on the jury list.[4] The appellants' statistics demonstrated that in the Eastern District only 30.03% of the food stamp recipients over twenty-one are registered to vote. In contrast 77.76% of the remainder of the voting age population have registered.[5] The district court felt these figures proved substantial underrepresentation, concluding, ". . . food stamp recipients, as a group, are 47.73% underrepresented on voters registration list." 370 F. Supp. at 304. The district court erred in concluding that this underrepresentation on the voters' registration list thereby established an underrepresentation on the grand jury list. The above statistics show only the registration rate differential between food stamp recipients of voting age and the voting age population at-large. That figure, however, does not accurately

---

3. Neither the absolute nor the comparative measure of underrepresentation can be considered determinative of the substantiality question. *See* Gewin, An Analysis of Jury Selection Decision, 506 F.2d 811, 834-35 (1975) (printed as an appendix to Foster v. Sparks, 506 F.2d 805 (5th Cir. 1975)).

4. Appellants defined "poor people" for the purpose of their challenge as "food stamp recipients." They took a statistically significant sample of food stamp recipients and checked to see how many of those names appeared on the voter registration lists in Eastern District of Louisiana parishes. The measure of economic status, then, was whether a prospective juror qualified for food stamp assistance rather than an absolute dollar ceiling on income.

5. Appellants point out that their figures probably understate the differential because not all poor people register for food stamps. But in this case, there is no indication of the steps taken to purge the voter registration list. Failure to do so would tend to offset the understatement.

disclose the impact of that registration differential on the jury list. We think the district court should have proceeded to assess the impact of this registration rate differential on a typical grand jury, just as it did in the case of blacks. The figures showed that food stamp recipients constitute 10.51% of the voting age population, and 4.34% of those on the jury list. Thus a twenty-three person grand jury that mirrored the jury list would contain 1.0 food stamp recipients, while one that mirrored the voting age population would contain 2.4 food stamp recipients. The registration rate differential has the same impact as in the case of blacks. We find this underrepresentation is not so substantial as to require supplementation of the voter registration list. Thus we express no view on the cognizability of the class of food stamp recipients or poor people under 28 U.S.C. §1862.

Our result comports with the underlying purposes of the Jury Selection Act of 1968. The Act primarily sought to eliminate the "key man" system in the federal jury selection process. Congress felt that utilization of voter registration lists as the primary source of names for the master jury list would provide a more representative cross section of the community. That body recognized that in some instances, failure of particular groups in a community to register would mean that the voter registration list would not accurately represent a fair cross section of the community. But where, as in this case, the impact of the underrepresentation does not substantially affect the composition of the average grand jury, the Act does not require the district to incur the substantial expense and administrative inconvenience necessary to supplement the voter registration list.

Affirmed.

Notes and Problems

1. The criminal defendants seeking reversal in United States v. Goff do not, as in *Castaneda*, present statistical evidence as to the actual number of blacks or poor people on the jury list in the Eastern District of Louisana. Instead, they attack the selection procedure directly, suggesting that the procedure cannot be free of bias if the underlying list from which jurors are selected, the voter registration list, underrepresents the black and poor populations. The basic statistical issue in *Goff* was whether the degree of underrepresentation was "substantial." *Goff* was decided before the Supreme Court established the two-or-three-standard-deviations guideline in *Castaneda*. Reflect on what numbers are needed to calculate the standard deviation. In *Castaneda* the "group of suspect composition" was the collection of

individual jurors actually selected to serve on grand juries. Here the "group of suspect composition" is the subset of the voting-age population actually registered to vote. Presuming that the selection from the registration list was nondiscriminatory, the defendants allege that the list itself was discriminatory because blacks were underrepresented on that list by 5.27% and the poor by 47.73%.

If you know the number of registered voters, you can calculate the number of standard deviations the observed number of each minority group is from the expected number. Assume that a million people are registered to vote in the Eastern District of Louisiana. We are given the percentages of blacks and poor in the voting-age population and on the registration list, and from this can infer the percentages of nonblacks and nonpoor in the relevant population. From this data we can also calculate the expected and observed numbers of black and poor on the voter registration lists. Calculate the number of standard deviations that the expected number of blacks (or poor) is from the observed number. Would *Goff* have come out differently after *Castaneda*?

2. Both the district and appellate courts examined the impact of the alleged statistical underrepresentation in the voter registration list on the make-up of a single grand jury. The appellate court found that a 23-person grand jury (an alternative "group of suspect composition") that statistically mirrored the voter registration list would contain 21.06% or 4.6 black persons, and 4.34% or 1.0 poor person; while one mirroring the voting-age population would contain 26.33% or 6.0 black persons, and 10.51% or 2.4 poor persons. Actually, Circuit Judge Thornberry's calculation is wrong with respect to the number of blacks on a jury statistically mirroring the voter registration list. A grand jury of 23 would have 4.8 blacks. Using expected numbers obtained from the populations figures, calculate the deviation of expected from observed numbers of black people in a jury drawn from the voter registration list and apply the two-to-three-standard-deviations rule. Then do the same for poor people. Assume that the actual jury contained 4.8 black persons and 1 poor person to get your observed values. Is your result different from that calculated in Note 1 above? Part of the explanation is due to the difference in size of the groups of suspect composition in Notes 1 and 2. The case that follows these notes, Inmates of Nebraska Penal & Correctional Complex v. Greenholtz, discusses this problem. The small numbers problem comes up many times in these materials and is the subject of additional scrutiny in the final chapter of the book.

3. In order to estimate the number of poor people on the voter registration list, the defendants took what the court refers to in foot-

note 3 as "a statistically significant sample" of food-stamp recipients and checked to see how many were registered. The process of sampling saves time and money, compared to the alternative of checking all food-stamp recipients. Scientific sampling, which makes sure the sample is large enough to give a reliable result, was encountered in Craig v. Boren in Chapter III and is also the subject of Chapter VI of this book.

4. The defendants' estimation of the number of poor people assumed that all poor people were registered for food stamps. The court and the defendants realize that the estimate probably understated the number of poor people because some may not have been food-stamp recipients. This underestimation means that P_1, the percentage of poor people in the population, is understated. What is the effect on defendants' statistical case of using a smaller-than-accurate P_1? To answer this question, recalculate the results for the problems in Notes 1 and 2 with higher values of P_1. Remember that a higher P_1 means a lower P_2, since $P_2 = 1.00 - P_1$.

INMATES OF THE NEBRASKA PENAL & CORRECTIONAL COMPLEX v. GREENHOLTZ
567 F.2d 1368 (8th Cir. 1977)

VAN OOSTERHOUT, Senior Circuit Judge.

On this appeal a subclass of plaintiffs-appellants consisting of native American and Mexican-American inmates of the Nebraska Penal and Correctional Complex renew their claim that defendant members of the Nebraska Board of Parole have denied discretionary parole to subclass members on grounds which are racially and ethnically discriminatory, in violation of the fourteenth amendment. . . .

I

Plaintiffs' case at trial consisted of four primary components: (1) a statistical analysis and expert testimony in support thereof; (2) testimony elicited from Board members and from expert witnesses concerning the differences in norms and values of white and minority cultures; (3) a number of individual cases of alleged discrimination; and (4) a number of alleged racial slurs or epithets by Board members.

1. Plaintiffs' Exhibit 18, introduced and received in the rebuttal portion of plaintiffs' case, is a statistical analysis of data extracted from Board records of essentially all of the approximately twenty-two

hundred male inmates confined in the penal complex in 1972 and 1973. It purports to demonstrate that in those years native American and Mexican-American inmates who were eligible for discretionary parole received a substantially lower percentage of discretionary paroles than did eligible white and black inmates. The exhibit contains the following breakdown, by racial and ethnic group, of the number of inmates eligible for discretionary parole and the number of discretionary paroles received:

	White	Black	Native American	Mexican- American	Total
Eligible for release by discretionary parole	590	235	59	18	902
Received discretionary parole	358	148	24	5	535
Per cent that received discretionary parole	60.7	63.0	40.7	27.8	59.3

Exhibit 18, prepared by a witness whom the district court accepted as an expert, recited that the data utilized in arriving at the above statistics were gathered in a manner generally recognized in the field of statistics and that the results were statistically accurate. The witness corroborated this recitation at trial. He concluded: "The data indicate that there is a substantial relationship between race and whether or not an eligible inmate receives a discretionary parole.". . .

2. . . . The Board members uniformly denied that racial or ethnic identity played any role in the decisionmaking process and steadfastly maintained that all inmates were evaluated by the same criteria. . . .

4. . . . The comments by Board members contained in [three exhibits submitted by plaintiffs] are: (1) one native American was asked by a Board member whether his tribe was among the "bad Sioux" or "fighting Sioux"; (2) a second native American was asked whether his tribe was "friendly"; (3) following his dismissal from the hearing, the second native American was referred to as a "friendly Indian that's a parasite on society" and as a "Chippewa off the old block"; and (4) during a discussion of alcohol use by native Americans, one Board member commented: "They are professionally immune to snake bite." The exhibits indicate that a number of these comments were followed by laughter.

Our attention is also drawn to . . . testimony from plaintiffs' case-in-chief concerning the use of alcohol by native Americans. Chairman Greenholtz acknowledged having previously made the following statements with reference to a particular inmate:

> Indians are children of nature, really. They have a certain culture ingrained in their psychological makeup. Most of them, I mean the Reservation type like he and he learns and he knows he's done time before. . . . And drinking is a status symbol with those kind of guys. . . .

II

However difficult of application, the controlling legal principles in this case are relatively clear. The central purpose of the fourteenth amendment's equal protection clause is of course the prevention of official conduct discriminating on the basis of race. Washington v. Davis, 426 U.S. 229, 239 (1976). Yet the Supreme Court has made abundantly clear that official conduct is not always discriminatory merely because it adversely affects a greater proportion of one race than another:

> Disproportionate impact is not irrelevant, but it is not the sole touchstone of an invidious racial discrimination forbidden by the Constitution. Standing alone, it does not trigger the rule that racial classifications are to be subjected to the strictest scrutiny and are justifiable only by the weightiest of considerations.

Id. at 242, 96 S. Ct. at 2049 (citation omitted). It is incumbent upon plaintiffs to establish that a racially disproportionate impact, if there is one, was occasioned by a racially motivated purpose.

This is not to say that evidence of disproportionate impact, or statistical evidence in particular, is unimportant. Necessarily, an invidious discriminatory purpose may often be inferred from the totality of the relevant facts, including the fact, if it is true, that the challenged conduct bears more heavily upon one race than another. The determination ultimately required demands a sensitive inquiry into such circumstantial and direct evidence of intent as may be available, including the impact of the challenged action, its historical background and its legislative or administrative history. Moreover, because legislative and administrative bodies operating under broad mandates rarely make decisions motivated by a single concern, it is not necessary that plaintiffs prove the challenged conduct rested solely on racially motivated purposes.

The Supreme Court has made it "unmistakably clear that '[s]tatistical analyses have served and will continue to serve an important role' in cases in which the existence of discrimination is a disputed issue." International Brotherhood of Teamsters v. United States, 431 U.S. 324, 339 (1977). "Statistics showing racial or ethnic imbalance are probative . . . [because and] only because such imbalance is often a telltale sign of purposeful discrimination." Id.n. 20. Indeed, the statistical disparity shown may occasionally be so "gross" or "stark" or "dramatic" that it alone will constitute prima facie proof of purposeful discrimination. We are only cautioned that such extreme disparities are rare, that statistical evidence, like any other kind of evidence, may be rebutted, and that the probative worth of statistical evidence depends on all of the surrounding facts and circumstances.

III

We examine plaintiffs' statistical evidence first. According to Exhibit 18, whites, blacks, native Americans and Mexican-Americans were discretionarily paroled in 1972 and 1973 at the respective percentage rates of 60.7, 63.0, 40.7 and 27.8. The district court, following its own reanalysis of the figures contained in the exhibit, was unconvinced that the statistics demonstrated any significant disparities in the treatment of the several racial and ethnic groups. It reasoned:

> [C]aucasian inmates comprise 65.4 per cent of the population eligible for a discretionary parole; caucasians received 66.9 per cent of the discretionary paroles awarded. Black inmates comprise 26.1 per cent of the population eligible for a discretionary parole; blacks received 27.7 per cent of the discretionary paroles awarded. Indian inmates comprised 6.5 per cent of the population eligible for a discretionary parole; Indians received 4.5 per cent of the discretionary paroles awarded. Mexican-American inmates comprise 2 per cent of the population eligible for a discretionary parole; Mexican-Americans received .9 per cent of the discretionary paroles awarded. These figures show statistical disparities of two per cent or less which the Court considers to be insignificant.

436 F. Supp. at 441-42.

Before discussing our view of the significance or insignificance of the statistical disparities shown, we point out that plaintiffs' statistics are assailed on other grounds as well. Contrary to assertions in plaintiffs' brief, the district court did not find plaintiffs' method of study reliable and their results accurate, although there was expert testimony to this effect and the court did find defendants had produced no substantial evidence to rebut it. The court expressed concern that

F. Standard Deviations as Statistical Proof   101

the size of the Mexican-American group (18) on which the study was based might be too small to yield a reliable result, a concern which we share but need not expressly resolve. The court also expressed concern that racial and ethnic identity was the only variable controlled by the study, while a number of obviously important factors were left uncontrolled. Defendants additionally urge that the 1972-1973 time period chosen by plaintiffs is too "stale" to support an inference that Board practices are now, or were at the time of trial, discriminatory. In short plaintiffs' statistics provide much grist for argument, most of it not resolved by the district court and most of it ill-suited to resolution for the first time on appeal. Accordingly, we bypass this entire phase of the controversy and assume without deciding that plaintiffs' statistics accurately and reliably reflect what they purport to reflect.

So assuming, then, we consider whether the statistical disparities shown are sufficiently significant substantially to assist plaintiffs in establishing a prima facie case of racial and ethnic discrimination. We are mindful of the Supreme Court's recent admonition that the primary responsibility for evaluating statistical evidence rests with the trial court.

The district court's conclusion that the disparities are insignificant rests on its comparison of the percentage of eligible inmates who were native American and Mexican-American with the percentage of discretionary paroles awarded which were awarded to native Americans and Mexican-Americans. Plaintiffs contend that this method of analysis is deceptive and the conclusion reached by the district court illusory. Plaintiffs further contend the disparities shown are indeed significant.

First, we find nothing inherently deceptive in the district court's method of analysis. A similar approach has proved useful in a variety of contexts. For example, in Castaneda v. Partida, [430 U.S. 482 (1977),] the Supreme Court assessed a claim of discrimination in the selection of grand jurors by comparing the percentage of a county's population which was Mexican-American with the percentage of grand jurors selected to serve in the county who were Mexican-American. And in Hazelwood School District v. United States, [433 U.S. 299 (1977),] the Supreme Court, in assessing a claim of discriminatory employment practices, indicated the relevance of comparing the percentage of the relevant labor market which was black with the percentage of persons employed who were black. The district court's analysis in this case was not fundamentally different.

Second, we nonetheless agree with plaintiffs that the district court's conclusion does not follow from that court's analysis. Specifically, we do not think "disparities of two per cent or less" are always insignificant. Although the method of analysis employed by the district court is indeed often useful, it is not invariably a sure guide to a correct result. Like all statistical evidence, the district court's figures must be considered in light of all the surrounding facts and circumstances. International Brotherhood of Teamsters v. United States, supra 431 U.S. at 339, 97 S. Ct. 1843. Necessarily, as a minority group's comparative size in a relevant population (in this case those eligible for discretionary parole) diminishes, increasingly smaller differences between the two percentage figures become significant. This is so because the use of such percentages considers the disparate impact upon a minority group as a function of the minority group's comparative size in the larger population. It follows that the disparate impact is diluted in cases, like this one, where the group is comparatively small.

[The point is aptly illustrated by the Mexican-American subclass members in this case, who comprise only 2.0 per cent of the relevant population. Even if no Mexican-Americans had been discretionarily paroled, the statistical disparity would have been only 2.0 per cent and therefore, under the district court's reasoning, insignificant.

We additionally point out that our comments in this regard bear no necessary relation to the problem of small population size. We may suppose that plaintiffs' study spanned twenty rather than two years and that all nonpercentage figures in Exhibit 18 were precisely ten times larger. If that were so, the population sizes would clearly be large enough, but the pertinent percentage figures, and therefore the district court's analysis, would remain unaltered. Moreover, if 5350 of 9020 eligible inmates, but among them only 240 of 590 native Americans and 50 of 180 Mexican-Americans, had received discretionary paroles, standard deviation calculations of the type explained in Castaneda v. Partida, supra 430 U.S. at 496-97 n. 17, would quickly reveal the considerable unlikelihood that racial and ethnic identity played no role in the receipt of a discretionary parole. (fn.18)]

Third and finally, although we decline to agree with the district court that the disparities here shown are *entirely* without significance, we conclude that they are not sufficiently significant substantially to assist plaintiffs in establishing a prima facie case of racial and ethnic discrimination. To be sure, Exhibit 18 reveals some disparity. Eligible whites and blacks, over a two-year period, were discretionarily paroled at the respective rates of 60.7 and 63.0 per cent, while eligible native Americans and Mexican-Americans, over the same period, were discretionarily paroled at the respective rates of 40.7 and 27.8 per cent. This of course means that whites and blacks were discretionarily paroled at rates roughly one-and-one-half times that for native Americans and more than twice that for Mexican-Americans. These figures, like the district court's, are unquestionably relevant ones.

Although these percentages are on first glance somewhat disturbing, we think the bare numbers in Exhibit 18 tell the most revealing story. The 40.7 per cent figure relative to native Americans represents but 24 individuals, and the 27.8 per cent figure relative to Mexican-Americans represents but 5. If the 535 discretionary paroles awarded to all inmates had been distributed proportionately among all racial and ethnic groups, native Americans would have received only 35 (6.5 per cent of 535) and Mexican-Americans only 11 (2.0 per cent of 535) discretionary paroles. Thus, for native Americans the difference between "expected" (35) and observed (24) number of discretionary paroles received was only 11, and for Mexican-Americans the difference between "expected" (11) and observed (5) number of discretionary paroles received was only 6. In each instance the difference represents less than two standard deviations. [In computing a standard deviation, we use the method described in Castaneda v. Partida, supra at 496-97 n. 17, 97 S. Ct. 1272. For native Americans one standard deviation equals the square root of the product of the total number of individuals who received discretionary paroles (535) times the probability of randomly selecting a native American from the eligible inmate population (.065) times the probability of randomly selecting someone other than a native American from the same population (.935). This computes to approximately 5.7. A similar computation for Mexican-Americans yields approximately 3.2.   (fn. 19)] Even with substantially larger populations, the Supreme Court has recently been unwilling to conclude that racial or ethnic identity plays a substantial role in challenged conduct unless observed and expected numbers differ by more than "two or three standard deviations." Hazelwood School Dist. v. United States, supra 433 U.S. at 309, 97 S. Ct. 2736 n. 14; Castaneda v. Partida, supra 430 U.S. at 496-97 n. 17, 97 S. Ct. 1272.

The specific analysis in Hazelwood School Dist. v. United States is instructive. The Supreme Court there considered a claim that the school district had engaged in a pattern or practice of employment discrimination against blacks, allegedly in violation of Title VII. A majority of this court, in part because the percentage of black teachers employed by the district was very low in comparison with the percentage of black teachers available in the St. Louis County and City area, had previously concluded the Government had established an unrebutted prima facie pattern or practice case. 534 F.2d 805 (8th Cir. 1976). The Supreme Court vacated our judgment. Since Title VII

Chapter IV. Standard Deviations

was not applicable to the district until 1972, the Court explained, the district could have rebutted the Government's prima facie case by showing racially neutral hiring practices after 1972. 433 U.S. at 309, 97 S. Ct. 2736. Statistical data of record revealed that 3.7 per cent (15 of 405) of the teachers hired in the 1972-73 and 1973-74 school years were black. This figure, the Court noted, should be compared with the percentage of blacks in the relevant labor market. Id. The Court did not make the comparison, however, because the relevant labor market area was disputed, the Government contending it included both St. Louis County and St. Louis City (15.4 per cent black) and the district contending it included St. Louis County but not St. Louis City (5.7 per cent black). The Court did comment, however:

> The difference between these figures may well be important; *the disparity between 3.7%* (the percentage of Negro teachers hired by Hazelwood in 1972-1973 and 1973-1974) and *5.7% may be sufficiently small to weaken the Government's other proof*, while the disparity between 3.7% and 15.4% may be sufficiently large to reinforce it.[17]
>
> [17]Indeed, under the statistical methodology explained in Castaneda v. Partida, involving the calculation of the standard deviation as a measure of predicted fluctuations, the difference between using 15.4% and 5.7% as the area-wide figure *would be significant*. If the 15.4% figure is taken as the basis for comparison . . . the difference between the observed number of 15 Negro teachers hired (of a total of 405) would vary from the expected number of 62 by more than six standard deviations. . . . *If, however, the 5.7% areawide figures is used . . . the expected value of 23 would be less than two standard deviations from the observed total of 15.*

Id. (emphasis supplied).

The emphasized portions of the above quote are highly pertinent to our analysis because the statistics there referred to are qualitatively and quantitatively comparable to the statistics here. Specifically, the 5.7 per cent/3.7 per cent disparity is comparable to the 6.5 per cent/4.5 per cent (native American) and the 2.0 per cent/.9 per cent (Mexican-American) disparities noted by the district court below. In addition, the 15 and 23 observed and expected numbers are not far afield from the 24 and 35 (native American) and the 5 and 11 (Mexican-American) observed and expected numbers here; in each case the variance is slightly less than two standard deviations.

Plaintiffs place considerable reliance on our decision in Green v. Missouri Pacific Railroad, 523 F.2d 1290 (8th Cir. 1975). In that case, applying the rule of Griggs v. Duke Power Co., 401 U.S. 424, 91 S. Ct. 849, 28 L. Ed. 2d 158 (1971), that employment practices having a racially disproportionate impact are prima facie discriminatory under

F. Standard Deviations as Statistical Proof 105

Title VII, we accepted a statistical analysis purporting to demonstrate that a rigid policy of refusing to hire persons previously convicted of a crime (other than a minor traffic offense) disqualified a substantially higher rate of black applicants than white applicants. Plainly, however, the statistical showing made there was stronger than the one made here. From a total job applicant pool of 8488, the challenged policy had disqualified 292 individuals, 174 black and 118 white. Since the applicant pool was 39 per cent black and 61 per cent white, an expected random distribution of the 292 disqualifications would have been 114 black and 178 white. Observed and expected values in *Green* thus differed by 60, more than seven standard deviations.

[The statistical showing in Murrah v. Arkansas, 532 F.2d 105 (8th Cir. 1976), a jury discrimination case also cited by plaintiffs, was similarly much stronger than that here. In a county 19.7 per cent black, jury commissioners had selected 800 names to be placed in a jury wheel, no more than 58 of them black. Had 19.7 per cent of the 800 been black, the number of blacks placed in the jury wheel would have been 158. Observed and expected values thus differed by 100, nearly nine standard deviations. (fn. 21)]

The above analysis leads us to the conclusion that the statistical showing in this case is not quite sufficient to permit a confident determination that the disparities shown are related to racial and ethnic factors and is therefore not quite sufficient substantially to assist plaintiffs in establishing a prima facie case of racial and ethnic discrimination. Although the Supreme Court has not in terms established a two standard deviation "floor" for statistical significance, *Castaneda* and *Hazelwood* certainly imply that a confident showing of statistical significance begins to arise at that level.

We recognize, of course, that the statistical disparities in this case *approach* the level seemingly deemed significant in *Castaneda* and *Hazelwood*. For that reason we decline to go the distance with the district court and hold the disparities utterly without significance.[22] If the remainder of plaintiffs' evidence came close to establishing a prima facie case, we might agree that their statistics would tip the balance.

---

22. Even so, the district court may ultimately prove to be correct. The Supreme court majority in *Hazelwood* went so far as to say that a disparity comparable to those here "may be sufficiently small to weaken" the plaintiffs' other proof, 433 U.S. at 311, 97 S. Ct. at 2743, presumably because classwide purposeful discrimination would ordinarily result in a greater disparity. The use of the word "may", however, and the fact that Mr. Justice Brennan, concurring, 433 U.S. at 313, 97 S. Ct. 2736, and Mr. Justice Stevens, dissenting, 433 U.S. at 317, 97 S. Ct. 2736, appear to regard the *Hazelwood* figures in a somewhat different light, perhaps suggest that the Supreme Court has not yet reached a definitive resolution. We accordingly go only as far as the facts of this case require and hold that the statistical disparities here are not of *substantial* assistance to plaintiffs.

Insofar as plaintiffs rely on their statistics as a primary component of their prima facie case, however, we are in disagreement.

[It is a foregone conclusion that plaintiffs' statistics do not exhibit the sort of "gross" or "stark" or "dramatic" disparities which in rare cases can alone establish a prima facie case. This case is a far cry from *Castaneda,* where over an eleven-year period Mexican-Americans constituted 79 per cent of a county's population but only 39 per cent of its grand jurors; the difference between observed and expected number of Mexican-American grand jurors equaled approximately twenty-nine standard deviations. 430 U.S. at 497 n.17.

Plaintiffs also suggest that defendants' decision-making on parole matters, like the method of grand juror selection in *Castaneda,* is "highly subjective" and "susceptible to abuse as applied." See 430 U.S. at 497. Even if we assume that the analogy is an apt one, it would only be useful where the disparities themselves were substantial. (fn. 23)]

IV

Our conclusion that the statistical disparities shown are not sufficiently significant substantially to support an inference of purposeful discrimination does not end the judicial inquiry. Under Washington v. Davis, supra, the inference can arise from any probative evidentiary sources, including but not limited to statistics. Of course, as the significance of the statistical disparities lessens, the quality and quantity of nonstatistical evidence required to raise the inference become correspondingly greater. We accordingly turn to a consideration of plaintiffs' nonstatistical evidence. . . .

Plaintiffs' experts isolated a number of problems which native Americans and to a lesser extent Mexican-Americans commonly encounter when considered for parole. . . . Factors such as these could well provide a noninvidious explanation for such variance in treatment as may exist. There is accordingly no necessary or even likely inference of discriminatory purpose, and the district court was not compelled to find one. . . .

Finally, we consider alleged racial slurs made by defendant Board members. The district court found these statements "subject to varying interpretations" and afforded them "little or no weight." 436 F. Supp. at 439. Viewing these alleged slurs in context, we agree. . . .

V

Our discussion has embraced those portions of plaintiffs' evidence which we find to be of the most arguable relevance in establishing purposeful discrimination. Further elucidation of the evidence and of our reasons for rejecting it would unduly prolong this opinion and serve no useful purpose. Although we cannot fully subscribe to the views expressed by the district court, particularly with reference to plaintiffs' statistics, a careful review of plaintiffs' entire case convinces us that the district court correctly found no substantial support for an inference of discriminatory purpose. We accordingly agree with the district court that plaintiffs failed to establish a prima facie equal protection case.

The judgment appealed from, insofar as it concerns the subclass of plaintiffs alleging denial of discretionary parole on grounds of racial and ethnic discrimination, is affirmed.

Notes and Problems

1. As in the previous cases, the plaintiff's description in Inmates of the Nebraska Penal & Correctional Complex v. Greenholtz of the procedural stage at which discrimination occurs is critical in determining N, the size of the group of suspect composition. The size of N is important in making the standard-deviation calculations. Discrimination might have occurred in selecting from the prison population of 2200 persons those 902 who were eligible for discretionary parole, or it might have occurred in deciding which of those 902 eligible prisoners were to be paroled. To determine which is the group of suspect composition one must decide where the discrimination took place. If it is where eligibility for discretionary parole is established, then N is 902. If it is where the board exercises its discretion, then N is 535. Where did discrimination allegedly occur here?

2. In Section III of the opinion, the court notes the significance of choosing the measure of the suspect group once it has been identified. Had plaintiffs presented a longer history of similar treatment, they might have prevailed on statistical grounds, even though the percentage of those minorities denied discretionary parole was unchanged. The court calculated a disparity between expected and observed values of less than two standard deviations for Mexican-Americans and Native Americans (1.88 and 1.93, respectively). Had the Mexican-American and Native American plaintiffs presented 20 years of such discretionary behavior instead of just two, the resulting number of standard deviations would be 5.6 and 6.1, respectively, because N, the expected number of parolees, and the observed number of parolees would all have been larger. You can check this conclusion yourself by calculating the disparities, assuming that the size of the group of suspect composition and the expected and observed numbers are all ten times larger. The implications for advocacy are fairly straightforward. It would still be necessary, of course, to establish that ten years is the relevant time period.

3. The relevance of the distinction between absolute disparities (the approach followed by the district court) and standard deviations (the approach followed by the court of appeals) is easily demonstrated by comparing the absolute deviation of 2%, which the trial court found insignificant, to the number of standard deviations this represents. As an exercise, apply the two-or-three-standard-deviations rule to the court's example of a 2% Mexican-American eligible population with no discretionary parolees.

4. The standard-deviation approach can give what may be termed an advantage to small groups. Compare the disparity calculated in Note 3 to that calculated for a group representing 26% of the population (as blacks did in this case) of whom only 24% received discretionary parole.

5. Note this court's characterization of the criterion of two standard deviations as being not a "floor" for statistical significance, but rather a level at which a "confident showing of statistical significance begins to arise." In reading the following cases, note the extent to which this level of two standard deviations has been adopted as a floor.

BOARD OF EDUCATION v. CALIFANO
584 F.2d 576 (2d Cir. 1978)

OAKES, C.J. . . .

[The Board of Education for the City of New York sued to enjoin the Department of Health, Education, and Welfare from holding them ineligible for assistance under the Emergency School Aid Act (ESAA). This Act is a program designed to aid in desegregating schools and support quality integrated schools. Applicant school systems must demonstrate that their method of assigning teachers to schools does not identify such schools as intended for students of a particular race, color, or national origin. The Chancellor of the Central School Board in New York City has ultimate control over teacher assignment. — Ed.]

The ESAA applications here at issue were for grants in the 1977-78 school year. To analyze whether there was compliance with the statute and regulations, HEW used 1975-76 data. Racial and ethnic statistics demonstrated that in school year 1975-76 62.6% of high school students were minority students whereas 8.2% of high school teachers were minority teachers. Seventy per cent of minority high school teachers were assigned to high schools in which minority student enrollment exceeded 70%, even though these high schools employed only 48% of the system's high school teachers. Conversely, in high schools in which there were proportionately a low number of minority teachers, minority student enrollments were below 40%.

[The high schools with proportionately a high or low number of minority teachers are as follows:

High Schools with Minority Student Enrollments over 90%		% Minority Teachers
Harlem	100%	70.0%
Ben Franklin	98.3	27.9
Park East	93.8	40.0
Harlem Prep	98.4	69.2
Lower East Side	100	63.2
M. L. King, Jr.	96.0	25.0
Satellite Acad.	92.7	25.0
Jane Addams	98.7	34.3
Boys & Girls	99.9	20.9
Eastern District	97.0	18.0
Bushwick	94.2	20.4
Pacific	99.8	37.5
Redirection	97.7	47.6
August Martin	97.6	16.7

Source: Board of Education v. Califano, fn. 25.

High Schools with Minority Student Enrollments under 40%		% Minority Teachers
Stuyvesant	31.0%	2.8%
Bronx H.S. of Science	31.3	2.8
Lafayette	29.2	0.6
Midwood	32.6	1.7
Abraham Lincoln	37.0	0.7
James Madison	35.6	0.9
New Utrecht	22.5	0.0
Fort Hamilton	30.0	3.7
Sheepshead Bay	32.5	3.7
F.D. Roosevelt	29.4	1.8
South Shore	36.9	2.4
William Grady	22.2	0.0
Benjamin Cardozo	38.5	3.2
Francis Lewis	36.9	1.7
Forest Hills	38.8	0.8
Long Island City	30.2	2.8
Richmond Hill	28.5	3.4
Bayside	30.4	1.3
New Dorp	4.3	0.0
Curtis	32.1	3.0
Tottenville	3.7	1.9
Susan E. Wagner	13.0	2.5
Ralph McKee	19.1	3.1

Source: Board of Education v. Califano, fn. 25]

Chapter IV. Standard Deviations

Similar correlations between the racial/ethnic composition of the faculty of community school districts and the racial/ethnic composition of the student bodies within those school districts exist. For the same school year, 14.3% of the teachers and 69.7% of the students in elementary schools were minority, and 16.7% of the teachers and 70.1% of the junior high school students were minority. Quite clearly, the schools with minority student enrollments over 90% identifiably had the highest percentage of minority faculty by a substantial margin.

CSDs with Minority Student Enrollments over 90%		% Minority Teachers
CSD # 1	93.6%	10.4%
4	98.8	24.6
5	99.2	56.7
7	99.0	27.9
9	97.7	26.9
12	98.3	26.7
13	97.0	34.7
14	90.3	14.6
16	99.6	39.0
17	96.1	16.8
19	91.5	12.2
23	99.6	30.0

Source: Board of Education v. Califano, fn. 26.

Similarly, community school districts with minority student enrollments under 50% contained a disproportionately low percentage of minority faculty.

CSDs with Minority Student Enrollments under 50%		% Minority Faculty
CSD 20	31.5	0.4%
21	34.9	2.5
22	29.1	1.7
24	44.0	5.9
25	29.5	2.6
26	25.8	2.7
27	48.4	7.3
30	48.8	6.2
31	16.3	3.1

Source: Board of Education v. Califano, fn. 27.

F. Standard Deviations as Statistical Proof 111

[Other information indicates that ten of the 32 CSDs in New York City employ minority faculty members in excess of 20%. Those CSDs have minority student enrollments varying from 86.7 to 99.6 percent. (fn. 27)]

Upon the "remand" to HEW, HEW found that the racial assignment of faculty in the central school district was, as HEW put it, "strikingly illustrated by the absence of minority teachers" at certain academic, i.e., nonvocational high schools. Ten of these were demonstrated to have a disproportionately low number of full-time minority teachers in the 1975-76 school year. All ten of these schools were among the thirteen academic high schools with full-time faculties having a percentage of black teachers at or below two standard deviations, which was 1.2%; the mean of full-time black teachers in academic high schools systemwide was then 5.2%.

Academic High Schools	% Black	Total Teachers
Stuyvesant	.9	109
Lafayette	.6	166
Midwood	.9	115
Abraham Lincoln	.7	137
James Madison	.9	117
New Utrecht	0.0	163
Fort Hamilton	.6	163
F. D. Roosevelt	1.2	169
Beach Channel	.7	137
Francis Lewis	.8	121
Forest Hills	0.0	125
Bayside	.7	150
New Dorp	0.0	105

Source: Board of Education v. Califano, fn. 28.

[In Castaneda v. Partida, 430 U.S. 482, 496-97 & n.17 (1977), a grand jury discrimination case, the Court adopted a statistical methodology used in the social sciences for the prediction of fluctuations from an expected value, known as the standard deviation, defined for the binomial distribution as the square root of the product of the total number in the sample (n) times the probability of selecting a minority (p) times the probability of selecting a non-minority (q), thus $\sqrt{npq}$. To express the standard deviation in proportionate terms, the formula is $\sqrt{pq/n}$. The statistical approach was also utilized by the Court in a school segregation case, Hazelwood School Dist. v. United States, 433 U.S. 299, 308-09 & n.14 (1977):

> A precise method of measuring the significance of such statistical disparities was explained in Castaneda v. Partida, 430 U.S. 482, 496-497, n.17. It involves calculation of the "standard deviation" as a measure of predicted fluctuations from the expected value of a sample. Using the 5.7% figure as the basis for calculating the expected value, the expected number of Negroes on the Hazelwood teaching staff would be roughly 63 in 1972-1973 and 70 in 1973-1974. The observed number in those years was 16 and 22, respectively. The difference between the observed and expected values was more than six standard deviations in 1972-1973 and more than five standard deviations in 1973-1974. The Court in *Castaneda* noted that "[a]s a general rule for such large samples, if the difference between the expected value and the observed number is greater than two or three standard deviations," then the hypothesis that teachers were hired without regard to race would be suspect. 430 U.S., at 497, n.17.

See id. at 311-12 n. 17, 97 S. Ct. 1272. In this case the standard deviation is 1.94% above or below 5.1%, or p, since the average size of academic high school faculties is 128 teachers, and 94.9% is the nonblack force of teachers at those schools. The square root of $\sqrt{pq/n}$ is 1.94%. Thus, in reference to the schools listed in note 28 supra, all have a standard deviation of two or more: 5.1% (minority teachers in all academic high schools) minus 1.94% (one deviation), 1.94% (a second deviation) = 1.2%. If the same calculations are made for the schools referred to in note 28 supra with respect to minorities in the teaching population systemwide (8.2%), the difference between the expected percentage of minority teachers to the actual percentage is of course higher. In the case of Lafayette, it would exceed three standard deviations. (fn. 29)]

To take another example for the same school year, 8.2% of academic high school teachers in the Central Board's employ were members of minority groups, black or Hispanic. Lafayette High School, for one, with a total of 166 teachers had only one minority teacher, even though it could have been expected based on systemwide statistics to have had fourteen minority teachers. Lafayette's proportion of minority students was 29.2%. In contrast, for the same year Boys High School in Brooklyn had more than two and one-half times the number of full-time minority teachers than the expected rate; its student body was 99.9% minority.

These substantial disproportions are not contested by the appellants, nor do they deny that the schools were statistically "racially identifiable" as a result of the significant disparities in staff assign-

High School Lafayette (Brooklyn)

School Year	Total Full Time Teachers	Expected Percent of Minority Teachers	Expected Number of Minority Teachers	Actual Number of Minority Teachers	Extent of Deviation		Percent of Minority Students
					Percent of Actual to Expected Number of Minority Teachers	Ratio of Expected to Actual Number of Minority Teachers	
1971-72	241	6.4%	15	1	6.7%	15:1	20.1%
1972-73	216	6.6%	14	1	7.1%	14:1	23.9%
1973-74	219	7.2%	16	1	6.2%	16:1	25.6%
1974-75	196	7.7%	15	1	6.7%	15:1	27.8%
1975-76	166	8.2%	14	1	7.1%	14:1	29.2%

Source: Board of Education v. Califano, fn. 30.

ments. The claim pressed below and on this appeal has been limited to the argument that the statute and regulation must be construed to require HEW to establish that the disparities resulted from purposeful or intentional discrimination in the constitutional sense. . . .

IV. DISCUSSION

A. CONSTITUTIONAL STANDARD VERSUS IMPACT STANDARD

The principal argument raised by appellants is that in evaluating the distribution of teachers throughout the New York City schools HEW should have employed the constitutional test of intentional discrimination. To find a violation of the Fourteenth Amendment, the constitutional standard requires a showing not only of disparate impact, but also of illicit motive.

While appellants argue that HEW's decision to deny ESAA funds relies solely on statistical evidence of disparate impact, contrary to Supreme Court cases construing the Fourteenth Amendment, we need not reach the question whether the evidence supports a finding of purposive segregative intent. Because we are dealing with an act of Congress, as amplified by HEW regulations, and not with a judicial determination whether certain acts have produced a Fourteenth Amendment violation, it is permissible for Congress to establish a higher standard, more protective of minority rights, than constitutional minimums require. For example, Title VII cases have not required proof of discriminatory motive, at least where the employer is unable to demonstrate that requirements causing a disparate impact are sufficiently related to the job.

Here, Congress intended to permit grant disqualification not only for purposeful discrimination but also for discrimination evidenced simply by an unjustified disparity in staff assignments. This conclusion seems clear from the statute which expressly requires that all ESAA "guidelines and criteria . . . be applied uniformly . . . without regard to the origin or cause of such segregation." 20 U.S.C. §1602(a). . . .

In disregarding "the origin or cause of segregation," 20 U.S.C. §1602(a), HEW determined that the Central Board failed to present a sufficient justification for the racial disparities in teacher and staff assignments. The proffered justifications for the substantial disparities in the predominantly ten nonminority academic high schools included (1) restrictions on the transfer of teachers written into the collective bargaining agreement, (2) the desirability of teaching assignments in those schools, (3) the unwillingness of many non-minority teachers to teach in predominantly minority schools and (4) the unequal distribution of licenses in specific areas. None of these explanations is adequate to justify the racial disparities in staff assignments. . . .

[The judgment of the district court affirming the denial of the grant applications is affirmed.]

Notes and Problems

1. As in many cases where the legal issues arise out of administrative decisions, Board of Education v. Califano presents a mountain of data susceptible to statistical analysis. Unlike previous cases in this chapter, there is no single standard-deviation calculation that proves the substantiality of the alleged disparate treatment. There is, for instance, no single group of suspect composition. Instead, the racial composition of each school's faculty might impermissibly "identify . . . such school as intended for students of a particular race, color, or national origin." The two-or-three-standard-deviation rule can be applied to the faculty composition of each school. Or such an analysis can be used to describe a larger picture — the racial job assignment pattern in the entire city school system. The court refers, for instance, to an HEW finding that for each of ten selected academic, i.e., nonvocational, schools the observed percentage of black teachers was equal to or greater than two standard deviations from the expected percentage. The court compiles examples of statistical disparities to lead to its overall conclusion of disparate impact of teaching assignments. In such cases, where a tremendous number of statistical calculations could be presented to the court, the lawyer-statistician-advocate's task is to emphasize those examples of disparity that are particularly compelling.

2. Note that the court's statistical analysis is wholly in percentage terms, rather than in the absolute numbers we saw in previous cases. For instance, the court calculates one standard deviation as 1.94% of the faculty rather than, for instance, 12 faculty members. Using percentages is sometimes more convenient than using the raw numbers. It may eliminate the need to calculate expected numbers, for instance. It makes the textual discussion of disparities clearer.

Working with percentages requires using a new formula for standard deviations. This alternative formula gives the same result but in percentage terms. The proportional formula for standard deviation is:

$$SD = \sqrt{\frac{P_1 \times P_2}{N}}$$

where P_1, P_2, and N, as before, represent the percentage of the population belonging to one category, the percentage of the population not in that category, and the number of people in the group of suspect composition, respectively.

Apply this formula to the now-familiar *Castaneda* calculation. In that case, 79.1% of the county's population was Mexican-American, 20.9% was not, and the number of people summoned to serve as grand jurors during the 11-year period 1962-1972 was 870. One standard deviation is:

$$SD = \sqrt{\frac{.791 \times .209}{870}} = .014 \text{ or } 1.4\%$$

To measure the absolute disparity in terms of percentage points, subtract the observed percentage of Mexican-American jurors, 39%, from the expected, 79.1%, which gives you 40.1%. In terms of standard deviations, this percentage disparity is 40.1% ÷ 1.4%, or nearly 29 standard deviations away from the expected percentage. Recall that 29 standard deviations was also the result using numbers of individuals. The Court in *Castaneda* calculated an expected number of 688 (.791 × 870), an observed number of 399, a disparity of 349 (688 − 399), and one standard deviation as 12 people. In numerical terms the observed number was 349/12 or 29 standard deviations from the expected value. The result of applying the two-or-three standard deviations rule will *always* be the same, no matter whether the percentage or numerical calculation method is used.

3. To test your understanding of the percentage approach to calculating statistical disparities, calculate the number of standard deviations between the observed percentage of black teachers at Stuyvesant High School and the expected percentage of 5.1% in all academic high schools. Assume that .9% of 109 teachers at Stuyvesant are black.

4. The legal issue in Board of Education v. Califano is whether the disparate-impact standard or the higher, tougher to prove, impact-plus-illicit-discriminatory-motive test must be met. An observed value that is two or three standard deviations away from the expected value may be enough to show discriminatory impact. How many standard deviations away from the expected value should the observed value be to allow one validly to infer intent? If the probability of finding a particular observed number purely by chance is one in

a trillion, should that evidence alone be sufficient to prove that the defendant intended to discriminate? Should statistical evidence alone ever be enough to allow us to infer an intention to discriminate? We will confront this issue again in subsequent cases.

EEOC v. UNITED VIRGINIA BANK/SEABOARD NATIONAL
615 F.2d 147 (4th Cir. 1980)

WIDENER, Circuit Judge:

On April 15, 1975, following a lengthy investigation, the Equal Employment Opportunity Commission (EEOC) filed this action against United Virginia Bank (bank or UVB) alleging that from July 2, 1965 to the date the complaint was filed UVB had discriminated against black applicants in its hiring procedures. . . .

Hiring and its attendant initial job assignment is the only issue in this case. Significantly, there is no claim of other racial discrimination in such things as promotions, transfers, pay, etc., which, as often as not, appear in litigation of this nature.

At trial, the EEOC presented the following as evidence of discrimination: (1) The principal part of its case was a statistical comparison of black employees at UVB with black people in the total area work force; (2) a statistical comparison of black and white applicant to hire ratios; (3) specific policies which allegedly discriminate against blacks, e.g., credit checks and a high school education requirement; and (4) individual instances of discrimination, mainly relating to an alleged failure to hire qualified black applicants when openings were available.

The centerpiece and keystone of the EEOC's case, both in the district court and on appeal, is that the proper statistical comparison in this case is between the percentage of black employees working in various job classifications at UVB and the percentage of black people in the local labor force. Black people represent 27 percent of the local labor force and 9.5 percent (32/336) of UVB's 1974 office/clerical workers, 78.6 percent of its service workers (11/14), 33.3% (1/3) of its operative employees, and 1.8 percent (2/111) of its managers.

The EEOC also introduced evidence it deemed showed discriminatory distribution of blacks in UVB's sixty job categories and that 43 of the 60 categories had no black employees.

The fundamental problem with the EEOC's statistical evidence lies in the fact that UVB's work force was compared with the work force as a whole. As the district court correctly recognized, this comparison was improper. It is clear that:

> When special qualifications are required to fill particular jobs, comparisons to the general population (rather than to the smaller group of individuals who possess the necessary qualifications) may have little probative value.

Hazelwood School District v. United States, 433 U.S. 299, 308 n.13 (1977). Indeed, we spelled out this proposition in our opinion on the first appeal in this case. EEOC v. United Virginia Bank/Seaboard National, 555 F.2d 403, 406 n.7 (1977). The district court followed our opinion on remand and held that the proper comparison was between UVB's work force and those individuals with the requisite qualifications for the particular job.

The EEOC, however, rigidly continues to argue that all the black local labor force is qualified for the office and clerical positions at UVB. This is simply not true. Tellers must be able to deal with the public, handle and account for money, and operate adding machines, typewriters and other office machines. The district court found that the entire percentage of black people in the local labor force would not provide an appropriate statistical group for comparison with UVB black employees. Since this determination was a factual one, it will not be disturbed unless clearly erroneous. FRCP 52(a). We, therefore, need not engage in a detailed re-examination of the job qualifications in this case for the district court's decision was obviously not clearly erroneous. The above discussion applies a fortiori to the EEOC's attempt to compare the number of black managers at UVB with the number of black people in the local labor force.

The EEOC failed to present any evidence as to the percentage of persons in the labor force qualified to hold the various positions at UVB. The bank did introduce a report entitled "Manpower Information for Affirmative Action Programs" prepared by the Virginia Employment Commission. The report, based on U.S. Labor Statistics in part, shows the percentage of black and other minorities in various occupations in 1975 for the Norfolk, Virginia Beach, Portsmouth Standard Metropolitan Statistical Area (SMSA). The district court took the data in this report and converted it into percentages which could be compared to the major job categories at UVB, and summarized these percentages in a table. The district court may have mistakenly thought, however, that the Virginia Employment Commission report was for 1970 rather than 1975. If so, it compared the SMSA percentages with UVB's 1970 work force when it should have compared them to the 1974 work force (the last full year before this action was initiated).

F. Standard Deviations as Statistical Proof 119

Category of Employment	% Blacks Employed in SMSA in 1975*	% Blacks Employed by UVB in 1970	% Blacks Employed by UVB in 1974
Official/Management	4.8	0	1.8
Office/Clerical	14.0	6.5	9.5
Operative	45.4	20.0	33.3
Service	49.6	55.0	78.6

* Mistakenly labeled 1970 in district court's opinion.
Source: EEOC v. United Virginia Bank/Seaboard National, fn. 2.

We do not intimate that the SMSA table necessarily demonstrates the percentage of black people in the labor force qualified to perform clerical work. However, the burden was on the EEOC to prove discrimination and to produce evidence to support its position. The SMSA report constitutes the only evidence which even remotely speaks to qualifications. Without these figures, the EEOC's case is virtually without any statistical evidence to support it.

Further problems emerge when an attempt is made to compare the SMSA percentages with the UVB employees. When the EEOC prepared its figures for black employees and applicants, it made no effort to exclude employees hired prior to the effective date of Title VII (July 2, 1965). Thus, the figures EEOC presents are weighted against UVB to the extent that white employees hired prior to the time Title VII was in effect were included in the employment figures. The Supreme Court has clearly stated that an employer who, after the effective date of the Act, "made all its employment decisions in a wholly nondiscriminatory way would not violate Title VII even if it had formerly maintained an all-white work force by purposefully excluding Negroes." Hazelwood School District v. United States, 433 U.S. 299, 309 (1976). *Hazelwood* involved a public employer, but it is clear that private employers are equally included within its rationale. Thus, in order to be valid, a statistical analysis such as the one presented here must exclude those persons hired prior to July 2, 1965 (the effective date of Title VII for private employers). To include pre-Act hires in the statistical analysis in this case would improperly weight the evidence and would tend to show present discrimination by an employer if it had discriminated prior to the effective date of the Act but had not discriminated after the Act took effect. It is therefore clear that the June 9, 1975 employment list as a whole was improperly used against UVB because the EEOC made no attempt to factor out the pre-Act hires. Our own examination of the employment records (the June 1975 list) shows at least 92 1975 employees who were hired prior to July 2, 1965 and are almost certainly bound to have been employed in June 1974. How many other pre-Act

TABLE 1.

Category	% Black Employees in SMSA in 1975[a]	Total No. Employed by UVB in 1974	Expected No. of Black Employees	Observed No. of Black Employees	Standard Deviation[b]	No. of Std. Deviations between Observed and Expected Values[c]
Official/Managerial	4.8	111	5.33	2	2.25	1.48
Office/Clerical	14.0	336	47.04	32	6.36	2.36
Operative	45.4	3	1.36	1	.87	.41[d]
Service	49.6	14	6.94	11	1.87	(−)2.14

b. Standard Deviation = $\sqrt{(S)(A)(B)}$

where S = sample size
  A = proportion of black people employed in population group
  B = proportion of non-black people employed in population group

c. No. Standard Deviations is calculated by dividing the Standard Deviation into the difference between the expected and observed no. of black people employed in the various categories.

d. Sample size probably too small to be statistically significant. However, because this category was included by the district court in its analysis, we have included it here. The same may be said in our recomputation below in Table II of the service category.

a. The figures for the percentage of black employees in the SMSA were apparently computed as the district court states from raw data appearing in the Virginia Employment Commission publication we have before referred to. At this point, it is well to emphasize that the use of these figures by the district court is authorized only in lieu of more reliable evidence going to the qualifications of black applicants to fill the various positions at UVB. In another case, *Patterson*, [615 F.2d 275,] with a similar situation facing us, we did not use the same statistical comparison employed by the district court here. Rather, we used a percentage measured by the percent of the total number of black employees who held given positions. These are the tabulated percentages shown in the Virginia Employment Commission publication. If those percentages were employed here, rather than the ones used by the district court, the result would be more favorable to UVB. For example, in the managerial category, the publications tables show 1.5% black, while the percentage at UVB is 1.8%, the comparison in the district court's opinion being 4.8 to 1.8; similarly, in the office/clerical category, the tabulated percentage is 10.1%, while UVB has 9.5%, and the district court's opinion compared 14.0 with 9.5; for operatives, the district court compared 45.4% with 33.3% employed at UVB, while the tables would compare a maximum of 13.5% with 33.3% employed; and in the service category the district court compared 49.6 with 78.6% at UVB, while the tabular value would be 20.6% to compare with UVB's 78.6%. It is apparent that UVB fares far better in three of the four categories by using the same standard we used in *Patterson*, but we think the method used by the district court in this case is probably better than the one we adopted in the *Patterson* case because the validity of the statistics used in *Patterson* depends on how closely the job distribution of the employer compares with that of the work force as a whole. A bank can hardly be said to be a typical employer with its emphasis on books, bookkeeping, and business machines, not to mention the considerable skill in dealing with money matters required of many of its employees.

F. Standard Deviations as Statistical Proof 121

employees quit, retired, or died during the year 1974-75 is not argued, but a brief reference to a disputed termination list indicates that the few pre-Act hires terminated during the period of June 1974-June 1975 were white, which of course would cause the *Castaneda-Hazelwood* analysis of standard deviations to be weighted further toward the defendant than it already is shown to be by the analysis which follows.

Furthermore, even if the pre-Act hires are not excluded, a comparison of the number of black employees of UVB with the number of black employees in similar jobs in the SMSA demonstrates that the difference between the two figures is probably not statistically significant. [Table 1, on the opposite page,] first prepared by the defendant from the district court's table, which we have recalculated, shows the [relevant] comparisons. In Castaneda v. Partida, 430 U.S. 482, 496 (1977), the Supreme Court indicated that figures falling outside two or three standard deviations from the expected values constitute a statistically significant difference. The number of operative workers involved here probably was too small for a statistically significant comparison, and in all events UVB was well over the mark in that category. We are left then with the official/managerial, office/clerical, and service categories which show that the observed numbers of black employees were 1.48, 2.36 and (−)2.14 standard deviations from the expected value on a particular day in June 1975. This difference is borderline at best in view of the Supreme Court ruling that a variation of two or three standard deviations would normally be considered significant, and in any event the numbers in the case at bar are far different than those in *Castaneda* where the observed value was more than 29 standard deviations from the expected value over an 11-year period.

Furthermore, the statistical significance which might be attached to these figures falls out when it is remembered that this table includes at least 92 pre-Act hires. Eight of these (5 black and 3 white) were in the service department. Thirty-five (33 white and 2 blacks) were in the office/clerical category, and 49, all white, were in the official/managerial category. None were operatives. When these pre-Act hires are removed from Table 1, . . . the results [are as shown in Table 2].

When the employees who were hired prior to the effective date of the statute are excluded from the table, we see that what doubtful statistical significance the statistical evidence had is reduced still further. The numbers of standard deviations in the two largest categories, comprising 354 out of 363 employees, were reduced in the official/managerial category from 1.48 to .35, and in the office/clerical category from 2.36 to 2.02. The standard deviations for the operative category remained the same, while in the service category the

TABLE 2.

Category	% Black Employees in SMSA	Total No. Employed by UVB in 1974	Expected No. Black Employees	Observed No. Black Employees	Standard Deviation	No. Standard Deviations
Official/ Managerial	4.8	53	2.55	2	1.56	.35
Office/ Clerical	14.0	301	42.14	30	6.01	2.02
Operative	45.4	3	1.36	1	.87	.41
Service	49.6	6	2.98	6	1.25	(−)2.41

standard deviations increased from -2.14 to -2.41. But in the service category, which is yet less than 3 standard deviations, the number of employees in the statistical sample was reduced to 6, probably too small a number for statistical comparison. Even if the number of 6 employees, however, should be considered as a large enough statistical sample, the EEOC offered no evidence at all that any of these 6 employees who, being employees, must be presumed to be readily available as witnesses, had applied for any position other than in the service category. Even more importantly, neither did it offer any evidence at all that any of the 6 had the qualifications to hold positions in another category, especially in the office/clerical or official/managerial categories. The operative category seems not to have been seriously considered by anyone.

Taking all of these things into consideration, we are of opinion the statistical evidence we have just discussed does not suffice to prove a prima facie case of discrimination.

Having shown that the district court incorrectly considered certain parts of the statistical evidence, its ruling that a prima facie case was established, if correct, must rely on the balance of the statistical information it considered of consequence. It found that there were no black employees in 43 of 60 specific job assignments and that those were largely the positions most highly paid; that black employees were in only 2 of 109 managerial positions; that in 1975 no black employees were employed in 5 specific job assignments in a total of 84 positions, and all 14 custodial positions were held by black employees. The district court considered this evidence as "the failure of UVB to place any blacks in the majority of positions and job categories." The question we have is whether or not this will support its ruling that a prima facie case was established, having shown that a part of the statistics which it relied on were improperly considered.

Even if the entire local labor force were the appropriate group for statistical comparison, there are still serious problems with these statistics. It is grossly misleading to say that 43 of the 60 job positions

F. Standard Deviations as Statistical Proof   123

were occupied only by white employees. The bank points out, without contradiction, that 22 of these categories contained only one employee, and 41 of the categories contained less than 5 employees. Absent other evidence, it is ordinarily improper to draw statistically significant comparisons from such small samples. While some statistical significance might be attached to such figures in some cases, in the case before us we feel that they lack statistical significance because, for practical purposes, they stand alone. With the exception of the individual employees mentioned elsewhere in the opinion, out of what must have been thousands of black applicants, the EEOC offered no evidence to show that any black applicant had the qualifications to hold one of these positions. Neither did it offer any evidence to show the seniority of the people in the positions, it being uncontradicted that the bank prefers to promote from within. It is significant to note here there is no claim of discrimination in promotions or transfers. Neither is there any claim of unequal pay because of race for people performing the same work. There not being statistical proof of discrimination within the larger group taken as a whole, we think it is not proper to break the group down into sub-components with so few members as one and to find a prima facie case within a sub-group when such finding is not permissible for the larger group of which the sub-group is a part. Accordingly, we are of opinion the EEOC did not prove a prima facie case of discrimination by statistical evidence. . . .

EEOC also complains that the bank's standards for hiring were subjective in some aspects, particularly in granting employment interviews and in the remarks made by a receptionist prior to such interviews. Out of the thousands of applicants received by the bank during the nine years in question, EEOC introduced evidence that 12 of them contained comments made by a receptionist, not by the person conducting the employment interviews, which it claims indicate racial discrimination in its hiring policy. More accurately, any intended derogation from some of the comments we think might be characterized as racial slurs. The district court found that the comments were not racially discriminatory, and we cannot say its findings are clearly erroneous. . . .

Accordingly, we are of opinion the EEOC failed to prove its case of discrimination against UVB with respect to hiring.

The judgment of the district court is accordingly affirmed.

Notes and Problems

1. One of the primary statistical issues in EEOC v. United Virginia Bank/Seaboard National is the proper choice of expected

and observed percentages of blacks in the UVB workforce, i.e., the choice of expected and observed values of P_1. Recall that P_1 is defined as the percentage of minority people in the population. In the problems following *Castaneda* we first confronted the issue of which minority people are to be included in this percentage. Justice Burger argued for including only those eligible to vote by reason of literacy, age, and nonfelonious behavior. The Fourth Circuit in the *UVB* case describes the EEOC's contention that the entire black local work force be included in the expected percentage (P_1) as the "centerpiece and keystone" of its case. The court insisted on some measure of "eligibility" for the UVB job openings, specifically, "that the proper comparison was between UVB's workforce and those individuals with the requisite qualifications for the particular job." This reduced the expected percentage of blacks from 27% to 4.8% for the official/managerial positions and from 27% to 14% for the office/clerical positions. What difference did these reductions make to the statistical finding? Note that for two other categories of positions, operative and service, the expected percentages of blacks rose following the qualifications readjustment. What effect did this have on the statistical findings?

2. The court also rejected the EEOC measure of the *observed* percentage of black employees in the various categories. Only discriminatory acts occurring after the effective date of Title VII in 1965 are actionable under its provisions. For this reason, UVB's employees hired before July 2, 1965, had to be ignored in calculating what observed percentage of employees were black. Deleting from factual consideration pre-Title VII acts is a frequent difficulty in employment discrimination cases and has a major impact on statistical conclusions. What would be the effect on the standard-deviation calculation in *UVB* of deleting pre-Title VII hiring if, for instance, everyone hired before the effective date of Title VII were white?

The calculation of standard deviations is a mere mathematical trick. The art of advocacy lies in choosing, presenting, and arguing for the numbers that best prove one's case.

UNITED STATES v. MASKENY
609 F.2d 183 (5th Cir. 1980)

COLEMAN, Circuit Judge.

Somewhere around 10 o'clock, P.M., March 6, 1978, Pilots (appellants) Darwin and Maskeny, accompanied by defendant Perkins, landed a twin engined airplane at the Sylvania, Georgia public airport, loaded with 3,623 pounds of marijuana. They were met with an entirely unexpected welcome from United States Customs Agents and Georgia state officers who had been awaiting their arrival.

The other defendants (appellants), also expecting the arrival, had stationed themselves and their trucks in the adjacent woods. They rushed to the plane and quickly transferred the cargo, after which they set out for Washington, D.C., only to be intercepted by other officers who had been waiting down the road. No ambush was ever more successfully laid or executed.

This happened because from the outset some of the participants unwittingly had been dealing with an undercover government agent who had been kept informed of everything that went on. Indeed, these conspirators were so gullible that an agent had no difficulty in leading them to believe that *he* was the owner of the publicly owned Sylvania airfield. While most everybody has heard of wooden nutmegs and of the Brooklyn Bridge, two of the conspirators actually paid the agent $2,000, cash in advance, for the privilege of landing their clandestine cargo on "his" airfield.

The defendant who had initiated the scheme with the government agent, and who handed over the $2,000 supplied by one of the others, quickly saw that he was irretrievably caught in the jaws of the lion, so he pleaded guilty. He testified for the prosecution, corroborated everything that the agent had sworn to, and wobbled only as to whether it was in November or December that he had received his first contact from one of the other conspirators, an immaterial matter in light of the evidence in the case. Defendant Kraince also pleaded guilty. The proof was annihilating but the remaining ten men stood trial to a jury. . . .

All defendants were found guilty on all charges. . . .

A. THE CONSTITUTIONAL CHALLENGE

In Duren v. Missouri, 439 U.S. 357, 99 S. Ct. 664, 58 L. Ed. 2d 579 (1979), the Supreme Court stated that a defendant's sixth amendment right to a petit jury selected from a fair cross-section of the community is violated when there is a systematic disproportion between the percentage of a "distinctive" group in the community and its representation in venires from which juries are selected, unless the state shows that the aspects of the process that result in the disproportion manifestly and primarily advance a significant state interest, id. 99 S. Ct. at 668-71 and 670 n. 26. The Court clearly set out the elements of a prima facie violation of the fair cross-section requirement: the defendant must show "(1) that the group alleged to be excluded is a 'distinctive' group in the community; (2) that the representation of this group in venires from which juries are selected is not fair and reasonable in relation to the number of such persons in the community; and (3) that this underrepresentation is due to systematic exclusion of the group in the jury-selection process." Id. at 668.

Appellant has alleged that the proportion of various groups responding to a jury service questionnaire and being placed on the qualified wheel varies to a constitutionally impermissible degree from the proportion of these groups in the community. We need not decide whether each of these groups is "distinctive" for purposes of a jury challenge[4] as we find that appellant has failed to make out a case of a constitutionally impermissible disproportion.

According to appellant's statistics, the disparity between the percentage of each allegedly "distinctive" group in the community and the percentage of that group either returning questionnaires or ending up on the qualified wheel is less than ten percent. The Supreme Court in Swain v. Alabama, 380 U.S. 202, 208-09, 85 S. Ct. 824, 13 L. Ed. 2d 759 (1965), held that underrepresentation by as much as ten percent did not show purposeful discrimination based on race. We recognize, however, that *Swain* has an equal protection case where purposeful discrimination must be shown and that the Court in *Duren* stated that a defendant need not show discriminatory purpose for a sixth amendment violation. The Court in *Duren,* however, discussed the statistical discrepancy needed to make out an equal protection violation along with its discussion of the disproportion that demonstrates a sixth amendment violation, 99 S. Ct. 670 n. 26. Thus, while the Court stated that statistical evidence is used to prove *different elements* in equal protection and sixth amendment claims, it did not indicate that the necessary amount of disparity itself would differ. This Court in Thompson v. Sheppard, 490 F.2d 830 (5th Cir. 1974), *cert. denied,* 420 U.S. 984, 95 S. Ct. 1415, 43 L. Ed. 2d 666 (1975), upheld a system that resulted in an eleven percent disparity between the percentage of black people in the population and black people on the jury list. Appellant urges that we either find that an absolute disparity below ten percent violates the constitution or that we base our decision on data derived from other statistical methods, specifically comparative disparity or disparity in standard deviations. It is true that the Supreme Court in Castaneda v. Partida, 430 U.S. 482, 496 n. 17, 97 S. Ct. 1272, 51 L. Ed. 2d 498 (1977), discussed disparity in standard deviations, but it seems clear to us that the Court in that case based its holding on an

---

4. Appellant argues, for example, that persons under the age of thirty who are eligible for jury service constitute such a group.

F. Standard Deviations as Statistical Proof 127

absolute disparity of 40 percent between the population of Mexican-Americans in the community and the percentage summoned for jury service, id. at 495, 97 S. Ct. 1272. In fact, the Court specifically noted that the actual disparities earlier accepted by the Court as adequate for a prima facie case[5] had all been within the range presented in *Castaneda,* id. at 496, 97 S. Ct. 1272. This Court as well has referred to statistical methods other than absolute disparity, but has never found a constitutional violation based on the data produced by such methods. Finally, the Supreme Court focused on absolute disparities in *Duren,* supra. Appellant argues that reliance on absolute disparity could lead to approving the total exclusion from juries of a minority that comprised less than ten percent of the population of the community. We need not, however, speculate here on how we would treat such a situation for all the groups analyzed in appellant's statistics comprise more than ten percent of the community.

In sum, we decline appellant's invitation to focus on comparative or standard deviation disparity and find that the absolute disparities shown do not make out a constitutional violation. . . .

Notes and Problems

1. United States v. Maskeny, decided in 1980, illustrates the continuing vitality of the absolute-disparity approach. In this case, the court rejected a disparity between the percentage in the community and the percentage in the jury of less than ten percentage points for each of the classes raised by defendants. We could easily check the implications of using the standard-deviations approach, but we are not told the size of the jury venire. Nonetheless, you can easily demonstrate that the effect of using standard deviations in such a case depends on the size of this group of suspect composition. Consider communities of varying size with populations of which 20% are eligible jurors under 30 years old (one class suggested by defendants). For communities with jury venires of sizes 24, 240, and 2400, the disparities from a ten-percentage-point underrepresentation are 1.22, 3.89, and 12.25 standard deviations respectively. It may be surprising to see increasing disparities in terms of numbers of standard deviations associated with the same percentage under-

---

5. By our calculations, the actual disparities in the cases cited by the Court were 27.1%, 24.4%, and 19.7%.

representation, 10% in this case, as the size of the group of suspect composition increases. This property of the standard-deviation approach contrasts sharply with the absolute-disparity approach, because the absolute disparity remains at 10% as the group's size increases. A disparity too small to be significant in absolute terms may rise to significance under the two-or-three-standard-deviation rule.

The increasing disparity associated with increasing group size is in accord with intuition. If two randomly selected people were kidnapped from downtown offices during business hours, it would not be surprising if neither were a lawyer. If 20 people were randomly selected, it would still not be surprising if no lawyer were among them, even given the supposed glut of attorneys. If 300 offices were randomly raided, however, we would casually and unscientifically expect that one would be a law office and that we could demand lawyer's ransom. More scientifically, we would say that there is a lower probability of finding no lawyer as the group size expands. And while the standard-deviation approach appropriately takes group size into account, the absolute-deviation approach does not. Judges' willingness to use the approach varies within and among jurisdictions.

2. Courts frequently confuse a "difference between percents" with a "percentage difference." Judge Coleman reports that the disparity between the percentage of each allegedly distinctive group in the community and the percentage of that group on the jury venire is less than ten percent. The judge is simply subtracting one percentage result, the observed percentage, from another, the expected percentage. It would be more accurate to say that there is less than a ten percentage-point difference between these values because this is a difference between percents. This confusion also arises in the Judge's discussion of Swain v. Alabama, 380 U.S. 202, 208-209, in which he reports the Supreme Court holding that underrepresentation by as much as 10% does not show purposeful discrimination. A 10% underrepresentation means that the observed value was 10% less than the expected value, a percentage difference between the two values, not that there was a ten percentage-point difference between observed and expected percentages.

Consider the following example. Thirty percent of the community is black. Only 20% of the jury list is black. The jury list contains 600 names. The difference between the observed and expected percents is ten percentage points, yet this is a 33% underrepresentation of blacks. We have one-third fewer blacks than we expect. This percentage difference can be calculated from the observed and expected percentages or numbers. If we use numbers, the expected number of blacks is 30% × 600, or 180, and observed is 20% × 600, or 120. The percentage underrepresentation is (180 − 120)/180, or 60/180, which equals 1/3 or 33%. If we use percentages, the percentage underrepresentation is (30% − 20%)/30%, which equals 1/3 or 33%. In Swain v. Alabama, the expected and observed percentages were 25% and 15% respectively. This is a ten percentage-point difference, but it represents a percentage underrepresentation of (25% − 15%)/25%, which equals .4 or 40% — not 10% as reported by Judge Coleman. This error frequently appears in opinions. In fact the Supreme Court in Swain v. Alabama made the same mistake when they held "We cannot say that purposeful discrimination based on race alone is satisfactorily proved by showing that an identifiable group in a community is underrepresented by as much as 10%." The Supreme Court cited Akins v. Texas, 325 U.S. 398, 405 (1944), where the expected percentage of black grand jurors was 15% and the observed percentage was only 8.35%, a 6.65 percentage-point difference but a 42% underrepresentation ((15% − 8.35%)/15%). The Court also cited Cassell v. Texas, 339 U.S. 282, 284-285 (1949), where the difference between expected and observed percentages was 8.8 percentage points (15.5% − 6.7% = 8.8%), but the blacks were underrepresented on grand jury panels by 57% ((15.5% − 6.7%)/15.5%).

In order to appreciate the difference between a "percentage difference" and a percentage-point "difference between percentages," perform the following exercises. The eligible population of Hidalgo County is 79.1% Spanish-surnamed.

(a) The percentage underrepresentation of Spanish-surnamed people on a jury venire of 870 people was 50.7%. How many standard deviations does a disparity of this magnitude represent?
(b) If the percentage-point disparity between expected and observed was 50.7 percentage points, how many standard deviations was the observed from the expected value?
(c) If the percentage-point disparity is 40.1%, how many standard deviations was the observed number from expected?
(d) If the percentage underrepresentation was 40.1%, how many standard deviations was observed from expected?

3. What arguments and evidence would have been necessary to win this case for defendants using a standard-deviations approach?

Note: In re Coca-Cola Co.

While reviewing complaints alleging deceptive or misleading advertising, the Federal Trade Commission frequently encounters statistical evidence. In re Coca-Cola Co., 83 F.T.C. 746 (1975), presents a slightly different context in which the standard deviation was used. The complaint had alleged that advertisements promoting Hi-C, a fruit drink sold by Coca-Cola, "were and are misleading in material respects and constituted, and now constitute, 'false advertisements' as that term is defined in the Federal Trade Commission Act . . . and are false, misleading, and deceptive." One issue was the advertiser's claim of nutritional benefit, particularly the large vitamin C content. The FTC considered the following evidence on this issue:

44. The frame of reference used by the nutritionists who testified in this proceeding for assessing the substantiality of the stipulated vitamin C content of Hi-C was the Recommended Dietary Allowance for vitamin C adopted by the Food and Nutrition Board, a division of the National Research Council/National Academy of Science. This organization consists of many of this country's leading medical and nutritional experts and advises various governmental agencies on matters of health and nutrition.

45. The Recommended Dietary Allowances are designed to measure the nutrient intake of a given nutrient from all food sources on a daily basis. The allowances are intended to provide goals for providing adequate nutrient intakes for practically all normal people in the United States. As such, the RDA's are set at a level in excess of average physiological need in order to take into account differences in individual requirements. Thus, in establishing the RDA levels for vitamin C, the Food and Nutrition Board relied on studies showing that the mean utilization of that nutrient by healthy adult males was 21.5 mg. per day, with a standard deviation of ± 8 mg., whereas the RDA for vitamin C for adult males is 60 mg. The Food and Nutrition Board noted that the RDA level for vitamin C "provides a generous increment" above actual utilization in order to take into account variability in individual needs and to provide a "surplus" to compensate for the fact that some of the vitamin C found in certain foods is lost during the cooking and preparation of these foods. The RDA for vitamin C is considered "exceedingly generous" for all normal conditions.

46. The RDA for children between the ages of 2 and 12, who are the primary consumers of Hi-C, is 40 mg. A single serving of Hi-C, containing 44 mg. of vitamin C, provides 110 percent of the RDA for vitamin C for children. Thus, one serving of Hi-C provides more vitamin C than the amount recommended by the Food and Nutrition Board as the appropriate daily dietary intake for healthy children from all food sources. Such a product is "high" in vitamin C, or an excellent source of vitamin C, in relation to the nutritional needs of children.

83 F.T.C., at 776. Because a single serving of Hi-C contains 110% of the RDA for vitamin C for children, there is no question of its nutri-

tional adequacy. The statistical issue might have been more interesting if the single serving content of vitamin C had been one standard deviation above the "mean utilization of that nutrient" by healthy children but less than the RDA. What would this have meant for Coca-Cola's claims?

G. THE COEFFICIENT OF VARIATION

IN RE FORTE-FAIRBAIRN, INC.
62 F.T.C. 1146 (1963)

INITIAL DECISION BY WILLIAM K. JACKSON,
HEARING EXAMINER. . . .

[The complaint charged Forte-Fairbairn, Inc., with unfair and deceptive acts and practices and unfair methods of competition in that fiber stocks sold as "Baby Llama" fibers were in fact composed of fibers other than baby llama, specifically, baby alpaca fibers. In an effort to identify the fibers, the hearing examiner traced in detail the paper trail, following the fibers from Peru to the ultimate purchaser. He then considered the following evidence. — ED.]

45. Prior to 1960, no scientist ever had had occasion to develop reliable diagnostic criteria for distinguishing baby alpaca from baby llama. The first known work in this area was a study by Dr. Von Bergen in 1960 involving alpaca, llama and vicuna, including the baby fibers. The absence of earlier scientific work in this area is explained by the fact that there was no commercial interest in the baby fibers until recent years.

46. In developing reliable diagnostic criteria for distinguishing the baby fibers, Dr. Von Bergen gathered samples of known origin of both baby llama and baby alpaca skins. These were obtained from the Peruvian Department of Agriculture's Animal Experimental Farm. Although Dr. Golub had examined a few known samples of baby llama prior to 1962, he had not made an extensive study of the matter. Dr. Golub obtained known samples of baby llama from the following sources: The San Diego Zoological Park; the National Zoological Park, the Catskill Game Farm, the Franklin Park Zoo and the York Animal Farm. Although he attempted to obtain known samples of baby alpaca in the United States, he was unable to find any and secured his samples of known baby alpaca from the Animal Experimental Farm in Peru. Both Dr. Von Bergen and Dr. Golub had

studied adult alpaca and llama fibers previously and were familiar with their characteristics.

47. In their research to discover what criteria are diagnostic in distinguishing baby alpaca from baby llama, Drs. Von Bergen and Golub subjected the known samples to the various standard identification techniques which appear in the scientific literature. . . .

48. Analytical criteria expressed statistically which are sometimes useful in identifying fibers include the calculation of the standard deviation (*S.D.*) and the coefficient of variation (*C.V.*). The *S.D.* is a statistical criterion based upon the fact that in plotting a normal distributional curve of a fiber population, one observes a curve which is in the shape of a bell. The *S.D.* is one-sixth of the base line across the curve so plotted. The *C.V.* is an arbitrary number derived by dividing the *S.D.* by the average diameter of the fiber population. In effect, the *C.V.* is the percentage which the *S.D.* is of the average diameter. It is a convenient form for expressing the distribution of the fiber population from the smallest to the largest in relation to the average diameter.

49. Fiber fineness, which must be measured in plotting the distributional curve of a fiber population and in calculating the *S.D.* and *C.V.*, is sometimes helpful for diagnostic purposes. Fiber fineness is measured by the wedge projection method whereby the fiber is observed at a magnification of 500 times. The widths of the fibers are recorded by a wedge ruler which covers a range of 10 to 70 microns, which is the normal average range of all animal fibers used in the textile industry. While visual observation of itself is sometimes helpful in fiber identification, both experts agreed that it would be of little or no value in distinguishing baby alpaca from baby llama.

50. Of the foregoing techniques and observations some were considered to be diagnostic in distinguishing baby alpaca from baby llama, whereas others were not. For example, the average diameter of the 16 fiber samples permitted the experts to conclude that they came from baby animals. This fact, however, was not considered diagnostic in differentiating baby alpaca and llama because the baby fibers of both animals average about the same. Similarly, the length of the fibers and the presence of fiber tips and roots clearly pointed to the conclusion that the 16 fiber samples were composed of baby fibers, but was not considered diagnostic in distinguishing baby alpaca from baby llama.

51. The two experts agreed that there exist both morphological and statistical criteria which permit distinguishing baby alpaca from baby llama fibers. Both the samples of known origin and the samples

G. The Coefficient of Variation

of unknown origin were subjected to these tests. The diagnostic factors are as follows:

A. Upon microscopic examination of the cross-section of a population of baby llama fibers, the presence of a relatively high number of coarse fibers is apparent. This is a characteristic of known samples of baby llama, which even in infancy exhibit the double coated fleece, which is a pronounced characteristic of that animal in adulthood. The presence of coarse fibers, however, is not observed in known samples of baby alpaca. That animal has a single coated fleece, both in adulthood and as a baby. The presence of coarse hairs was observed in the 16 fiber samples as well as in the known samples of baby llama.

B. The shape of the medulla was also considered diagnostic. The medulla of a baby alpaca fiber is quite round and shows little or no tendency toward irregularity. The medulla of a llama fiber, on the other hand, exhibits a pronounced tendency toward such irregularity. This characteristic was observed in the 16 fiber samples as well as in known samples of baby llama.

C. While Dr. Von Bergen noted that some differences between baby alpaca and llama exist insofar as the appearance of the scale structure is concerned, he did not rely on this factor in reaching his conclusions. Dr. Golub, on the other hand, pointed out that if an examination is made of the lower two-thirds portion of fibers of medium size, a difference in the scale margins can be observed. He testified that while the literature makes no reference to the baby animals in this regard, he made a special study of baby alpaca and baby llama to see if this feature was observable, and he found that it was. Thus, he testified that in fibers of intermediate size the scale margins of baby alpaca are very jagged and irregular, whereas those of baby llama are smooth. In this respect, he found that the 16 fiber samples corresponded with the known samples of baby llama which he had studied.

D. Both experts agreed that the $C.V.$ of the fiber population provided a reliable criterion for distinguishing baby alpaca from baby llama. The $C.V.$ of the samples of baby alpaca studied by Dr. Von Bergen was between 18% to 25%, which reflects the relatively uniform staple of the fleece. Known samples of baby llama, on the other hand, range between 28% to 35%. Known samples of baby alpaca studied by Dr. Golub had an average $C.V.$ of 22%, whereas the known samples of baby llama studied by him were consistently above 30%. In examining the 16 fiber samples, Dr. Von Bergen found that the $C.V.$ averaged in the neighborhood of 30.5% to 31.5%, whereas Dr. Golub found a range between slightly below 30% to as high as 39%. Both experts agreed that their observations of the 16 fiber samples in terms of *C.V.* agreed with known samples of baby llama, but did not agree with their observations of known samples of baby alpaca.

E. The *S.D.* of the 16 fiber samples was approximately 6 to 7 microns, which corresponds to the *S.D.* of the known baby llama samples tested.

52. Both Drs. Von Bergen and Golub testified that by using these various diagnostic factors they found a complete correlation between the morphological and microscopic features of the baby llama samples of known origin and the 16 samples of stock from Forte-Fairbairn. Both Drs. Von Bergen and Golub concluded, based upon their analyses of over 1000 fiber samples taken from each of the 16 fiber stocks, at issue in this proceeding, that the stocks were composed wholly of baby llama and that there was no evidence of the presence of fibers other than baby llama, such as baby alpaca, adult alpaca, adult llama or guanaco. Both experts were also in agreement that if any such other fibers had been present in any substantial quantity, their presence would have been detected.

DISCUSSION

The essential allegations of the complaint in this matter read as follows:

> PARAGRAPH FOUR: In the course and conduct of their business, as aforesaid, respondents have made representations concerning their said products on sales invoices. Among and typical of the representations made was the invoicing of their fiber stocks as "Baby Llama".
>
> PARAGRAPH FIVE: The aforesaid representations were false, misleading and deceptive. In truth and in fact, said fiber stocks were not composed wholly of "Baby Llama" but were composed of fibers other than baby llama.

As previously found, the representations offered in support of the above quoted allegations of the complaint were made in connection with a sale of fiber stocks invoiced as "Baby Llama" by respondents on November 25, 1959, to Northfield Mills. The burden of proof is on complaint counsel to establish by clear and convincing evidence that the fiber stocks involved in this sale were "not composed wholly of 'baby llama' but were composed of fibers other than baby llama". To do this, complaint counsel has relied almost exclusively on documentary evidence consisting of foreign supplier invoices and shipping memoranda originating in Peru which describe the fiber stocks as "baby alpaca". All subsequent documents identifying the fibers as "baby alpaca" were copied from these earlier supplier invoices and shipping memoranda originating in Peru. It is clear, therefore, that the later documents and actions of respondents in reliance thereon are in and of themselves entitled to no more weight than the original supplier invoices and shipping memoranda upon which they were based. . . .

There is no doubt that these documents are sufficient to sustain a prima facie case in support of the allegations of the complaint. To rebut this evidence, however, respondents introduced uncontradicted and unimpeached direct evidence based upon tests conducted by two eminently qualified experts demonstrating that the fiber stocks in issue in this proceeding were composed wholly of baby llama. The issue to be decided here, therefore, is whether the identity of the fibers based upon foreign supplier invoices and shipping memoranda is to prevail over direct scientific proof based upon expert testimony of fiber technologists who have analyzed samples of the fiber stocks involved herein.

In focusing on this issue, the hearing examiner has not overlooked the fact that at the time respondents issued the corrected invoices in January 1960, they had no scientific proof, but merely relied upon the information supplied by Mr. Michell. The hearing examiner has also considered respondents' failure to inform Mrs. Murphy in 1961 of their change in identification of the fiber stocks. Although these actions of respondents are not satisfactorily explained and may be inconsistent with their prior conduct, they are not determinative of the issue in this case. The question is not whether respondents had good cause to change the invoices in 1960 or whether their treatment of Mrs. Murphy was fair and above board, but rather of what in truth and in fact these fiber stocks are composed.

In the Matter of Alscap, Inc. [60 F.T.C. 275] (Docket No. 8292, February 14, 1962), the Commission recently adopted a hearing examiner's initial decision wherein he held in effect that blind reliance on foreign labels inaccurately describing fiber content is no defense to a charge of mislabeling, where the Commission adduced scientific evidence through expert witnesses establishing the correct fiber content. In view of the decision in this case, it would appear that the evidentiary weight to be accorded foreign supplier invoices and other overseas shipping advices cannot overcome direct scientific evidence on the issue of fiber identity or content. . . .

By admitting these documents into evidence, the hearing examiner in no way passed upon their trustworthiness, but on the contrary, indicated at that time that any weight to be given to them would depend upon the sources of information from which they were made and the method and circumstances of their preparation. (See Uniform Rules of Evidence, Rule 63(13).) . . . [T]he clerks who prepared these documents did not speak English; the clerks merely copied the pertinent information, including the description of the fibers from other documents; moreover, the clerks never personally inspected the fiber stocks and even if they had they would have been unable to determine the differences between baby alpaca and baby llama. Such lack of personal knowledge of the identity of the fiber stocks by the clerks making the invoices or shipping advices seriously impairs the weight to be given to these documents. Consequently, these documents and the entries and acts of respondents made in reliance thereon are not to be accorded great evidentiary weight in this proceeding.

Finally, the hearing examiner is not unmindful that Donald Forte's January 6, 1960, letter contains admissions that "a few Baby Alpaca Skins are used" and "occasionally, a skin of a Baby Alpaca is included". These statements were presumably based upon information received by Godsoe from Mr. Michell, who as already found had never inspected the fiber stocks in issue and who did not appear or testify. Moreover, the exact details of Michell's conversations with Godsoe are not contained in the record and the hearing examiner consequently accords them little weight. In any event, admissions of this character cannot overcome direct scientific proof to the contrary.

Upon the basis of the entire record and the previous decision of the Commission in the *Alscap* case, the hearing examiner concludes that complaint counsel has failed to sustain the burden of establishing that the fiber stocks in issue were composed of fibers other than baby llama.

CONCLUSIONS

1. The Federal Trade Commission has jurisdiction of and over the respondents and the subject matter of this proceeding.
2. The complaint herein states a cause of action, and this proceeding is in the public interest.
3. The reliable, probative and substantial evidence in this record does not sustain the allegations of the complaint that respondents have engaged in unfair acts and practices or unfair methods of competition in violation of the Federal Trade Commission Act by falsely identifying fiber stocks on invoices as "baby llama".

ORDER

Accordingly, it is ordered, that the complaint in this matter be, and hereby is, dismissed.

G. The Coefficient of Variation 137

DECISION OF THE COMMISSION

Pursuant to Section 4.19 of the Commission's Rules of Practice effective June 1, 1962, the initial decision of the hearing examiner shall, on the 13th day of April 1963, become the decision of the Commission.

Notes and Problems

1. In re Forte-Fairbairn, Inc., demonstrates the versatility of the standard deviation. Many Federal Trade Commission decisions employ statistics in a similar manner in order to evaluate manufacturers' claims regarding their products. Note the extent to which statistical and scientific evidence is given more weight than documentary evidence.

2. The coefficient of variation (C.V.) presented in this case describes in percentage terms how large the standard deviation of fiber diameters was relative to the average diameter of the fibers. The scientists were unable to distinguish between alpaca and llama fibers on the basis of the average diameters of those fibers alone, because the average diameters of the two fibers were equal. Recall that the standard deviation measures how much variation there is from the mean diameter, and that the experts found more diameter variation among llama fibers than among alpaca fibers. So they used the coefficient of variation to represent the standard deviations for the two fibers in percentage terms, dividing the standard deviation by the mean diameter for each type of fiber:

$$C.V. = \frac{SD}{M_{diameter}}$$

Since the mean diameter was the same for each type of fiber and the standard deviation for llama fibers was greater, the coefficient of variation for the llama fibers was greater. The experts reported $C.V.$'s of 30% to 39% for llamas and 18% to 25% for alpacas.

You might wonder why the experts needed to calculate the coefficient of variation when the difference between alpaca and llama fibers would have been apparent from the standard deviations alone. Whenever the averages for the groups being compared are identical, as they were in this case, the standard deviation alone would be sufficient for comparison. Because the coefficient of variation is equally useful where averages are identical and more useful than the standard deviation when averages are different, scientists and lawmakers frequently use the coefficient of variation as the summary measure of variability.

Note: B.F. Goodrich Co. v. Department of Transportation

In B.F. Goodrich Co. v. Department of Transportation, 541 F.2d 1178 (6th Cir. 1976), the court examined National Highway Traffic Safety Act standards for testing automobile tires under the Tire Quality Grading Regulation. The testing agency required that the manufacturing of tested tires be subject to quality controls with coefficients of variation below 5%. Although a tire is designed to certain specifications, due to inconsistencies in the manufacturing process there might be slight variations in such specifications as tire weight or circumference or tread depth. The coefficient of variation measures the percentage variation around that specification for each quality characterisic observed. From measurements of actual tread depth of newly manufactured tires, for instance, a standard deviation can be calculated. The testing agency required that this standard deviation, that is, the typical variation around the specification, be no more than 5% of the specification. Thus, if the specifications called for a tread depth of one-half inch, the standard deviation could be no greater than 5% of 0.5 inches or 0.025 inches. By stating the quality standard in terms of a single coefficient of variation, the testing agency avoided having to state different standard deviation requirements for each specification.

Problem: Taylor v. Weaver Oil & Gas Corp.

In Taylor v. Weaver Oil & Gas Corp., 18 Fair Empl. Prac. Cas. (BNA) 23 (S.D. Tex. 1978), plaintiff Carol Taylor alleged sex discrimination on the grounds that women employees were given smaller raises than their male counterparts. Recognizing that some variation among employees in salary increases was inevitable, given different skills and experience, the defendant's expert compared a summary measure of the variation in raises for a period in which there could have been no sex discrimination to the measure of variation in raises for the period at issue in the lawsuit. The defendant's theory was that if sex discrimination was a factor in setting raises during the second period, then the coefficient of variation would show greater variation in raises during the second period. The coefficients of variation were, in fact, the same for both periods, and defendant prevailed. Note that a comparison of standard deviations alone would not have worked in this case because the average raises were different.

The following are hypothetical data showing raises for two

periods. In which period is the variation in raises, as a percentage of average raise, greater?

EXHIBIT IV-1.
Hypothetical Data for Two Salary Raises

Employee Number	May 1977 Raise ($)	Employee Number	April 1982 Raise ($)
6	2000	23	8,200
17	1000	54	9,500
33	4000	57	11,200
71	2000	71	10,000
72	3500	87	7,200

H. THE NORMAL DISTRIBUTION

In In re Forte-Fairbairn, Inc., Hearing Examiner Jackson described the standard deviation by reference to the "*normal distribution curve* of a fiber population . . . which [when plotted] is in the shape of a bell." He was referring to a graph with the frequency on the ordinate or vertical axis and various fiber diameters on the abscissa. The diameter of the fiber is the variable. The frequency is the measurement and indicates the number of times we observed a fiber with a given diameter. The "normal distribution curve" assumes that most fibers will have a diameter close to the mean value, which was 18.5 microns. That indicates that 18.5 is the most frequent measurement. So at 18.5 microns on our graph the frequency will be the highest. See the graph in Exhibit IV-2. Each x represents one fiber with a diameter equal to 18.5 microns. This hypothetical sample shows 15 fibers with a diameter of 18.5. Our assumption that most fibers will have a diameter close to the mean of 18.5 implies that as we get further away from 18.5 the frequency of other diameters will decrease. The y's in the figure represent other observations. Very seldom will we see a fiber with a diameter of 5 or 35 microns. A greater frequency of observations is seen as one approaches 18.5 microns from either side.

Recall that the standard deviation is a measure of variation from the mean. In this case, one standard deviation for baby llama fibers was 6.5 microns. One standard deviation on either side of the mean takes us up to 25 microns (18.5 + 6.5 microns) and down to 12 microns (18.5 − 6.5 microns). Two standard deviations take us up to

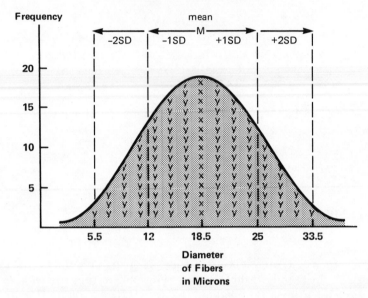

EXHIBIT IV-2.
Normal Curve for Llama Fibers

31.5 microns and down to 5.5 microns. In the next chapter we will see that whenever a sample is expected to be *normally distributed* around the mean, then 68% of the observations will be within one standard deviation of the mean, 95.4% of all observations are within two standard deviations, 99.7% of all observations are within three standard deviations of the mean, and more than 99.99% of all observations are within four standard deviations of the mean. These characteristics of a normal distribution make it a particularly useful tool. In fact, the assumption of a normal distribution underlies most of the statistical tests we will discuss in succeeding chapters. And while you need not understand the theory of a normal distribution to use the tools we will develop, the tests are based on the normal curve and statisticians regularly refer to it, so the well-informed lawyer should have some notion of what it means.

To understand the significance of the normal curve, look back at the *Castaneda* case (see pages 82-89 above). The majority opinion indicates that each jury will have an expected 79.1% Mexican-Americans, reflecting the population of Hidalgo County. This is the mean value on our normal distribution curve. Not all juries will have this percentage. Some will have more Mexican-Americans and some fewer. But we expect that most samples of people drawn from the population to serve on juries will have about 79.1% Mexican-

EXHIBIT IV-3.
Normal Curve for *Castaneda*

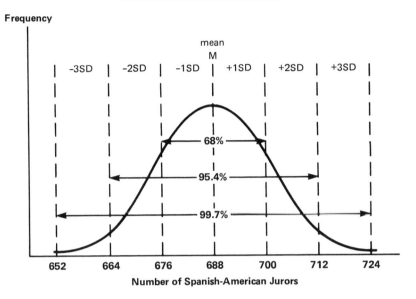

Americans. This implies that we expect very few juries to have as few as 10% or as many as 90% Mexican-Americans, although some undoubtedly will just as a matter of chance. In the 11-year period, 870 jurors were summoned. We would expect 79.1% or 688 of them to be Mexican-Americans, but we observed only 399. One standard deviation was 12 jurors. The graph in Exhibit IV-3 depicts the *Castaneda* normal curve. The mean or expected value is 688 with one, two, and three standard deviations above and below the mean indicated.

Note that 99.7% of the observations will be within three standard deviations of the expected value of 688. That means that only .3% of random drawings will be more than three standard deviations from the mean. The observed value in fact was 29 standard deviations from the mean. There is a negligibly small chance that such an event could occur by chance, which the Court calculated at less than 1 in 10^{24}. There must be something other than random drawing going on. Comparison of the normal curve with the observed frequency does not prove that discrimination was the cause; it only indicates that the drawings were not random.

CHAPTER V

SIGNIFICANCE TESTING

A. STATISTICAL VERSUS PRACTICAL SIGNIFICANCE

Whenever Walter Cronkite or Dan Rather or one of their ilk reports the results of a statistical study (usually a poll), he indicates the "margin of error" to which the conclusions are subject. A recent report indicated that 49% of voters surveyed favored tax cuts, while 27% did not. "These statistical conclusions," said Dan, "are subject to a margin of error of plus or minus 2%." Dan was reporting the significance of his statistics. Certainly if he had said that the margin of error was plus or minus 40%, the statistics would seem much less valuable. That would mean that between 9% and 89% of voters favored the cuts, while 0% to 67% did not. This is about as informative as Dan's coming on the screen with a special report to say, "San Francisco may or may not have been destroyed in the latest earthquake. More news at 11:00."

In statistics, significance does not have exactly the same meaning as in ordinary parlance. Ordinarily a "significant" number or event is one of importance, one of great magnitude, or a noteworthy occurrence. In statistics, the significance indicates, not the attention a number deserves or how important it is, but rather whether it is mathematically accurate enough to fit our purposes, whatever they may be. Thus, Dan Rather's statement that the margin of error was only 2% indicated that the numbers involved are more accurate than if the margin were 40%.

It is very important, then, not to confuse *statistical* significance with *practical* significance. A measurement with great statistical significance is one that is unlikely to have occurred by chance, while a measurement with great practical significance is one of great magnitude or is otherwise noteworthy. A scientific measurement of an earth tremor that is very precise and reliable, such as a 1.5 reading on the Richter scale, may be statistically significant in that it is accurate and the measurement did not occur due to a chance malfunction of the instrument. But a 1.5 reading may be of relatively little practical significance. The newspapers may never mention it. A 12.0 reading from an imprecise instrument that frequently gives random results would have little statistical significance even though, if it happened to be accurate this time, it would have great practical significance. You can see from this example that a high level of statistical significance does not necessarily imply great practical significance. On the other hand, a low level of statistical significance may cause one to doubt the practical significance of the measurement because it may not be accurate.

Statisticians typically assess the statistical significance of a result before examining its practical significance. A lawyer might find it easier to distinguish between the two by identifying statistical significance with the credibility of the evidence and practical significance with the relevance and materiality of the evidence. Evidence that is relevant and material, that is, practically significant, will be given very little weight if it is not credible.

The accuracy or reliability of a statistical measurement is an indication of the likelihood that we could get such a measurement just by chance. Since a measurement that could have occurred by chance is not reliable, tests of *statistical significance* allow us to calculate precisely the probability that our measurement could be the result of a random occurrence. We have already encountered one such test. In Castaneda v. Partida, the Supreme Court cited the general rule that "if the difference between the expected value and the observed number is greater than two or three standard deviations, then the hypothesis that the jury drawing was random would be suspect." Intuitively, having an observation of so many standard deviations from the mean indicates that this observed result is not at all what you would expect if there were no discrimination. The "hypothesis" to which the Court refers is the initial assumption that the jury system is not discriminatory, i.e., that race has nothing to do with being selected for jury duty (remember "innocent until proven guilty"?). Because the assumption is that one factor has *nothing* to do with another, this assumption, which we are testing statistically, is called the *null hypothesis*. Note what the court presents as the null hypothesis in the cases in this chapter. They are included as an introduction to formal mathematical tests that have been developed by statisticians to indicate how reliable a statistical conclusion is. Another, perhaps more descriptive, phrase for significance testing is *reliability testing* or *accuracy testing*. Naturally, courts will be concerned with the accuracy or reliability of your statistics, and cases in this and subsequent chapters reveal a fair degree of sophistication on the part of courts in using mathematical tests to evaluate the significance of statistical conclusions.

B. TYPE 1 AND TYPE 2 ERRORS

The null hypothesis is analogous to a restatement of the factual issues involved in a case in a form testable by statistics. Finders of fact occasionally make errors in resolving the factual issues before them. An error is described in statistics as being either a false rejection of the null hypothesis or a false failure to reject. A false rejection type of error occurs when you conclude, "No, this null hypothesis isn't true," and you are wrong because actually the null hypothesis is true. A false failure to reject occurs when you conclude, "This null hypothesis should not be rejected," but the hypothesis is, in fact, false. The false rejection is called, arbitrarily, a *type 1 error*. The false failure to reject is a *type 2 error*. The type 2 error, failure to reject a false hypothesis, sounds just like *accepting* a false hypothesis. But accepting and failing to reject a hypothesis are different. Statistical tests are designed to permit one to conclude that a statement is not likely to be true. They are not constructed to prove statements true. So one either rejects a null hypothesis or fails to do so.

Consider a statement of the null hypothesis involved in Ballew v. Georgia, above: "The size of the jury has *no* effect on the outcome of the trial." If we do not reject this null hypothesis, it is because we have not demonstrated to our satisfaction that jury size does have an effect on outcome or that different outcomes do result from jury size. There is some evidence (see Zeisel's article, above) that we would be wrong in failing to reject this hypothesis, i.e., there is some chance that we made a type 2 error (falsely failed to reject the hypothesis). The Court in Ballew v. Georgia, however, *rejected* the null hypothesis, in effect saying, "We would be wrong to say there is no effect on the outcome of jury size." And since jury size might have an effect on outcome, the Court concluded that Sixth Amendment rights required maintenance of jury size above a minimum level. When the Court rejected the null hypothesis, they may have been committing a type 1 error, a false rejection, by incorrectly concluding that the hypothesis isn't true. To summarize the characteristics of both types of statistical error:

Type 1 Errors. Incorrectly disagree with an assertion
Falsely reject the null hypothesis

	Conclude there is an effect or a relationship when there is none.
Type 2 Errors.	Incorrectly fail to disagree with an assertion Falsely fail to reject the null hypothesis Conclude there is no significant relationship when there is one.

Note again that the null hypothesis is stated as a negative, "there is *no* effect" or "there is *no* relationship." An incorrect rejection of this assertion is a type 1 error. An incorrect failure to reject is a type 2 error. The problem is, how to remember this? One way is to memorize it. A second way is to invent a mnemonic device that will stick in your mind. It has been suggested that a false rejection has one *f*, so is type 1, while a false failure has two *f*'s, so is type 2.

In criminal trials the null hypothesis is, "This defendant did not commit the crime." A type 1 error would send an innocent party to jail by falsely rejecting this hypothesis. A type 2 error would set a guilty person free by failing to prove beyond a reasonable doubt that the defendant did commit the crime. Historically, in law we have been more concerned with type 1 errors than with type 2. The beyond-a-reasonable-doubt test suggests a willingness to let some guilty people go free. This primary focus on type 1 errors is also true of statistics. As you will see, we will calculate the probability of type 1 errors but not of type 2 errors.

While reading United States v. Fatico, consider the extent to which the court is willing to accept error in the trial of various factual issues. When reading the following cases, establish what null hypotheses are being tested, that is, what factual issues are involved and what sources of error exist.

UNITED STATES v. FATICO
458 F. Supp. 388 (E.D.N.Y. 1978)

WEINSTEIN, District Judge.

In view of prior proceedings, see United States v. Fatico, 441 F. Supp. 1285, 1287 (E.D.N.Y. 1977), *reversed,* 579 F.2d 707 (2d Cir. 1978), the key question of law now presented is what burden of proof must the government meet in establishing a critical fact not proved at a criminal trial that may substantially enhance the sentence to be imposed upon a defendant. There are no precedents directly on point.

The critical factual issue is whether the defendant was a "made" member of an organized crime family. Clear, unequivocal and convincing evidence adduced by the government at the sentencing hearing establishes this proposition of fact. . . .

Defendant was indicted with others for receiving goods stolen from interstate commerce during three hijackings of trucks from Kennedy Airport. 76-CR-80, 76-CR-81 and 76-CR-218 (E.D.N.Y.). At his initial trial on indictment 76-CR-218 the jury failed to agree. Defendant then entered a guilty plea to the conspiracy charge in indictment 76-CR-81 in satisfaction of all charges in the three pending cases. He now faces a maximum penalty of five years imprisonment and a $10,000 fine. . . .

C. BURDEN OF PROOF

1. THE CONTINUUM

a. Burdens in General

We begin with the caution of Justice Brennan in Speiser v. Randall, 357 U.S. 513, 520-21, 78 S. Ct. 1332, 1339, 2 L. Ed. 2d 1460 (1958), about the crucial nature of fact finding procedures:

> To experienced lawyers it is commonplace that the outcome of a lawsuit — and hence the vindication of legal rights — depends more often on how the factfinder appraises the facts than on a disputed construction of a statute or interpretation of a line of precedents. Thus the procedures by which the facts of the case are determined assume an importance fully as great as the validity of the substantive rule of law to be applied. And *the more important the rights at stake the more important must be the procedural safeguards surrounding those rights.* (Emphasis supplied.)

The "question of what degree of proof is required . . . is the kind of question which has traditionally been left to the judiciary to resolve. . . ." Woodby v. Immigration & Naturalization Serv., 385 U.S. 276, 284, 87 S. Ct. 483, 487, 17 L. Ed. 2d 362 (1966).

> Broadly stated, the standard of proof reflects the risk of winning or losing a given adversary proceeding or, stated differently, the certainty with which the party bearing the burden of proof must convince the factfinder.

In re Ballay, 157 U.S. App. D.C. 59, 73, 482 F.2d 648, 662 (1973).

As Justice Harlan explained in his concurrence in In re Winship, 397 U.S. 358, 370, 90 S. Ct. 1068, 1075-76, 25 L. Ed. 2d 368 (1970), the choice of an appropriate burden of proof depends in large measure on society's assessment of the stakes involved in a judicial proceeding.

> [I]n a judicial proceeding in which there is a dispute about the facts of some earlier event, the factfinder cannot acquire unassailably accurate knowledge of what happened. Instead, all the factfinder can acquire is a belief of what *probably* happened. The intensity of this belief — the degree to which a factfinder is convinced that a given act actually occurred — can, of course, vary. In this regard, a standard of proof represents an attempt to instruct the factfinder concerning the degree of confidence our society thinks he should have in the correctness of factual conclusions for a particular type of adjudication. Although the phrases "preponderance of the evidence" and "proof beyond a reasonable doubt" are quantitatively imprecise, they do communicate to the finder of fact different notions concerning the degree of confidence he is expected to have in the correctness of his factual conclusions. (Emphasis in original.)

Thus, the burden of proof in any particular class of cases lies along a continuum from low probability to very high probability.

b. Preponderance of the Evidence

As a general rule, a "preponderance of the evidence" — more probable than not — standard is relied upon in civil suits where the law is indifferent as between plaintiffs and defendants, but seeks to minimize the probability of error.

> In a civil suit between two private parties for money damages, for example, we view it as no more serious in general for there to be an erroneous verdict in the defendant's favor than for there to be an erroneous verdict in the plaintiff's favor. A preponderance of the evidence standard therefore seems peculiarly appropriate for, as explained most sensibly, it simply requires the trier of fact "to believe that the existence of a fact is more probable than its nonexistence before [he] may find in favor of the party who has the burden to persuade the [judge] of the fact's existence."

In re Winship, 397 U.S. 358, 371-72, 90 S. Ct. 1068, 1076, 25 L. Ed. 2d 368 (1970) (Harlan concurring) (footnotes omitted). Quantified, the preponderance standard would be 50 + % probable. But cf. M. Finkelstein, Quantitative Methods in Law, 59-78 (1978) (equalization of errors between parties may require higher probability than minimization of errors — i.e., more than 50 + %).

The preponderance of the evidence test has also been used to determine the admissibility of evidence under the constitutional exclusionary rules. See Lego v. Twomey, 404 U.S. 477, 92 S. Ct. 619, 30 L. Ed. 2d 618 (1972) (plurality opinion) (voluntariness of a confession). In *Lego,* the Court explained that the procedures to determine the validity of a confession are "designed to safeguard the right of an individual, entirely apart from his guilt or innocence, not to be compelled to condemn himself by his own utterances." 404 U.S. at 486, 92 S. Ct. at 625. The jury must still determine the "accuracy or weight of confessions admitted into evidence." Id. The Court thus concluded that:

> Since the purpose that a voluntariness hearing is designed to serve has nothing whatever to do with improving the reliability of jury verdicts, we cannot accept the charge that judging the admissibility of a confession by a preponderance of the evidence undermines the mandate of In re Winship. 397 U.S. 358, 90 S. Ct. 1068, 25 L. Ed. 2d 368 (1970).

After sentencing, the defendant does not retain the opportunity to relitigate some questions that he has after an adverse pre-trial determination. In addition, in the case before us, the facts critical to sentencing are hardly collateral; they cut to the heart of the defendant's liberty. . . . Since the factual determination of the sentencing judge is final, the defendant deserves substantial protection, including a burden of proof higher than that used in negligence cases.

c. Clear and Convincing Evidence

In some civil proceedings where moral turpitude is implied the courts utilize the standard of "clear and convincing evidence" — a test somewhat stricter than preponderance of the evidence.

Where proof of another crime is being used as relevant evidence pursuant to Rules 401 to 404 of the Federal Rules of Evidence, the most common test articulated is some form of the "clear and convincing" standard. A panel of the Ninth Circuit has even suggested a beyond a reasonable doubt test. The Second Circuit applies a preponderance of the evidence test. These standards are designed to give defendants added protection not fully afforded by Rules 403 and 404. Since the crimes are merely evidence of intermediate propositions, not material elements of a crime being tried or of a sentence, there is theoretically no reason why any burden must be met as long as Rule 401's test of relevancy is satisfied — that is, the evidence has any tendency to make the material proposition "more probable or less probable than it would be without the evidence." See United States v. Schipani, 289 F. Supp. 43, 56 (E.D.N.Y. 1968), aff'd, 414 F.2d 1262 (2d Cir. 1969). The organized crime charge before us is more akin to a material proposition than to an intermediate evidentiary proposition. The line of cases dealing with other crime evidence is, therefore, not useful in determining an appropriate burden of proof on sentencing.

Quantified, the probabilities might be in the order of above 70% under a clear and convincing evidence burden.

d. Clear, Unequivocal and Convincing Evidence

"[I]n situations where the various interests of society are pitted against restrictions on the liberty of the individual, a more demanding standard is frequently imposed, such as proof by clear, unequivocal and convincing evidence." In re Balley, 157 U.S. App. D.C. 59, 73, 482 F.2d 648, 662 (1973). The Supreme Court has applied this stricter standard to deportation proceedings, denaturalization cases, and expatriation cases. In Woodby [v. Immigration & Naturalization Service, 385 U.S. 276, 285 (1960)], the Court explained:

> To be sure, a deportation proceeding is not a criminal prosecution. But it does not syllogistically follow that a person may be banished from this country upon no higher degree of proof than applies in a negligence case. This Court has not closed its eyes to the drastic deprivations that may follow when a resident of this country is compelled by our Government to forsake all the bonds formed here and go to a foreign land where he often has no contemporary identification.

In terms of percentages, the probabilities for clear, unequivocal and convincing evidence might be in the order of above 80% under this standard.

e. Proof beyond a Reasonable Doubt

The standard of "proof beyond a reasonable doubt" is constitutionally mandated for elements of a criminal offense. In re Winship, 397 U.S. 358, 364, 90 S. Ct. 1068, 25 L. Ed. 2d 368 (1970). Writing for the majority in *Winship,* Justice Brennan enumerated the "cogent reasons" why the "'reasonable-doubt' standard plays a vital role in the American scheme of criminal procedure" and "is a prime instrument for reducing the risk of convictions resting on factual error." Id. at 363, 90 S. Ct. at 1072.

> The accused during a criminal prosecution has at stake interest of immense importance, both because of the possibility that he may lose his liberty upon conviction and because of the certainty that he would be stigmatized by the conviction. Accordingly, a society that values the good name and freedom of every individual should not condemn a man for commission of a crime when there is reasonable doubt about his guilt. As we said in Speiser v. Randall, supra, 357 U.S. at 525-526, 78 S. Ct., at 1342; "There is always in litigation a margin of error, representing error in fact finding, which both parties must take into account. Where one party has at stake an interest of transcending value—as a criminal defendant his liberty—this margin of error is reduced as to

him by the process of placing on the other party the burden of . . . persuading the factfinder at the conclusion of the trial of his guilt beyond a reasonable doubt. Due process commands that no man shall lose his liberty unless the Government has borne the burden of . . . convincing the factfinder of his guilt." . . .

Moreover, use of the reasonable-doubt standard is indispensable to command the respect and confidence of the community in applications of the criminal law. It is critical that the moral force of the criminal law not be diluted by a standard of proof that leaves people in doubt whether innocent men are being condemned.

Id. at 363-64, 90 S.Ct. at 1072-73.

In capital cases, the beyond a reasonable doubt standard has been utilized for findings of fact necessary to impose the death penalty after a finding of guilt.

Many state courts, in interpreting state recidivism statutes, have held that proof of past crimes must be established beyond a reasonable doubt.

In civil commitment cases, where the stakes most resemble those at risk in a criminal trial, some courts have held that the beyond a reasonable doubt standard is required.

If quantified, the beyond a reasonable doubt standard might be in the range of 95 + % probable.

. . . In Hollis v. Smith, the Second Circuit determined that due process requires proof of the critical fact at issue by "clear, unequivocal and convincing evidence." 571 F.2d 685, 695-96 (2d Cir. 1978). It found the evidence relied upon for the longer sentence did not measure up to that standard and granted a writ of habeas corpus.

In the instant case, proof by the Government that the defendant is a member of organized crime was not established in the criminal trial. As in *Hollis,* proof of this critical fact will result in a substantially longer period of incarceration. But, unlike *Hollis,* proof of the fact is not a previously defined prerequisite to a longer sentence. This difference, however, is of little consequence and Judge Friendly did not base his holding on it. "[T]he potential unfairness to defendant may be equally as great when an increase in sentence is based on facts not specified by the legislature as when the legislature has specifically delineated standards." Note, The Constitutionality of Statutes Permitting Increased Sentences for Habitual or Dangerous Criminals, 89 Harv. L. Rev. 356, 375 (1975).

Following what we believe to be the letter and spirit of *Hollis,* and the need to protect critical rights of liberty, we hold that when the fact of membership in organized crime will result in a much longer and harsher sentence, it must be established by "clear, unequivocal and convincing evidence." . . .

It is important to note that we do not hold that this standard of proof is fixed for all possible disputed facts at sentencing. Where the sentencing judge will give a matter only slight weight, a preponderance standard might be suitable. In some instances, for example, a dispute may arise about how much support a defendant gave an estranged wife and, since the matter might require an extensive and bitter hearing, some rough approximation based on 50+% probabilities will normally satisfy everyone.

If the defendant challenges what the judge regards as a peripheral issue, the normal practice is for the court to state that it will assume defendant's version for the purposes of sentencing. At the other end of the spectrum, where there is a dispute about a recent serious felony conviction, ease of proof suggests that the court should require proof beyond a reasonable doubt if its existence will enhance the sentence.

Flexibility is even reflected in the standard charge on reasonable doubt — "a doubt sufficient to cause a prudent person to hesitate to act in the most important affairs of his life." Holt v. United States, 218 U.S. 245, 254, 31 S. Ct. 2, 6-7, 54 L. Ed. 1021 (1910). The charge gives the trier considerable freedom to require greater probability for more important issues. As Professor Friedman sensibly observed, confirming what judges see happening in the courtroom:

> [J]udges and jurors alike must be "satisfied" of the truth of allegations or denials of fact. What amounts to satisfaction will vary with the issues involved. The more trivial the question, the more easily and swiftly will satisfaction materialize. The more momentous and serious its consequences, the greater the caution and deliberation demanded, that is, the greater the amount of cogent evidence before there can be any "satisfaction" about where the truth lies.

Friedman, Standards of Proof, 33 Can. Bar Rev. 665, 670 (1955).

The issue of membership in an organized crime family may be even more important than a prior conviction — and problems of proof are much more difficult. Considering the need to avoid extended sentencing hearings, the standard suggested by the Second Circuit in *Hollis* is appropriate. As indicated below, most judges in this district would place the probabilities of a "beyond a reasonable doubt" standard lower than would this court and would not find the *Hollis* test particularly high. For this and other reasons, this court believes a "beyond a reasonable doubt" burden more consonant with the tradition of American due process. Based on cases in this Circuit, however, *Hollis* probably articulates the highest burden acceptable to the Court of Appeals. Any lower standard under the circumstances would be imprudent. . . .

Professor Underwood, in a recent article cites a number of studies suggesting that judges, as well as laymen, will not always make the fine distinctions between preponderance, clear and convincing, clear unequivocal and convincing, and beyond a reasonable doubt described in this and other opinions. The Thumb on the Scales of Justice: Burdens of Persuasion in Criminal Cases, 86 Yale L.J. 1299, 1311 (1977). Those interviewed placed the probability standard higher than would be expected on theoretical grounds for a preponderance and somewhat lower than might be expected for beyond a reasonable doubt, indicating a narrower range in which to insert the two intermediate burdens.

> [A]lmost a third of the responding judges put "beyond a reasonable doubt" at 100%, another third put it at 90% or 95%, and most of the rest put it at 80% or 85%. For the preponderance standard, by contrast, over half put it between 60% and 75%. Questionnaires sent to jurors and students produced slightly lower results for the reasonable doubt instruction, and rather higher results for the preponderance standard; still, for most people the distinction was clear.

A survey of district judges in the Eastern District of New York indicates the following assessment of probabilities:

Probabilities Associated with Standards of Proof, Judges, Eastern District of New York

Judge	Preponderance	Clear and Convincing	Clear, Unequivocal and Convincing	Beyond a Reasonable Doubt
1	50+%	60-70%	65-75%	80%
2	50+%	67%	70%	76%
3	50+%	60%	70%	85%
4	51%	65%	67%	90%
5	50+%	Standard is Elusive and Unhelpful		90%
6	50+%	70+%	70+%	85%
7	50+%	70+%	80+%	95%
8	50.1%	75%	75%	85%
9	50+%	60%	90%	85%
10	51%	Cannot Estimate Numerically		

. . . The high probability required in criminal cases, however, does not mean that most guilty people who are tried are acquitted. In almost all cases the guilt is so clear or the doubt so great that precise quantification is of no moment. In some few instances — which this court would roughly estimate on the basis of experience as no more than one in ten cases — it may make a difference whether the trier's perception of the standard is 80, 90, 95, or 99%.

The standard can never be set at certainty or 100% probability, because

> Time is irreversible, events unique, and any reconstruction of the past at best an approximation. As a result of this lack of certainty about what happened, it is inescapable that the trier's conclusions be based on probabilities.

Maguire, et al., Cases and Materials on Evidence 1 (6th ed. 1973). Setting the standard at 100% in order to avoid any chance of convicting the innocent would thus result in a zero conviction rate and acquittal of all the guilty. As Professor Posner points out:

> If the standard of proof is set at so high a level that the probability of an innocent person's being convicted is zero, the conviction rate for guilty people will also be zero, since only with a zero conviction rate can all possibility of an innocent person's being convicted be eliminated.

Posner, An Economic Approach to Legal Procedure and Judicial Administration, 2 J. of Legal Studies 399, 411 (1973).

Quantification of these standards has not been well developed for reasons not now relevant. Nevertheless, there is little doubt that utilizing one rather than the other of the four burdens discussed in this opinion will make a difference in the results in some cases.

. . . While we must remain dubious of any conclusions based upon hearsay, the Government's proof here meets the rigorous burden of "clear, unequivocal and convincing evidence." The probability is at least 80% that defendant is an active member of an organized crime family. . . .

Notes and Problems

1. In United States v. Fatico, Judge Weinstein wrestled with the problem of finding a constitutionally acceptable size for errors in a sentencing hearing. What was the null hypothesis being tested in this hearing? What type of error might Judge Weinstein have made in deciding to reject the null hypothesis? What is the probability that, by rejecting the null, the judge made this error? Does this probability of error seem high or low? This last question is the statistical equivalent of asking whether the burden of proof established by the court is appropriate for the sentencing context.

2. In *Castaneda* and subsequent discrimination cases, the courts followed a two-or-three-standard-deviations guideline. From our discussion of the normal distribution curve, you know that an observa-

tion more than two standard deviations from the expected or mean value is likely to occur by chance only 4.6% of the time. This follows from the fact that 95.4% of observations are less than two standard deviations from the mean. Does this mean that courts in discrimination cases are willing to accept no more than a 4.6% probability of a type 1 error? Is this an appropriate standard, given Judge Weinstein's discussion in *Fatico*? Or is the standard deviation guideline being applied to issues that are different from the ultimate issues addressed by Judge Weinstein?

CERTIFIED COLOR MANUFACTURERS ASSOCIATION v. MATHEWS
543 F.2d 284 (D.C. Cir. 1976)

WILKEY, Circuit Judge:

This appeal is a challenge to the action of the Commissioner of Food and Drugs in terminating provisional approval of the color additive FD&C Red No. 2 ("Red No. 2") pursuant to the Transitional Provisions of the 1960 Color Additives Amendments to the Federal Food, Drug, and Cosmetic Act. The Certified Color Manufacturers Association and other parties (collectively referred to as "CCMA" or "appellants") are appealing from the order of the District Court granting summary judgment to the defendants-appellees and dismissing CCMA's complaint with prejudice, and thus frustrating appellants' effort to enjoin the publication of the Commissioner's order in the Federal Register. Appellants seek reversal of the District Court order, and a remand with directions to enter judgment for them and to enter an order directing the Commissioner to restore Red No. 2 to the provisional list.

Appellants urge that the District Court erred in holding that (1) the Commissioner had not acted arbitrarily or capriciously, or in excess of statutory authority, in terminating the provisional listing of Red No. 2; and (2) such action need not be preceded by notice and opportunity for comment.

We affirm the order of the District Court.

I. BACKGROUND

A. STATUTORY FRAMEWORK

To provide a proper perspective from which to analyze the regulatory action here challenged, it is helpful first to describe the statutory framework within which the action was taken. The Color Additive Amendments of 1960 reflect a Congressional and administrative response to the need in contemporary society for a scientifically and administratively sound basis for determining the safety of artificial color additives, widely used for coloring food, drugs, and cosmetics. The Amendments reflect a general unwillingness to allow widespread use of such products in the absence of scientific information on the effect of these products on the human body. The previously used system had some glaring deficiencies, and the 1960 Amendments were designed to overcome them. This was accomplished by the establishment of a dual system of registration: a permanent listing and a provisional listing. A color additive would be permanently listed if those desirous of producing it had proven to the satisfaction of the Commissioner that it was safe for its intended use. Generally speaking, until such listing, certain color additives were provisionally listed for a reasonable interim period pending completion of the scientific investigations necessary to make a safety determination.

The statute required that a color additive be *"deemed unsafe"* within the meaning of various sections of the Federal Food, Drug, and Cosmetic Act for use in food, drugs or cosmetics *unless* the Commissioner had issued a *regulation* which stated that the *additive had been found* to be suitable and safe for general use, or for a particular use. The burden of establishing safety was placed on the industry. The Commissioner was required to exercise his scientific judgment in determining whether safety had been proven, but was expressly precluded from listing any additive for any use if it was found to induce cancer in a human or animal.

In order to avoid a statutory presumption that all additives not permanently listed at the date of enactment were unsafe, Title II of the Amendments ("Transitional Provisions") provided for the continued use of commercially established additives—such as Red No. 2—on an interim basis. This was allowed only "to the extent consistent with the public health, pending completion of the scientific investigations needed as a basis for making determinations" on the safety of the additive for permanent approval. This provisional listing was to expire on a "closing date" which was established as either 12 January 1963 ("initial closing date"), or such later date as the Commissioner determined. The original closing date could be postponed "if in the [Commissioner's] judgment such action [was] consistent with the objective of carrying to completion in good faith, as soon as reasonably practicable" the investigation needed for making the safety determination.

Throughout this transitional period, while the permanent safety determinations were being made, the Commissioner was granted broad discretionary authority summarily to terminate a provisional listing "forthwith whenever in his judgment such action [was] necessary to protect the public health." During the period following postponement of the closing date, the Commissioner additionally was given broad authority to terminate that postponement "at any time" if he found that, inter alia, the basis for the postponement no longer existed because of a change in circumstances.

It is thus apparent that after 12 January 1963 the exercise of either one of these discretionary powers would achieve the same result: the affected color additive would be removed from the marketplace; no regulation listing it for permanent approval would be in effect, and it therefore would be "deemed unsafe."

In the instant case, the Commissioner exercised both of these powers. He terminated the postponement of the closing date of Red No. 2, and additionally terminated its provisional listing.

B. AGENCY ACTION

Red No. 2 is a petroleum derived color additive widely used in this country artificially to create or brighten the white, brown, purple, or red colors of various foods, durgs, and cosmetics. It is used alone or in combination with other compounds to attain the desired color. It has no apparent nutrient value and its primary function is to enhance the eye appeal of a particular product. It has been widely used in this country under federal regulatory control for nearly seventy years. Since the initiation of more sophisticated scientific investigations following enactment of the 1960 Color Additive Amendments, it has failed to receive FDA permanent approval for safety.

Acting pursuant to his authority under the Transitional Provisions, the Commissioner has postponed the closing date a total of fifteen times, either at the request of interested parties or on his own initiative. The most recent postponement occurred 5 January 1976 and was to expire 30 September 1976. A petition for permanent listing pursuant to 21 U.S.C. §376 has been under active consideration since late 1968, but had not been acted on prior to the action here challenged. At one point the agency was on the verge of affirmatively acting on the petition; before such action could be taken, however, information became available on several studies undertaken in the Soviet Union which raised new concerns over the safety of Red No. 2. One study concluded that amaranth, the generic name for Red No. 2, was poisonous to the reproductive organs of laboratory rats; the remaining studies concluded that amaranth was carcinogenic, that is, it caused cancer.

158   Chapter V. Significance Testing

While FDA scientists questioned the validity of the Soviet tests, the new studies pointed out the absence of information on the effects of Red No. 2 on reproductive physiology and raised concern over its possible carcinogenicity. Accordingly, the FDA undertook its own chronic feeding study over a two-and-one-half year period in which Osborne-Mandel rats were fed Red No. 2 in their diet at concentrations of 3%, 0.3%, 0.003%, and 0%. The duration of that study and the follow-up evaluations have been the primary reason for postponing the closing date of Red No. 2's provisional listing for the last several years.

The most recent postponement came after a meeting of the Toxicological Advisory Committee ("TAC") established by the Commissioner pursuant to the statute to make recommendations on the safety of Red No. 2. At its November 1975 meeting the TAC saw no reason to recommend against postponing the closing date, thus continuing the period for evaluation. It recognized that the feeding study was severely flawed because the dosage levels of Red No. 2 in the various rat groups had not been maintained and because there had been substantial tissue loss due to autolysis. It nevertheless believed that portions of the study could still be used to establish safety and recommended continuing review, including a biostatistical study of the data. Accordingly, on 14 November 1975 the Commissioner published notice of his intent to postpone the closing date of the provisional listing of Red No. 2, along with all other provisionally listed additives, stating as his reason that "[r]eview of the data submitted to establish [safety] is underway. . . ." After receiving no adverse comments, the closing date was formally postponed on 5 January 1976.

In the interim, Dr. D.W. Gaylor had performed the biostatistical analysis requested by the TAC. In a memorandum to the Chairman of the TAC, dated December 1975, Dr. Gaylor presented his analysis and concluded that "it appears that feeding FD&C Red No. 2 at a high dosage results in a statistically significant increase in a variety of malignant neoplasms among aged Osborne-Mandel female rats." In preparation for his statistical analysis, Dr. Gaylor had isolated malignant neoplasms found in 44 rats in the low dosage group and in 44 rats in the high dosage group. His analysis resulted in findings that (1) for those animals dying before the end of the study, that is, before the passage of 131 weeks, the proportion of those dying with cancer in the high dosage group was not significantly higher statistically than those dying with cancer in the low dosage group;[30] (2) for those rats

---

30. Of the animals dying before the end of the study, 7 of 23 from the high dosage group were found to have cancer, while from the low dosage group, 4 out of 30 had cancer.

sacrificed at the end of the study and found to have malignant neoplasms, it was determined that there was less than a 5%, or a 4%, probability that the higher frequency of malignant neoplasms found in the high dosage group would be due to random occurrence;[31] and (3) in evaluating the experiment for all animals (interim deaths and terminal sacrifice), it was found that there was less than a 2% probability that the higher frequency of malignant neoplasms found in the high dosage group would be due to random occurrence.[32]

The statistical relationship thus uncovered was viewed as a matter of "great concern," and a special ad hoc "Working Group" was convened on 14 January 1976 to review the study. It submitted a report to the Commissioner on 16 January 1976 in which it concluded: (1) given the deficiencies in the chronic feeding study, "inferences drawn by Dr. Gaylor represent the minimum possible observable differences," (2) the chronic feeding study "could never be used as a demonstration of safety," and (3) the principles of the statistical analysis were accepted, but the strength of any conclusions would depend on reanalysis "using a slight redefinition of tumor types."

[The report summarized the statistical evidence as follows:

The evidence from this experiment demonstrating that FD&C Red No. 2 *is* a carcinogen is twofold:
1. A statistically significant increase in malignant tumors in females in the high-dose group versus low dose ("controls").
2. A possible dose-response relationship in females using all dose groups.

The evidence from this experiment not supporting the carcinogenicity of FD&C Red No. 2 is fourfold:
1. No statistically significant increase in malignant tumors was demonstrated in male animals.
2. No statistically significant events are observed if one aggregates all tumor types.
3. No unusual or unique tumor types were observed in the test animals.
4. No single tumor type showed an increase over that observed in the control animals. . . . (fn. 35]

Following receipt of this report, on 19 January the Commissioner prepared, and gave notice of intent to publish, a regulation to termi-

---

31. At terminal sacrifice, 14 rats in the low dosage group and an equal number in the high dosage group were found not to have cancer; however, seven rats in the high dosage group and no rats in the low dosage group were found to have cancer. The one degree of freedom chi-squared value corrected for continuity was found to be statistically significant at the $P < .05$ level. Application of "Fisher's exact test" resulted in a two-sided significance level of $P < .04$.

32. The rates were statistically significant at the $P < .02$ level.

nate the provisional listing of Red No. 2. The order was signed on 23 January, to become effective on publication in the Federal Register. Prior to publication, CCMA filed its complaint in the District Court and was granted a temporary restraining order the next day. After consideration of the papers filed, the administrative record, and oral argument of the parties, the District Court vacated the temporary restraining order, denied CCMA's request for a preliminary injunction, granted summary judgment to the defendants-appellees, and dismissed the complaint with prejudice.

Following CCMA's unsuccessful attempts to obtain a stay from this court and the Supreme Court, the regulation was published on 10 February 1976. The Commissioner therein reviewed the general progression of events leading to the most recent postponement of the closing date, the Gaylor analysis and the Working Group's report, and concluded that provisional listing of Red No. 2 was "no longer appropriate." Accordingly, he terminated the provisional listing and additionally terminated the postponement of the closing date.

He concluded that termination of the listing was "necessary to protect the public health" because of the concern engendered by the Gaylor analysis and because the available data did not overcome that concern. The Commissioner stated:

> ... The provisional listing of FD&C Red No. 2 should be terminated because such action is necessary to protect the public health, in that questions have been raised about the safety of the color additive and the available data do not permit a determination of safety.

He terminated the postponement of the closing date on a finding that the basis for the last postponement no longer existed because of the change in circumstances brought about by the Gaylor analysis and the conclusion of the Working Group that the feeding study could not be used to demonstrate safety. He further stated:

> This postponement was based on the assumption that the studies being evaluated *could* establish the safety of the color additive. Since it has now become clear that *the studies cannot satisfactorily resolve* this issue, the Commissioner finds that *the basis for the postponement no longer exists* and hereby terminates the postponement of the closing date in accordance with section 203(a)(2) of the transitional provisions of the Color Additive Amendments of 1960.

"Cessat ratione, cessat" postponement.

Each action resulted in the termination of the provisional listing of Red No. 2. If we find, therefore, that the District Court was correct in determining that the Commissioner acted properly in taking either action, the effect is to sanction the general conclusion of the Commissioner. . . .

III. THE ISSUES ON APPEAL. . .

B. TERMINATION OF THE PROVISIONAL LISTING

The principal issue raised by appellants is that the District Court erred in failing to find that the Commissioner had acted arbitrarily and capriciously, and in excess of statutory authority in terminating the provisional listing of Red No. 2. We shall first consider the statutory argument and then the evidentiary considerations.

The primary thrust of appellants' statutory argument is that the standard authorizing the Commissioner to act "forthwith whenever in his judgment such action is necessary to protect the public health" requires that there be an actual threat to the public, amounting for all practical purposes to an imminent health hazard. We believe this argument is without foundation. When Congress has intended so to restrict the discretion available to an agency to take regulatory action similar to that involved here, it has not hesitated to indicate that intent by the inclusion of appropriately restrictive language. Here Congress attached no such "qualifiers" and chose instead to provide for a much broader standard, designed, we think, to allow for precautionary or prophylactic responses to perceived risks. There is nothing in the statute or its legislative history to indicate that Congress intended to restrict regulatory action to situations where there was a substantial likelihood of serious harm to the public. The overall Congressional intent appears to be to the contrary. . . .

Our review of the record, in light of the broad discretion conferred on the Commissioner, convinces us that there was no abuse of discretion in terminating the provisional listing of Red No. 2. The information available to him indicated a statistically significant relationship between high dosages of Red No. 2 and the occurrence of cancer in aged female rats. That relationship concededly did not establish conclusive proof that Red No. 2 was a carcinogen, but it was at least suggestive of it, and that statistical analysis was later confirmed, albeit at lower levels of significance. Moreover — and this was the angle from which he was to view these findings — he had been advised that the principal study under evaluation *could not* be used to *establish* safety. Based on this information, he determined in the exercise of his scientific judgment that a question as to the safety of the additive existed sufficient to warrant termination. We cannot say that this was an irrational choice, nor can we say that it was beyond his statutory authority. . . .

162   Chapter V. Significance Testing

Therefore, in light of the broad discretionary authority available to the Commissioner, and our necessarily limited role as judges, we affirm the District Court's holding on this issue. . . .

Notes and Problems

1.   In Certified Color Manufacturers Association v. Mathews, the circuit court refers regularly to the statistical significance of the biostatistical analysis performed by Dr. Gaylor. In footnotes 31 and 32 of the opinion, reference is made to a specific measure of statistical significance, a chi-square value. "Chi" is pronounced "kī" to rhyme with "sky," and chi-square is symbolized by χ^2. This summary statistic will be discussed in detail below. Briefly, however, the findings of statistical significance at the $P < .05$, $P < .04$, and $P < .02$ levels indicate that the court can be 95%, 96%, and 98% certain, respectively, that the null hypotheses involved in the specific tests carried out on rats should be rejected.

2.   Several null hypotheses are examined in this case. The null hypothesis considered by the circuit court and the Commissioner of Food and Drugs is different from that tested by Dr. Gaylor. An appreciation of the procedural posture of this case and of the statutory framework is required to state the null hypothesis that confronted the Commissioner. What is that null hypothesis?

3.   In the excerpt describing Dr. Gaylor's statistical analysis, you will find his three statistical conclusions (see pages 158–159 and footnotes 30-32). What null hypothesis was associated with each? For two of these three you can determine the probability of a type 1 error associated with each conclusion.

4.   What scientific evidence would have been the most convincing proof that the dye is unsafe for humans? Does Dr. Gaylor's study of rats address the scientific issues relevant to this case? The study does not prove that Red No. 2 *causes* cancer in rats. It does show a relationship between increased dosage and cancer rates in rats. What does it show about humans? The Commissioner's termination of the postponement and provisional listing are based on the statistical conclusions of Dr. Gaylor and the Working Group. Given what we know of Dr. Gaylor's probabilities of error, can we determine the Commissioner's probability of error? Can we at least guess whether the Commissioner's probability of error is larger or smaller than Dr. Gaylor's?

C. CHI-SQUARE TESTING

CHINESE FOR AFFIRMATIVE ACTION v. FCC
595 F.2d 621 (D.C. Cir. 1978)

[Organizations concerned with employment opportunities for Chinese-Americans appealed from Federal Communications Commission orders renewing broadcasters' licenses. The Court of Appeals, Wilkey, Circuit Judge, held that: where facts before the Federal Communications Commission established that the licensee had employed a substantial number of Asian-Americans during the expiring term, that the licensee's employment figures for minorities in general fell within the zone of reasonableness and that the licensee's affirmative action program was effective, and where there was no allegation of overt discrimination against Asian-Americans, the Commission properly concluded that no substantial and material questions of fact had been raised and that license renewal without prospective remedies was consistent with the public interest. The companion case, Bilateral Bicultural Coalition on Mass Media v. FCC, referred to in this opinion as *Bilingual II,* was remanded upon a finding that the licensee's employment of Mexican-Americans was outside the zone of reasonableness and raised issues of intentional discrimination. — ED.]

SPOTTSWOOD W. ROBINSON, III, Circuit Judge, dissenting in part: . . .

My difficulty with today's decision stems solely from the court's failure to utilize the occasion as an *en banc* body to review thoroughly the Commission's present philosophy toward employment discrimination by its licensees. This default is especially troubling in light of the court's professed endeavor "to state definitively the position of this Court on the issues raised and to govern related cases in the future." Although I concur to the result in *Bilingual II,* I think the court errs grievously in rejecting out of hand the allegations of employment bias in *Chinese for Affirmative Action,* and in placing its imprimatur on the license renewal granted therein.

The statistical showing in *Chinese,* while not so compelling as that in *Bilingual II,* did establish a prima facie case of intentional racial discrimination when measured by normal evidentiary standards, from which neither the Commission nor the court has advanced any cause for departure. The Commission, however, gave those statistics only cursory examination — on the thesis that the minority group involved is not dominant in the broadcast community — and ultimately it disposed of the challenger's presentation on grounds that I can regard only as an effective abandonment of its equal-employment policies in the recent past. The court, in turn, ignores the Commission's stated rationale and substitutes another, and, even if it were not our duty to refrain from a displacement of that sort, I could not accept the court's unexplained conclusions. The court disdains both a hard look at the Commission's decision, and a rigorous study and elucidation of its own reasoning.

I must, then, dissent from the court's endorsement of the Commission's limitation of the coverage of its anti-discrimination rules to groups dominant in the licensee's service area. I must also demur to the court's condonation of Commission acceptance — without even the seeds of a logical explanation — of statistical disparities that naturally give rise to an inference of purposeful racial bias. . . .

II

It is now far beyond dispute that broadcasters must not intentionally discriminate, and that past activity of that nature will justify, if not demand, nonrenewal of a communications license. Still we are left with the problem of determining what showing is necessary to generate a "substantial and material [issue] of fact" statutorily requiring — absent Commission efforts otherwise resolving the dispute — a hearing before the Commission properly may "find that grant of the application would be consistent with" the public interest.

The type of showing most commonly attempted — and that with which we deal here — is statistical in nature. The Commission initially expressed distrust of statistics but, before that misgiving was subjected to judicial scrutiny, the Commission purported to come around to the view that "'[i]n the problem of racial discrimination, statistics often tell much, and Courts listen.'" Quite recently the Commission has declared that "a statistical prima facie case of employment discrimination may be made out by comparing the composition, by race and sex, of the company's staff with the city's labor force composition in general." Indeed, because a statistical presentation is perfectly capable of supporting an inference of prohibited conduct, it would be irrational to ignore it totally. Moreover, in the field of discrimination, it is my experience that statistics are recognized as a uniquely valuable type of evidence: "statistics are frequently the best available evidence of . . . discrimination" because they are often "the only available avenue of proof." Consequently, the pertinent question is not whether statistics will ever suffice, for they often will, but what level of statistically-indicated disparity will trigger an administrative investigation. . . .

C. Chi-Square Testing   165

The Communications Commission must allocate its limited resources to a broad range of important responsibilities, of which the fight against job-bias is but one — although certainly a vital one. Because any statistical showing leaves open the possibility, slight though it might be, that the disparity is due to chance, the Commission may adopt a level of statistical significance[54] reasonably calculated to avoid exorbitant efforts over coincidental differentials — what statisticians call "type [I] errors." And more than that, because the Commission is concerned with long-term effects on the use of airwaves in the public interest and not with remedying isolated incidents, it could perhaps ignore some differences that are significant but of small magnitude.[57] Disparities may often be attributable to the temporarily-undetected character flaw of a single personnel officer or employment agency, and not to licensee-approved or -condoned employment practices.

More broadly, all that statistics can ever show is that the null hypothesis — that the composition of the licensee's labor force is the result of random distribution and that race and sex had no relationship to employment opportunity — should be rejected. Even though a disparity is statistically significant, it does not irrebuttably prove the reason for the difference, which could be any number of things — intentional racism by employment officials, a disinclination among minority-group members to seek media jobs, unintentional use of culturally-biased employment tests and standards or the lingering effects of past discrimination by educational institutions. Thus, when the magnitude of disparity is relatively low in comparison with a convincing refutation of calculated bias, the Commission might conclude that the probability that a hearing would reveal a discriminatory intent is too insubstantial to justify an expenditure of precious resources. But, "absent explanation, it is ordinarily to be expected that nondiscriminatory hiring practices will in time result in a work force more or less representative of the racial and ethnic composition of the population in the community from which employees are hired." The corollary is that where a long-term disparity is established and no satisfactory explanation is given, it ordinarily can be expected that intentional manipulation has been worked. . . .

54. Social scientists commonly accept 1 in 20 as the maximum chance of erroneously rejecting random distribution, and a statistical showing less likely to mislead is considered to be significant. A 1 in 40 chance of error has been suggested as the standard for "overwhelming" statistical showings. Bogen & Falcon, The Use of Racial Statistics in Fair Housing Cases, 35 Md. L. Rev. 59, 77 (1974).
57. See H. Blalock, [Social Statistics (2d ed. 1972), at 163,] (statistical significance is not the same as practical significance because in isolation it tells nothing about the importance or magnitude of the differences).

166   Chapter V.   Significance Testing

We can — and indeed we must — indulge the Commission great latitude in deciding whether to designate a renewal application for investigation or hearing. But it goes without saying that the Commission may not yield to licensees the freedom to engage in willful misconduct. Thus the Commission must develop — obediently to its responsibilities as guardian of the public interest in broadcasting — its own framework for analyzing statistics, and as a court we must appraise its effort to determine whether it comports with the Commission's statutory duties.

The Commission has informed us that it will at least make further inquiry when licensee employment of minorities is less than 50 percent of parity with their percentages in the area workforce,[73] or when employment in upper-level positions[74] is less than 25 percent of parity. Because this proposal is not at issue here, I have no need to discuss it, but I must note that it arrived unadorned with any explanation. This is the more discomforting because figures showing 50 percent of parity — not to mention those indicating only 25 percent of parity — would, both as a matter of common sense and of statistical science, often be significant and relatively large in magnitude.

---

73. FCC responses [to Questions Posed by Court in D.C. Cir. Nos. 75-1855 and 75-2181, at 1 (Sept. 16, 1972)]. A guideline beginning at 50% of parity and increasing 5% per year to a maximum of 90% of parity was proposed in Note, [53 B.U.L. Rev. 657, 658 (1973)]. Although we have no call to discuss particular figures, such a stepped criterion seems a reasonable effort to combine the requirements of the antidiscrimination and affirmative-action rules. The Commission on Civil Rights argues for a test of 80% of parity. United States Commission on Civil Rights, [Windowdressing on the Set: Women and Minorities in Television 151 (1977)].

Licensees with fewer than ten employees are not rigidly subject to the Commission's 50% guideline because the Commission believes that the sample size is too small to give rise to any statistically reliable finding. FCC Responses, [supra note 73,] at 1-2. This would often but not always be true.

74. The Commission currently requires licensees to report on employment practices in their "upper-four" job categories. Licensees as a whole have reported that fully 77.2% of all broadcast employees occupied such positions in 1975, United States Commission on Civil Rights, supra note 73, at 143, and we are in the dark as to why this may be so. Without elucidation, I can accept this startling statistic only as reflecting a distorted view of what jobs are important, making a mockery of any figures purporting to show that a licensee has an appreciable number of women or minority-group members in critical positions. Put another way, the most mundane jobs must often be classified as upper-level.

For instance, in this very case CBS asserts that women are proportionally represented in important positions because in 1975 half of all females occupied upper-four positions. This claim ignores the telling fact that of only 14 employees not in the highest categories all but one were women. This is hardly an "exemplary" performance. Contra, Brief for Intervenor CBS, at 12. Because statistics showing overall proportional employment may easily mask a design to keep minorities and women out of truly influential jobs and because those positions are the more likely to have an impact on programming, the Commission needs reliable information on upper-level employment.

C. Chi-Square Testing   167

[For instance, the Supreme Court recently held that a showing that the percentage of Mexican-Americans selected for grand juries over a ten-year period in a Texas county was only 50% of parity not only gave rise to an inference of intentional discrimination but made out a prima facie case in that regard. Castaneda v. Partida, 430 U.S. 482, 97 S. Ct. 1272, 50 L. Ed. 2d 498 (1977); see Turner v. Fouche, 396 U.S. 346, 90 S. Ct. 532, 24 L. Ed. 2d 567 (1970) (violation found although 62% of parity); Whitus v. Georgia, 385 U.S. 545, 87 S. Ct. 643, 17 L. Ed. 2d 599 (1967) (violation found on 33% of parity). See also Washington v. Davis, 426 U.S. 229, 241, 96 S. Ct. 2040, 2048, 48 L. Ed. 2d 597, 608 (1976) (showing of intentional discrimination is a necessary element in establishing that Equal Protection Clause was breached in constituting jury).

Particularly characteristic of the Commission's reasoning is its counsel's statement in oral argument that the agency uses the "zone of reasonableness" only to find out whether the figures promote an inference that something is wrong with a licensee's affirmative-action plan, and not to ascertain whether there is an indication of purposeful discrimination. See, e.g., Swanson Broadcasting, Inc., 42 Rad. Reg. 2d (P&F) 1002 (Renewal Branch 1978) (although licensee's performance fails under 50/25 standard, renewal will be granted without hearing but with imposition of reporting conditions). The Commission will infer intentional misconduct only when the statistics unveil a situation so far outside the zone that it may be classified as egregious. Because statistics that do not seem sufficiently outrageous to the Commission often are probative of intentional discrimination, see, e.g., notes 88-92 infra and accompanying text, the Commission's decision to assume in all such cases that ineffective affirmative action is the real culprit is totally arbitrary. The Commission has advanced this theory to justify prospective orientation in *Bilingual II,* but the court pays it the deference it truly deserves—absolutely none. . . .   (fn. 76)]

I can only assume that when the Commission first makes a decision under its newly-revealed template it will fully elucidate it and its application. It may not simply invoke talismanically the fact that it is not the Equal Employment Opportunity Commission to reject out of hand a statistical showing that in analogous areas of the law would indicate "substantial under-representation" and erect a prima facie case of intentional discrimination. Because it is not EEOC, the Commission may justifiably be guided by consideration other than those central to EEOC's responsibilities. But if the question is what demonstrates intentional discrimination—without regard to the factors that go into deciding what to do about it—the Commission cannot reasonably decide that statistics that have been accepted as *prov-*

ing disparate treatment do not so much as raise a substantial *issue* of such conduct in its eyes.

III

My central concern in this opinion is the Commission's handling of the statistical attack by Chinese for Affirmative Action on Columbia Broadcasting System's employment of Asian-Americans at KCBS, its San Francisco AM station. An initial but easily surmountable objection to CAA's challenge is the Commission's apparent position that it need not examine statistics reflecting job-treatment of Asian-Americans — a minority group not "clearly the predominate minority in a licensee's SMSA." To begin with, this unquestionably was a major — and unexplained — departure from past Commission policy. In 1974, for instance, the Commission decreed that "[t]he non-discrimination provision applies to *all* persons, whether or not the individual is a member of a conventionally defined minority group," and many other manifestations of this view might be cited.

Despite the court's approval today of this abrupt shift, I see no reason whatever for countenancing purposeful discrimination merely because it is aimed at only one small group. The Commission's anti-discrimination rules were promulgated in large part to assure that programming needs of minority groups in the licensees' audiences are met. In other programming-related areas such as ascertainment, the Commission has not limited the licensee's responsibility to one of only serving "dominant" groups. Even if a justification for a divergent view on job-bias were possible, the Commission has not yet expressed it, and an explanation — by the Commission, not the court — is necessary. [On the Commission's behalf, the court posits the rationale that "such small percentages generally will provide no meaningful basis for comparison with the workforces of most radio stations, which often employ only 25 to 75 workers." Maj. Op. at note 7. Yet the sample size is actually much larger, for over the three-year license term many employees come and go. Moreover, we do not even know where the "25 to 75" figure came from. Even if the statistics for such stations would often be unreliable, that would be no reason to reject those statistical presentations, like the one in this case, that undeniably are meaningful. And even if one accepts the court's rationale, it does not rescue the Commission in this case, since KCBS employed more than 75 workers.

As a general matter, the value of a statistical showing depends not merely upon the size of the sample but also upon the extent of the expected observation and the degree of disparity—elements unknowable until the Commission looks at the particular group's statistics. In short, as the application of the logic of statistics that it purports to be, the court's conclusion that statistics for small groups are always unreliable is simply wrong.

Curiously, the court somehow later concludes that Asian-Americans constitute "a significant minority group." Maj. Op., —U.S. App. D.C. at—, 595 F.2d at 631. I am not sure how a significant group differs from a dominant one and why statistical evidence as to this significant group should not have been accepted even under the theory the court supplies for the Commission. (fn. 86)]. Since the Commission has articulated no rationale for ignoring the complaints of Asian-Americans merely because of the size of their group, I must proceed to the question whether the Commission properly can reject on the merits the statistical showing they proffered.

I am satisfied that CAA's figures, unless somehow successfully rebutted, generated an inference of calculated employment discrimination. With 6.9% of the Standard Metropolitan Statistical Area labor force consisting of Asian-Americans, the expected number within the licensee's 249 employees during the license term was 17. [Due to the inadequacies of the Commission's reporting form, we do not know how many employees worked for KCBS during the license term. We do know that about 83 workers were employed on the average at any one time, that the average Asian-American left after nine months and that the turnover rate for Asian-Americans was only "relatively high." See Maj. Op. text at note 43. Even assuming that the tenure for all employees averaged a third again longer than that for Asian-Americans, we have a total sample of 249 employees, 6.9% of which is roughly 17. (fn. 90)] As a result of a preliminary Commission inquiry, we know that KCBS employed a total of ten, and perhaps only nine, Asian-Americans for various lengths of time during the license term under review. Although this performance exceeds the 50-percent-of-parity range now proposed by the Commission, the unlikelihood that so great a disparity occurred solely by chance is only 3.67 percent — or about one in 27 — which is within the region normally considered significant.

[In the absence of Commission adoption and justification of some other standard of statistical significance, we are left to rely upon what social statisticians traditionally have accepted. See note 54 supra. The figure in text was arrived at by using the chi square test. See P. Hoel, [Introduction to Mathematical Statistics 298 (4th ed. 1971)]. See also F. Mosteller, R. Rourke, G. Thomas, Probability with Statistical Applications 494 (2d ed. 1970) (binomial distribution chart). This result is close to the upper boundary of normally acceptable type I error, see note 54, supra, but I am unwilling to ignore it until the Commission provides a reasoned explanation for why it is in the public interest to do so. And if the Commission should decide to exclude the Filipino, the result would become even more compelling.

Just looking at the 1974 figures alone, as the court challenges me to do, the evidence still indicates discrimination. Accepting KCBS's corrections as true, still only one Asian-American was employed in May of 1974 when the 1974 report was submitted. This is barely more than 1% of KCBS's employees and should be compared to an expected (in the absence of discrimination) figure of nearly six.

The computations could also be made on the basis of person/months. See CBS, Inc., supra note 29, 56 F.C.C.2d at 302; quency of no more than 94. These figures are arrived at without too much difficulty from the information available to the Commission. The license term in question ran from December 1, 1971, through November 30, 1974. The eight unreported Asian-Americans were employed during that time for a total of approximately 34 person/months. See CBS, Inc., supra note 29, 56 F.C.C.2d at 302; Appendix to Letter from Eleanor S. Applewaite, General Attorney, CBS, Inc., to Richard J. Shiben, Chief, FCC Renewal and Transfer Division (Oct. 3, 1975), J. App. 208. The one reported Asian-American could not have worked more than 18 months during the license term. Compare Memorandum of CBS in Response to a Question Raised in the Department of Justice Amicus Brief, appendix at 6-7 with id., appendix at 11-12. The Filipino employee may have worked as many as 42 months.

Using the test employed by the Supreme Court in Castaneda v. Partida, supra note 76, 430 U.S. at 496-497 n. 17, 97 S. Ct. at 1281 n. 17, 51 L. Ed. 2d at 512 n. 17, and in Hazelwood School Dist. v. United States, [433 U.S. 299 (1977)], we arrive at a standard deviation of about 13. The gross disparity is thus approximately 9 standard deviations — and nearly 12 if we exclude the Filipino, see note 91 supra — and "the hypothesis that the [employment practices were] random would be suspect to a social scientist." Castaneda v. Partida, supra note 76, 430 U.S. at 496-497 n. 17, 97 S. Ct. at 1281 n. 17, 51 L. Ed. 2d at 512 n. 17. Using a person/month as the sample measurement factors in the length of the employees' stay, the shortness of which might be due to intentional discrimination. The choice of months as the unit of time is somewhat arbitrary, however, and, if we are concerned primarily with hiring practices, person/month figures are misleading because each unit is not tied to an independent hiring decision. (fn. 92)] This requires rejection of the null hypothesis that the licensee's hands are unquestionably clean. It does not conclusively prove intentional discrimination, of course, but it certainly does summon the Commission to ascertain whether, in light of all relevant information — on hand or procurable through further inquiry — the chances of an eventual finding of willful discrimination merit a hearing.

It was because Asian-Americans are not a dominant minority group in KCBS's service area that the Commission felt that an investigation into intentional discrimination was unwarranted. The court shuns the ground administratively assigned for rejecting CAA's contentions and instead detects in the Commission's opinion a finding that CAA's statistical showing was examined but found wanting. That opinion is hardly a model of precise analysis, but it is possible to discern four factors that the Commission could have deemed incompatible with an inference of purposeful discrimination against Asian-Americans had it thought itself bound to inquire into that subject. These are KCBS's in-term performance with respect to women and minorities overall, its affirmative action plan, its post-term performance with respect to Asian-Americans and the fact that it had employed some Asian-Americans during the term.

Even assuming that the Commission actually did consider these factors to the end the court supposes, it never ventured any rationale for an implicit conclusion that they served to so dispel the inference of past purposeful discrimination naturally flowing from the statistical disparity demonstrated by CAA. And it is by now a basic tenet of administrative law that agency explication of the basis of its decision not only aids judicial review but also improves agency decisionmaking; a corollary is that judicial review of agency action is not to be undertaken in the dark. I would thus remand *Chinese,* as well as *Bilingual II,* so that the Commission can tell us just how it arrived at its decision. . . .

IV

To sum up, it seems to me that the Commission's performance in these two cases reflects its broad misapprehension of the value — indeed, the logical compulsion — of statistical evidence and of the methods by which statistical showings can be rebutted. That malaise has fatally infected the result in *Chinese.* It could be that the Commission ultimately should renew KCBS's license, but it has not yet satisfactorily explained why that should be so. The Commission cannot brush statistical challenges aside with flip rejoinders about the nature of its responsibilities. It is not the Equal Employment Opportunity Commission, and the Communications Act is not the Civil Rights Act, but evidence is evidence, and no less than other tribunals must the Commission give it due respect.

In both cases, the challengers' statistics indicated intentional discrimination, which the Commission admits is normally a compelling ground for denial of a license renewal. The Commission refused to consider the showing in *Chinese,* and even had it done so the fact remains that it has not elucidated what countervailing factors could and how they would overcome the evidentiary inference. And if they do outweigh the inference, it should be for the Commission, not the court, to explain how that is done. Accordingly, I concur in the court's disposition of *Bilingual II,* but respectfully dissent from the court's failure to reach the same outcome in *Chinese.*

1. Calculating the Chi-Square Value

Standard deviations are one method of comparing expected values of variables to observed values. In Chinese for Affirmative Action v. FCC (*CAA*), the court was comparing the expected number of Asian-American employees to the observed number (see page 169 above). The population mean, 6.9% Asian-Americans in the community, times 249 employees gives an expected number of Asian-American employees of 17 (0.069 × 249). Since an employee is either Asian-American or not, we can use the binomial formula for standard deviation. In this case the standard deviation, SD, is:

$$\sqrt{N \times P_{Asian-American} \times P_{not}} = \sqrt{249 \times .069 \times .931} = \sqrt{16} = 4$$

The observed number of Asian-American employees is nine. The statistical question is: how many standard deviations is the observed number, nine, from the expected number of employees, 17. The answer is two standard deviations: One standard deviation is four; two are eight; and 17 minus nine is also eight. Castaneda v. Partida indicates that an observation that is two or three standard deviations away from the mean (or expected value) is suspicious.

Using the *chi-square test,* we can make another test of the null hypothesis that race has no relationship to employment at the radio station involved. This test allows us to evaluate the likelihood that such an observed number could occur by chance, that is, that only nine Asian-American employees were hired but that there was no discrimination (or other nonrandom factor) involved.

The chi-square test is similar to, but not quite the same as, the calculation of variance. We are trying to determine whether proportionately more non-Asian-Americans are hired than Asian-Americans.

C. Chi-Square Testing

The chi-square test approaches this problem by comparing the observed numbers of employees of each type, symbol O_j (with O standing for observed and j standing for the type of employee), with the expected number of employees of each type, symbol E_j.

The formula for calculating the chi-square value, symbol χ^2, is

$$\chi^2 = \Sigma \frac{(O_j - E_j)^2}{E_j}$$

Applying this formula means that, for each type of employee, we must:

Step 1. Subtract the expected number from the observed number.
Step 2. Square the difference.
Step 3. Divide by the expected number (giving us one value for each employee type).
Step 4. Add these values together to get the chi-square value.

Of a total 249 employees, the observed number of Asian-Americans is nine, while expected number is 17. Simple subtraction indicates that the expected number of non-Asian-Americans is 249 minus 17, or 232, while the observed number is 249 minus 9, or 240. The following steps summarize the calculation of the chi-square value.

EXHIBIT V-1.
Calculation of the Chi-Square Value in *CAA*

Step	j = Asian-Americans	j = Non-Asian-Americans
E_j (Expected)	17	232
O_j (Observed)	9	240
(1) $O_j - E_j$	−8	+8
(2) $(O_j - E_j)^2$	64	64
(3) $(O_j - E_j)^2 \div E_j$	64 ÷ 17 = 3.76	64 ÷ 232 = .276
(4) $\Sigma[(O_j - E_j)^2 \div E_j] = 3.76 + .276 = 4.036 = \chi^2$		

Lines (1), (2), (3), and (4) show the calculations to get out χ^2 value of 4.036. Now that we have this number, what do we do with it?

Look back at the reason for doing this test. We want to know whether to reject the null hypothesis that race has no relationship to employment at the radio station. The chi-square value will tell us the probability that the null hypothesis is true, i.e., the probability of a type 1 error. To translate the chi-square value into a probability, we must refer to the chi-square table.

2. Degrees of Freedom

Before we can use the chi-square table, however, we must become somewhat familiar with the concept of degrees of freedom. Strictly speaking, the number of *degrees of freedom* is the number of items that can be freely chosen in forming a designated group. In the situation at hand, the group we are examining is the collection of employees at the radio station. The number of employees is fixed (historically) at 249 people. The number of items we are considering is "the number of j's," i.e., the number of types of employees, Asian-Americans and non-Asian-Americans. Degrees of freedom is a mathematical concept. The question of free choice is transposed into mathematical concept here and is formulated, "If we choose a number of one type of employee (e.g., 17 Asian-Americans), are we (mathematically) free to choose the number of the other type?" The answer is no. If the total is 249 and one type of employee constitutes 17, there must be, by mathematical necessity, 232 of the other type. We are "free to choose" the number of only *one* type. The degrees of freedom are the number of types that we are mathematically free to choose. In this example, we are free to choose one, therefore the degree of freedom equals one.

Contrast this example to an examination of all the racial types of employees at the radio station, e.g., blacks, Hispanics, Caucasians and Asian-Americans. In such a case we would say that we are mathematically free to choose the number in each of three categories, but the fourth would be mathematically determined because we have a fixed total number of employees.

In general, the number of degrees of freedom ($d.f.$) is equal to the number (n) of types (j) minus one. The chi-square table shows probabilities (along the top of the table) for various degrees of freedom (along the side of the table).

3. Using the Chi-Square Table

Once we have calculated the chi-square value as:

$$\chi^2 = \Sigma \frac{(O_j - E_j)^2}{E_j}$$

and the number of degrees of freedom as:

$$d.f. = n - 1$$

we are ready to use the table to determine the probability that the null hypothesis is true or false, i.e., whether we should reject the null hypothesis or not. We will also be able to determine the probability of making a type 1 error.

Consider our *CAA* example. For this particular null hypothesis (that race bears no relationship to employment at the radio station), the chi-square value is 4.036. Look at the row of probabilities along the top of the table. $P = .90$ means a 90% chance of a type 1 error or, alternatively, 90 chances out of 100 that a particular racial mix will occur by chance. $P = .80, .70, .50, .30$, etc., mean that there is an 80%, 70%, 50%, 30%, etc., chance that a particular result will occur by chance. What is that particular result? On this table, the result is a particular chi-square value (e.g., 4.036 in *CAA*) for a given number of degrees of freedom (e.g., 1, as in *CAA*) if the null hypothesis is correct.

We are testing to see whether to reject the null hypothesis or not. The employment records give us a chi-square value of 4.036. For one degree of freedom, we look to see what is the probability of getting a chi-square value of 4.036 by chance when there is no discrimination. Alas, the number 4.036 does not appear on the table. But we do see that there is a 5% chance of getting 3.841 (which is less than 4.036) and a 2% chance of getting 5.412 (which is greater than 4.036). So we can conclude that the probability of getting 4.036 just by chance is between 2% and 5% (and probably closer to 5%, because 4.036 is closer to 3.841 than to 5.412). The chance of a type 1 error is between 2% and 5% if we reject the null hypothesis.

If there is only a 5% chance that the null hypothesis is true, then it is rather unlikely that race has *no* relationship to employment. We can be about 95% certain that rejecting the null hypothesis is the right thing to do. But five times out of 100 we will incorrectly reject this null hypothesis (type 1 error). This means that five defendants out of 100 will be found guilty when they are innocent. Is this too many? Don't we need to be more certain before jumping to conclusions? The other option is to refuse to reject the null hypothesis and conclude that this is insufficient proof of discrimination. But then we may be permitting a guilty party to go free.

This is the ultimate statistical-legal question. How high a standard of proof should be required for different types of issues? Circuit Judge Spottswood Robinson, dissenting in *CAA*, is conscious of this difficulty; he recognizes that a 5% chance of a type 1 error is barely acceptable. Note his comment in footnote 54.

Work through another approach to testing this null hypothesis by considering additional data given by Judge Robinson on page 170. During the license term, Asian-Americans worked an observed number of 94 person/months compared to an expected 206 person/months, while non-Asian-Americans worked 2892

EXHIBIT V-2.
Critical Values for χ^2

Degrees of Freedom	.90	.80	.70	.50	.30	.20	Significance Level (P) .10	.05	.02	.01	.001
1	.0158	.0642	.148	.455	1.074	1.642	2.706	3.841	5.412	6.635	10.827
2	.211	.446	.713	1.386	2.408	3.219	4.605	5.991	7.824	9.210	13.815
3	.584	1.005	1.424	2.366	3.665	4.642	6.251	7.815	9.837	11.345	16.266
4	1.064	1.649	2.195	3.357	4.878	5.989	7.779	9.488	11.668	13.277	18.467
5	1.610	2.343	3.000	4.351	6.064	7.289	9.236	11.070	13.388	15.086	20.515
6	2.204	3.070	3.828	5.348	7.231	8.558	10.645	12.592	15.033	16.812	22.457
7	2.833	3.822	4.671	6.346	8.383	9.803	12.017	14.067	16.622	18.475	24.322
8	3.490	4.594	5.527	7.344	9.524	11.030	13.362	15.507	18.168	20.090	26.125
9	4.168	5.380	6.393	8.343	10.656	12.242	14.684	16.919	19.679	21.666	27.877
10	4.865	6.179	7.267	9.342	11.781	13.442	15.987	18.307	21.161	23.209	29.588
11	5.578	6.989	8.148	10.341	12.899	14.631	17.275	19.675	22.618	24.725	31.264
12	6.304	7.807	9.034	11.340	14.011	15.812	18.549	21.026	24.054	26.217	32.909
13	7.042	8.634	9.926	12.340	15.119	16.985	19.812	22.362	25.472	27.688	34.528
14	7.790	9.467	10.821	13.339	16.222	18.151	21.064	23.685	26.873	29.141	36.123
15	8.547	10.307	11.721	14.339	17.322	19.311	22.307	24.996	28.259	30.578	37.697

16	9.312	11.152	12.624	15.338	18.418	20.465	23.542	26.296	29.633	32.000	39.252
17	10.085	12.002	13.531	16.338	19.511	21.615	24.769	27.587	30.995	33.409	40.790
18	10.865	12.857	14.440	17.338	20.601	22.760	25.989	28.869	32.346	34.805	42.312
19	11.651	13.716	15.352	18.338	21.689	23.900	27.204	30.144	33.687	36.191	43.820
20	12.443	14.578	16.266	19.337	22.775	25.038	28.412	31.410	35.020	37.566	45.315
21	13.240	15.445	17.182	20.337	23.858	26.171	29.615	32.671	36.343	38.932	46.797
22	14.041	16.314	18.101	21.337	24.939	27.301	30.813	33.924	37.659	40.289	48.268
23	14.848	17.187	19.021	22.337	26.018	28.429	32.007	35.172	38.968	41.638	49.728
24	15.659	18.062	19.943	23.337	27.096	29.553	33.196	36.415	40.270	42.980	51.179
25	16.473	18.940	20.867	24.337	28.172	30.675	34.382	37.652	41.566	44.314	52.620
26	17.292	19.820	21.792	25.336	29.246	31.795	35.563	38.885	42.856	45.642	54.052
27	18.114	20.703	22.719	26.336	30.319	32.912	36.741	40.113	44.140	46.963	55.476
28	18.939	21.588	23.647	27.336	31.391	34.027	37.916	41.337	45.419	48.278	56.893
29	19.768	22.475	24.577	28.336	32.461	35.139	39.087	42.557	46.693	49.588	58.302
30	20.599	23.364	25.508	29.336	33.530	36.250	40.256	43.773	47.962	50.892	59.703

Source: R. Fisher and F. Yates. Statistical Tables for Biological, Agricultural and Medical Research 47 (Table IV) (6th ed. 1963). Reprinted with permission of Oliver and Boyd, Edinburgh.

person/months compared to an expected 2780.[1] How do we use a chi-square test on this data?

For this example, the table of O_j's and E_j's looks like this:

EXHIBIT V-3.
Alternative Calculation of the Chi-Square Value in CAA

Step	j = Asian-Americans	j = Non-Asian-Americans
E_j (Expected)	206	2780
O_j (Observed)	94	2892
(1) $(O_j - E_j)$	−112	+112
(2) $(O_j - E_j)^2$	12544	12544
(3) $(O_j - E_j)^2 \div E_j$	60.89	4.51
(4) $\Sigma[(O_j - E_j)^2 \div E_j] = 65.40 = \chi^2$		

Once again we have one degree of freedom, but we can't find anything close to 65.40 on our table for 1 d.f. We must conclude that the chances of the null hypothesis being true with this size χ^2 value are (much) less than one chance in 100.

Using the *Castaneda* test, we calculate how far such an observation, i.e., 94 person/months, is from the mean, i.e., 206. Since we have a binomial situation, we can use the formula:

$$SD = \sqrt{N \times P_A \times P_B}$$

with N measuring total available person/months. One standard deviation equals $\sqrt{2986 \times .069 \times .931}$, which equals $\sqrt{191}$, or 13.8. How many standard deviations is 94 from 206? Approximately nine (206 − 94 ÷ 13) standard deviations from the mean. This is far greater than the two or three required by *Castaneda* to make us suspicious.

Notes and Problems

1. Note how the choice of variables to be measured affects the statistical outcome. Using the facts of Chinese for Affirmative Action v. FCC, we performed two different tests of statistical significance on the data: the standard deviation test and the chi-square test. The first variable was number of employees, and two categories were used —

---

1. How do we get these latter figures? If 6.9% of the population is Asian-American, then they would be expected to work 6.9% of the total person/months. If 206 is 6.9% of the total person/months, then the total person/months is 206 divided by 0.069, which equals 2986 person/months worked by all employees. Non-Asian-Americans would be expected to work 2986 minus 206, or 2780 person/months. Asian-Americans actually worked 94, according to Judge Robinson, so non-Asian-Americans must have worked 2986 minus 94, which equals 2892 person/months.

minority employees were compared to nonminority employees. We found a two-standard-deviation difference between the observed and the expected number of minorities and a corresponding chi-square value of 4.036, which we concluded was somewhat less than a 5% chance of error in rejecting the null hypothesis. The second variable, chosen by Judge Robinson, gave more startling results. It was the number of person/months worked by minority members and by nonminorities. A nine-standard-deviation disparity and a chi-square value of 65.4, representing much less than a 1% chance of error, resulted from analysis of this data. The latter results are much more convincing. Proper selection of variables to be measured is an essential part of advocacy with statistical tools.

2. Recall that the biostatistical analysis in Certified Color Manufacturers' Association v. Mathews relied on a chi-square test. The results of biological experiments on rats are frequently considered by administrative agencies, and it is useful to know how they are analyzed statistically. Consider one piece of evidence from Dr. Gaylor's study. Of 53 rats dying before the end of the study, 11 (or 21%) had cancer. Seven were from the group of 23 rats receiving high doses of Red No. 2, and four were from the 30 rats in the low-dosage group. If increased doses of Red No. 2 are safe, then there should be no significant difference between the cancer rates in high- and low-dosage groups. Our null hypothesis might be, "There is no relationship between dosage of Red No. 2 and cancer rates among those rats dying before the end of the study."

The only difficult conceptual step in applying the chi-square test is determining the expected number in each category, E_j. One approach in this case is to test whether the seven of 23 high-dosage rats and four of 30 low dosage rats are a significant departure from the mean cancer rate of 21%. Using this approach, we expect 21% of the 23, (or 4.8) high-dosage rats, and 21% of the 30 (or 6.3) low-dosage rats to get cancer. The chi-square table for analyzing this evidence appears as follows:

EXHIBIT V-4.
Calculation of Chi-Square Value from Red-Dye No. 2 Data

Step	j = High-Dosage Rats	j = Low-Dosage Rats
E_j (Expected)	4.8	6.3
O_j (Observed)	7	4
(1) $(O_j - E_j)$	2.2	2.3
(2) $(O_j - E_j)^2$	4.84	5.29
(3) $(O_j - E_j)^2 \div E_j$	1.01	0.84
(4) $\Sigma[(O_j - E_j)^2 \div E_j] = 1.85 = \chi^2$		

For one degree of freedom, a χ^2 value of 1.85 will occur by chance between 10% and 20% of the time; such a value represents too large a probability of a type 1 error to reject the null hypothesis.

3. Using this approach, confirm the finding that there is a significant statistical relationship between dosage and cancer among those rats sacrificed at the termination of the experiment. Among the 21 surviving high-dosage rats, seven had cancer. None of the 14 low-dosage surviving rats had cancer. What is your null hypothesis? What is the probability of a type 1 error if you reject the null hypothesis?

4. The approach we used in Note 2 to calculate the expected number is not the only possible one. If our null hypothesis were, "Increasing the dosage has no effect on the cancer rate," we might hypothesize a cancer rate among high-dosage rats not significantly different from 13%, the rate for low-dosage rats. This assumption gives an E_j of 13% × 23 = 3 high-dosage rats, and 13% × 30 = 4 low-dosage rats. Calculate the chi-square value for this test. You will find that the probability of a type 1 error lies between 2% and 5%, small enough to reject the null hypothesis. Check these results. Again, you can see the power in cleverly choosing the design of the statistical test, and you can begin to appreciate how manipulators of numbers can "prove anything."

Notes: Problems with Small Numbers

1. *Combining Categories.* You should be aware that the chi-square test is not precisely accurate whenever the expected value in any category of the variable, any E_j, is less than five. This criticism applies to the tests we performed in Notes 2, 3, and 4 above. The general rule is that if a category has an expected number less than five, then that category must be combined with another category. But it is impossible to do so if we have only two categories, as we did in Notes 2, 3, and 4 above. The particular advantage of the chi-square test is that it is most useful when we have a larger number of categories. The steps for using the chi-square test for more than two categories remain the same; the only differences are that there are more values to sum up in step 4 and that the degree of freedom changes. Recall the rule, $d.f. = n - 1$.

If you are confronted with only two categories and an E_j less than five, and if you still want to use a chi-square test, there are more sophisticated ways to correct the bias that this "small sample" creates.

Among these tests are the Yates Correction and the Fisher's Exact Test. The application of the first of these to the chi-square test is described in the following note.

2. *The Yates Correction.* The Yates Correction for small expected numbers of observations is not accepted unquestioningly by all statisticians. But it is frequently used and is worth a passing familiarity. Three limitations to its use are crucial. The first is that the correction makes your calculated chi-square value more accurate only when the expected number in at least one category, the E_j's, is less than five. The second is that the correction can be used only when there are just two categories. It does not apply when the degrees of freedom are greater than one. The explanation of the third limitation must wait until the Yates corrected chi-square formula is explained.

For one degree of freedom, where the E_j in at least one category is less than five, the chi-square calculation is:

$$\chi^2 = \Sigma \frac{(|O_j - E_j| - 1/2)^2}{E_j}$$

for all the j's. The only difference between this formula and the steps on page 179 is step 1. Instead of:

(1) $O_j - E_j$

we calculate:

(1)' $|O_j - E_j| - 1/2$

The straight lines on either side of "$O_j - E_j$" mean "take the absolute value of," which simply means ignore any minus sign that might result from subtracting E_j from O_j. The correction comes from subtracting $1/2$ from the difference between O_j and E_j. Perform this correction on the E_j's and O_j's for both categories. Steps 2, 3, and 4 are essentially unchanged. They look like this:

(2)' $(|O_j - E_j| - 1/2)^2$

(3)' $(|O_j - E_j| - 1/2)^2 \div E_j$

(4)' $\Sigma[(|O_j - E_j| - 1/2)^2 \div E_j] = \chi^2$.

Then look up the χ^2 value to perform the test as usual.

The third restriction on using the Yates Correction is that if $|O_j - E_j|$ is less than $1/2$, the correction should not be used. Using the Yates Correction, recalculate the results in Problems 3 and 4 above.

All of these restrictions and limitations make one wonder just how useful and accurate the correction is. Don't worry about it. Statisticians have not yet resolved this question.

3. *Age-Discrimination Hypothetical.* In an age-discrimination suit, we find the following data. Of 87 applicants for the apartment complex in question, 20 applicants were accepted. Of these applicants, two were elderly (over 90 years old) and 18 were not elderly. Elderly people make up 20% of the population. Use a chi-square test to determine the probability of this selection's occurring by chance. What is the null hypothesis? Do you reject it or not? If you reject it, what is the probability of a type 1 error? What result do you get when you apply the two-or-three-standard-deviations rule in this case?

4. *Causation.* A large chi-square value with a low associated probability of a type 1 error does not necessarily mean that discrimination *caused* the observed outcome. It only means that it is unlikely that the disparity between applications accepted for different age groups in Note 2 above occurred by chance. There may be other reasons why a smaller proportion of elderly people were accepted—their lower income, their greater tendency to have pets, or their loud parties. If such explanations would excuse the discriminatory behavior of the apartment complex, the advocates for either side cannot ignore them.

CHANCE v. BOARD OF EXAMINERS
330 F. Supp. 230 (S.D.N.Y. 1971)

MANSFIELD, Circuit Judge.

The fairness and validity of competitive examinations, once described by Gilbert and Sullivan as the means of attaining "a Duke's exalted station," have frequently been challenged in courts and elsewhere. We are here called upon to decide whether those examinations which have been prescribed and administered by the Board of Examiners of the City of New York (the "Board" herein) to candidates seeking licenses for permanent appointment to supervisory positions in the City's school system (principals, assistant principals, administrative assistants, etc.) are unconstitutional. We conclude that a sufficient showing has been made of violation of the Equal Protection Clause of the Fourteenth Amendment to warrant the issuance of preliminary injunctive relief.

The two named plaintiffs, who are respectively Black and Puerto Rican, have brought this purported class action on behalf of themselves and all other persons similarly situated pursuant to federal civil rights laws, 42 U.S.C. §§1981[2] and 1983.[3] They allege that the competitive examinations, which must be passed by a candidate before he or she can qualify for licensing and appointment, discriminate against persons of Black and Puerto Rican race, and have not been validated or shown fairly to measure the skill, ability and fitness of applicants to perform the duties of the positions for which the examinations are given, with the result that success on the examination does not indicate in any way that the candidate will succeed as a supervisor. This racial discriminatory effect, coupled with lack of justification or predictive value as measurements of abilities required to perform the jobs involved, is alleged to violate not only plaintiffs' federal constitutional rights but also (based on pendent jurisdiction) Art. 5, §6 of the New York State Constitution, and §§2590-j(3)(a)(1), 2569(1), and 2573(10) of the New York Education Law.

Plaintiffs seek a preliminary injunction under Rule 65, F.R. Civ. P., prohibiting the alleged violations of these laws. They also seek declaratory relief pursuant to 28 U.S.C. §2201. We have jurisdiction under 28 U.S.C. §§1331 and 1343(3).

The Board of Education has not actively opposed the motion for preliminary injunction, and it agrees that plaintiffs have presented triable issues of fact. The Board of Examiners ("Board" herein), however, has vigorously opposed the motion. . . .

An applicant for permanent appointment to a supervisory position in the New York City School System must, in addition to meeting state requirements for the position, obtain a New York City license. First, each such candidate must have met minimum education and experience requirements established by the City's Board of Education and the Chancellor, Harvey B. Scribner, who is the Chief Administrator of the School District of the City of New York. For instance, a candidate for principal of a day elementary school must, among other things, have had (1) four years' experience teaching in day schools under regular license and appointment as a teacher, and (2) two

---

2. 42 U.S.C. §1981: "*§1981. Equal rights under the law.* All persons within the jurisdiction of the United States shall have the same right in every State and Territory to make and enforce contracts, to sue, be parties, give evidence, and to the full and equal benefit of all laws and proceedings for the security of persons and property as is enjoyed by white citizens, and shall be subject to like punishment, pains, penalties, taxes, licenses, and exactions of every kind, and to no other."

3. 42 U.S.C. §1983: "*§1983. Civil action for deprivation of rights.* Every person who, under color of any statute, ordinance, regulation, custom, or usage, of any State or Territory, subjects, or causes to be subjected, any citizen of the United States or other person within the jurisdiction thereof to the deprivation of any rights, privileges, or immunities secured by the Constitution and laws, shall be liable to the party injured in an action at law, suit in equity, or other proper proceeding for redress."

years' experience of supervision in day schools under license and appointment, or meet various alternative experience requirements.

Next the candidate must pass an examination procedure prepared and administered by the Board for the particular type or classification of supervisory post desired, which may take as long as two years to complete. If the candidate successfully completes the testing procedure, he or she is granted a license and placed on a list of those eligible for assignment to the type of supervisory position involved. The appropriate school governing authority — either a central board of education or a community school board under New York City's present decentralized system — then selects the person it wishes from the eligible list to fill an open position. Since appointments of permanent supervisory personnel in the New York City School System must be made from lists of eligibles who have passed examinations, the Board from time to time announces and conducts examinations for particular supervisory posts (of which there are more than 50 different types) following which the number of persons eligible for appointment are supplemented by promulgation of lists of those who passed the latest examination. If a successful candidate, after being listed as eligible for appointment, is not appointed within four years, he or she is dropped from the list and must again pass the qualifying examinations to be listed as eligible.

Only in the cities of Buffalo and New York does state law provide for examinations in addition to *state* certification, N.Y. Education Law §2573 (10-a), and only the New York City School District maintains a Board of Examiners and the specific examination and licensing procedure here under attack. The Board has described itself as "a highly select group with broad professional background in education and related fields chosen through the most objective and impartially searching examination given under civil service." (Ex. 10, Item 10, attached to 5/28/71 aff. of Richard S. Barrett)

Were it not for New York City's special examination and licensing procedure, plaintiffs Chance and Mercado would have been appointed permanent elementary school principals. Both have been certified by the *state* for that position, and both are specially trained to be principals, having graduated from a year-long Fordham University Instructional Administrators and Principals Internship Program in Urban Education.

Plaintiff Boston M. Chance has been employed in the New York City public school system for the last 15 years and is acting principal of P.S. 104, an elementary school in the Bronx. Chance, who is of the Black race, possesses all of the basic qualifications of education and experience established by law and by the Board of Education and the

Chancellor of the New York City School District for the position of principal of an elementary school. However, he lacks a *city* license as elementary school principal and therefore is barred at present from securing a permanent position as principal. In September, 1968, Chance took the examination given by the Board for the position of Assistant Principal, Junior High School, but he failed it and thus was not placed on the eligibility list and was not issued a license entitling him to permanent appointment.

Plaintiff Louis Mercado, a Puerto Rican who also holds a New York State license as a principal, has been serving the New York City school system for the last 12 years. He is presently acting principal of P.S. 75, an elementary school in Manhattan, but he is barred from permanent appointment because he does not have a New York City license as an elementary school principal. Mercado is in a somewhat different position from Chance in that he does not allege that he has ever taken the relevant Board of Examiners' Supervisory Examination. While the present motion was pending — and while the parties were collecting statistical information pursuant to our order — the Board conducted their November, 1970 series of examinations for elementary school principal. Mercado withdrew from this examination and refused to take it on the grounds that the "Board of Examiners is not the appropriate agency for qualifying school personnel" and "the examination is not relevant. . . ."

Both Chance and Mercado were selected for their present acting principalships by their respective community school boards, in accordance with New York City's decentralized system. In some instances such local school boards found, after interviewing licensed principals listed as eligible by the Board, that persons not so licensed were more qualified to serve as principals than those interviewed and that they performed their duties in a superior manner.

There are approximately 1,000 licensed Principals of New York public schools of varying levels (e.g., elementary day, junior high school, high school, etc.), of whom some act as the heads of schools and others function in administrative positions. Of the 1,000 only 11 (or approximately 1%) are Black and only 1 is Puerto Rican. Furthermore, of the 750 licensed Principals of New York elementary schools only 5 (or less than 1%) are Black, and none is Puerto Rican. Of the 180 high school administrative assistants, none is Black or Puerto Rican.

Of the 1,610 licensed Assistant Principals of New York City junior high and elementary schools, only 7% are Black and only .2% are Puerto Rican. When the list for the position of Principal, elementary school, was originally promulgated, only 6 out of the 340 persons (or about 1.8%) were Black and none was Puerto Rican; and when the list for Principal, high school, was promulgated, none of the 22 licensed people was Black or Puerto Rican. The promulgated list of licensed Assistant Principals for junior high schools reveals that only 55 out of 690 persons (or 8%) were Black and none was Puerto Rican.

Plaintiffs contend that the written and oral examinations of the Board are the major factor accounting for this extremely low percentage of Black and Puerto Rican supervisors in a school system where 55% of the students are Black or Puerto Rican. Plaintiffs summarize their basic argument as follows:

> [T]hese tests place a premium on familiarity with organizational peculiarities of the New York City school system which, while having little to do with educational needs, are largely gained through coaching and assistance from present, predominately white, supervisory personnel. . . .
>
> The testing procedures do not indicate a candidate's ability to do the job being tested for. There is no evidence that they measure merit or fitness, they have never been validated, and they are unreliable psychological instruments.

Amended Complaint §§22, 23.

Rather than risk the endless delay that would be encountered while the parties obtained this essential evidence through pretrial discovery procedures, we directed the parties, in view of the importance of the issue, to use their best efforts to agree on a procedure whereby the Board of Examiners and the Board of Education would compile the necessary racial statistics. All parties cooperated fully and at considerable effort in working out almost all of the details of the procedure to be followed. Such differences as existed were resolved by court order. The result has been that after months of research we have been presented with the pass-fail statistics for the relevant racial and ethnic groupings of candidates for 50 supervisory examinations given over the past few years. In view of plaintiffs' claims that the examinations had a "chilling effect" inhibiting Black and Puerto Ricans from becoming candidates, this statistical survey ("the Survey") also includes figures as to those candidates who "Did Not Appear" to take the written test, which commenced the examination process, or who "Withdrew," i.e., took the written test but did not appear for subsequent parts of the examination.

All parties and amici have submitted briefs as to the relevance of the statistics thus adduced and the inferences that may be drawn from them. The parties also submitted affidavits by statistical experts. A hearing was held, at which each side's expert testified and was subject to cross-examination; and we heard more oral argument on the statistical data. After declining the opportunity to examine and cross-examine any other witnesses, including those presented by affidavit, both sides rested on the record thus adduced.

Upon the evidence thus presented we find that the examinations and testing procedures prepared and administered by the Board for the purpose of determining which candidates will be licensed as supervisory personnel have the effect of discriminating against Black and Puerto Rican candidates.

The Survey reveals that out of 6,201 candidates taking most of the supervisory examinations given in the last seven (7) years, including all such examinations within the last three (3) years, 5,910 were identified by race. Of the 5,910 thus identified, 818 were Black or Puerto Rican and 5,092 were Caucasian. Analysis of the aggregate pass-fail statistics for the entire group reveals that only 31.4% of the 818 Black and Puerto Rican cnadidates passed as compared with 44.3% of the 5,092 white candidates.[17] Thus on an overall basis, white candidates passed at almost 1 1/2 times the rate of Black and Puerto Rican candidates. These overall figures, however, tell only part of the story. Of greater significance are the results of two examinations which had by far the largest number of candidates, those for Assistant Principal of Day Elementary School and Assistant Principal of Junior High School [see the table on the following page].

Thus white candidates passed the examination for Assistant Principal of Junior High School at almost double the rate of Black and Puerto Rican candidates, and passed the examination for Assistant Principal of Day Elementary School at a rate one-third greater than Black and Puerto Rican candidates. The gross disparity in passing rates on these two examinations is of particular significance not only because they were taken by far more candidates than those taking any other examinations conducted in at least the last seven years, resulting in licensing of the largest number of supervisors, but also because the assistant principalship has traditionally been the route to and prerequisite for the most important supervisory position, Principal. To the extent that Black and Puerto Ricans are screened out by the examination for Assistant Principal they are not only prevented from becoming Assistant Principals but are kept out of the pool of eligibles for future examinations for the position of Principal as well. The fact that the process involves a series of examinations and that to reach the

---

17. Of the candidates identified as Puerto Rican only, 27.59% passed; of the candidates identified as Black, 31.56% passed.

Examination	Caucasian		Black		Puerto Rican		Black and Puerto Rican		Probability of Chance Result Less Than —
	Total	% Pass	Total	% Pass	Total	% Pass	Total	% Pass	
Assistant Principal, Day Elementary Schools, 1965 Examination (PF-03)	1171	61.3%	278	45.68%	7	28.57%	285	45.26%	1/1 million
Assistant Principal, Junior High Schools, 1968 Examination (PF-43)	1319	48.82%	236	26.27%	14	14.29%	250	25.60%	1/1 million

Source: Table adapted from Affidavit of plaintiff's expert, Professor Jacob Cohen, May 6, 1971, ¶5. The computations of probability are his; the racial statistics come directly from the Survey.

top one must pass several examinations at different times in his or her career serves to magnify the statistical differences between the white and non-white pass-fail rates. For instance, if we take a group of 100 Black and Puerto Rican candidates, on the one hand, and 1,000 white candidates, on the other, and assume a passing rate of 25% for the former and of 50% for the latter on a given assistant principal's examination as was approximately the case in the examination for Assistant Principal of Junior High School), the results would be as follows:

Black & Puerto Rican   25% × 100 = 25 ⎱ Licensed Assistant
White   50% × 1000 = 500 ⎰ Principals

The group of 525 licensed assistant principals would then form the pool of eligibles for the related principal's examination. Assuming the same relative pass rates, we have the following results:

Black & Puerto Rican   25% × 25 = 6.25 ⎱ Principals
White   50% × 500 = 250 ⎰

Thus the true resulting difference between the Black and Puerto Rican versus the white pass rates would be even more substantial: only 6.25% of the Blacks and Puerto Ricans would pass the two successive examinations as against 25% of the whites.

When we look at all 50 examinations which were the subject of the Survey, we find that only 34 were taken by at least one member of both the white and Black-Puerto Rican racial groups. One of these examinations (Assistant Administrative Director, given Dec. 1967, PF-17) was passed by everyone taking it. Another (Director, Bureau for Children with Retarded Mental Development, given Jan. 1968, PF-18) was not completed successfully by anyone taking it.[18] There remain 32 examinations where one or the other of the two main racial groups — Black and Puerto Rican in one group and white in the other — had a larger percentage passing than did the other group. *Of these 32 examinations the white group had a larger percentage passing in 25 examinations and the non-white group had a larger percentage passing in only 7 examinations.* Thus the whites passed at a proportionately higher rate in *three times* as many examinations as the non-whites. The probability of these results occurring by chance is less than 1.05 in 1,000.

---

18. The one Black who took it failed; no Puerto Ricans took it; and 5 whites took it. Four failed and the last one withdrew, which in effect meant that he failed because he could not successfully complete the examination process.

Plaintiffs' Exhibit 2 to oral hearing of 5/21/71. See also transcript of oral hearing of 5/21/71 at 14-19.

Plaintiffs offered the testimony of Dr. Jacob Cohen, an expert in the field of statistics, in support of the validity and significance of the Survey results. The Board, in turn, adduced the testimony of Dr. Nathan Jaspen, an expert in the same field, in opposition. Both witnesses possess outstanding qualifications. After reviewing their testimony and appraising them as witnesses, we are more persuaded by the testimony of Dr. Cohen with respect to certain crucial matters affecting the significance of the figures for present purposes.

Turning first to the gross aggregate pass-fail statistics, which reveal that the overall pass rate of white candidates (44.3%) was almost half again as high as the non-white rate (31.4%), Dr. Cohen testified that on the basis of such a large sample (5,910 out of 6,201 candidates), the test results were especially valuable and formed a sound basis for drawing valid statistical conclusions as to the difference in passing rates between the ethnic groups involved. In analyzing the statistics he used the Chi-Square Test (Yates-corrected), which is a method using formulas generally accepted by statistical experts to determine whether an observed difference in any given sample is greater than that which would be expected on the basis of mere chance or probability. He found with respect to the aggregate test that by "the Chi-Square (Yates-corrected) statistical test, the probability of the difference being a chance result not related to the factor of race is determined as *less than one in one billion.*" (Emphasis added) . . .

Finally we are impressed with the revealing statistics comparing the percentage of Black and Puerto Rican Principals to white Principals in the five largest school systems in the country:

City	Total No. of Principals	% Black	% Puerto Rican	% Black and Puerto Rican
Detroit	281	16.7%	—	16.7%
Philadelphia	267	16.7%	—	16.7%
Los Angeles	1,012	8.0%	1.7%	9.7%
Chicago	479	6.9%	—	6.9%
NEW YORK	862	1.3%	0.1%	1.4%

Thus New York has by far the lowest percentage of minority representation. The next lowest city, Chicago, has almost *5 times* the percentage of minority principals found in New York City, and as the following table shows there is a similar imbalance of minority Assistant Principals:

C. Chi-Square Testing

City	Total No. of Asst. Principals	% Black	% Puerto Rican	% Black and Puerto Rican
Detroit	360	24.7%	0.2%	24.9%
Philadelphia	225	37.0%	———	37.0%
Los Angeles	———	———	———	———
Chicago	714	32.5%	———	32.5%
NEW YORK	1,610	7.0%	0.2%	7.2%

Plaintiffs also argue that discrimination may be inferred from the fact that the percentage of Black and Puerto Rican Principals and Assistant Principals in New York City schools (1.4% and 7.2%, respectively) is far below the percentage of the total student body who are Black and Puerto Rican (55.8%) and when compared with similar figures for the five largest school systems in the country (New York, Los Angeles, Chicago, Detroit, and Philadelphia) constitutes not only the lowest minority representation in the supervisory ranks, but also the lowest ratio of such minority group supervisors to minority group students. We reject this contention. Supervisors are drawn from the pool of qualified teachers, most of whom attended elementary and high school long ago, and not from present-day students. Undoubtedly the low number of minority teachers eligible to take the supervisory examinations prescribed by the Board has been due in part to the fact that the percentage of minority students who 10 or 15 years ago went on to college and qualified for a teaching career, and thus provided the source of today's minority teachers, was much smaller than the number of white students following such a course, with the result that a larger pool of qualified white graduates entered the teaching profession. In addition the minority student population in New York City has increased during the same period, with the effect of increasing the racial imbalance between teachers and students. Current efforts to promote higher educational opportunities for minority groups will not produce qualified teachers for some years. But statistics as to the current dearth of qualified minority teachers do not have probative value with respect to the question before us, which is whether New York City's examination system discriminates against *minority candidates who have already qualified as licensed teachers.*

For the same reasons we are unimpressed with plaintiffs' comparisons between (1) the percentage of Black and Puerto Rican members of the general population in New York City, and (2) the percentage of Black and Puerto Rican Principals and Assistant Principals found in the City's total school supervisory personnel. Statistical comparisons to the general racial population of the community may be relevant in determining whether there is discrimination in job opportunities that are supposed to be open to the general public, in the selection of teachers from a pool of those already qualified and eligible for appointment, or in qualification of voters or jurors. But we are here dealing with candidates who must meet preliminary eligibility requirements as to education and experience that are not possessed by most of the general population. Where the education of our children is at stake, such insistence upon the highest possible quality in our teachers is a salutary and lawful objective, provided it does not result in racial discrimination between candidates who are otherwise eligible, which is the case here.

Notwithstanding the introduction of some evidence thus found irrelevant, the evidence establishes to our satisfaction that the examinations prepared and administered by the Board for the licensing of supervisory personnel in New York City schools do have the de facto effect of discriminating significantly and substantially against qualified Black and Puerto Rican applicants. However, the existence of such discrimination, standing alone, would not necessarily entitle plaintiffs to relief. The Constitution does not require that minority group candidates be licensed as supervisors in the same proportion as white candidates. The goal of the examination procedures should be to provide the best qualified supervisors, regardless of their race, and if the examinations appear reasonably constructive to measure knowledge, skills and abilities essential to a particular position, they should not be nullified because of a de facto discriminatory impact. We accordingly pass on to the question of whether the examinations under attack can be validated as relevant to the requirements of the positions for which they are given, i.e., whether they are "job-related." . . .

Reluctant as we are to invade a profession characterized by an expertise not shared by us, we must conclude on the record before us that while the Board has adopted procedures designed for content validity, it does not appear in practice to have achieved this goal. Our conclusion, which is based upon our appraisal of affidavits of experts furnished by the parties, is confirmed by our own study of some of the examinations themselves. Turning first to the short-answer, multiple-choice tests, many of the questions strike us as having little relevance to the qualities expected of a school supervisor. Some examples are cited in the margin.[23]

23. *1961 Examination for Principal, Day Elementary School*
  "64. Of the following characters in the nursery rhyme, THE BURIAL OF POOR COCK ROBIN, the one who killed Cock Robin is the
  1. Lark

A review of the balance of a typical short-answer test and, indeed, of the essay type test reveals that the questions appear to be aimed at testing the candidate's ability to memorize rather than the qualities normally associated with a school administrator. . . .

CONCLUSION

The evidence reveals that the examinations prepared and administered by the Board of Examiners for the licensing of supervisory personnel, such as Principals and Assistant Principals, have the de facto effect of discriminating significantly and substantially against Black and Puerto Rican applicants.

Despite the fact that candidates for such positions are licensed teachers who have satisfied prerequisites as to education and experience established by the Board of Education for supervisory positions and have already been certified by the State of New York for the positions sought, a survey of the results of examinations taken by 5,910 applicants (of whom 818 were Black or Puerto Rican) reveals that white candidates have received passing grades at almost $1\,^{1}/_{2}$ times the rate of Black and Puerto Rican candidates and that on one important examination given in 1968 for the position of Assistant Principal, Junior High School, white candidates passed at almost double the rate

      2. Thrush
      3. Bull
      4. Sparrow"

1965 Examination for Assistant Principal, Junior High School
   "211. *I've Got a Little List,* from the *Mikado* is sung by
      1. Nanki-Poo   2. Pish-Tush   3. Ko-Ko   4. Pooh-Bah
   "212. Arthur Sullivan of the team of Gilbert and Sullivan, wrote
      1. Gypsy Love Song   2. The Lost Chord
      3. Ah, Sweet Mystery   4. A Kiss in the Dark
         of Life
   "218. Which one of the following violin makers is *NOT* of the great triumvirate of Cremona?
      1. Amati   2. Stradivarius   3. Guarnerius
      4. Maggini"

1968 Examination for Assistant Principal, Junior High School
   "63. The author of *Dodsworth* is also the author of:
      1. *Daisy Miller*
      2. *Elmer Gantry*
      3. *Henry Esmond*
      4. *Mrs. Dalloway*
   "74. The author of *Tender is the Night* also wrote:
      1. *The Great Gatsby*
      2. *Butterfield 8*
      3. *The Sun Also Rises*
      4. *Sanctuary*"

of Black and Puerto Rican candidates. The discriminatory effect in the latter case is aggravated by the fact that the Assistant Principalship has traditionally been an essential prerequisite to the more important supervisory position of Principal.

The existence of such de facto racial discrimination is further confirmed by the fact that only 1.4% of the Principals, and 7.2% of the Assistant Principals in New York City schools are Black or Puerto Rican, percentages which are far below those for the same positions in the four other largest city school systems in the United States. For example, the percentage of Black and Puerto Rican Principals in each of the cities of Detroit and Philadelphia is 16.7%, or 12 times as high as that in New York.

Such a discriminatory impact is constitutionally suspect and places the burden on the Board to show that the examinations can be justified as necessary to obtain Principals, Assistant Principals and supervisors possessing the skills and qualifications required for successful performance of the duties of these positions. The Board has failed to meet this burden. Although it has taken some steps towards securing content and predictive validity for the examinations and has been improving the examinations during the last two years, the Board has not in practice achieved the goals of constructing examination procedures that are truly job-related. Many objectionable features remain, with the result that some 37 minority Acting Principals and 131 minority Acting Assistant Principals, who are considered fully qualified and are desired for permanent appointment by the community school boards, are rendered ineligible for such permanent appointment. A study of the written examinations reveals that major portions of them call simply for regurgitation of memorized material. Furthermore, the oral examination procedure leaves open the question of whether white candidates are not being favored — albeit unconsciously — by committees of examination assistants who have been entirely or predominantly white.

There appears to be a strong likelihood that plaintiffs will prevail on the merits at trial. It further appears that plaintiffs would suffer greater harm from denial of preliminary injunctive relief than defendants would suffer from the granting of relief. Denial of relief would perpetuate existing racial discrimination, depriving plaintiffs and others similarly situated of an equal opportunity for permanent appointment and licensing as supervisors. During the long period before the case would finally be adjudicated on the merits, permanent appointments would be made from lists promulgated by the Board, which would have the effect of threatening the continued employment of those holding acting appointments, since New York law requires vacancies to be filled from the eligibility lists if such lists exist. N.Y. Education Law §2573(2); Board of Education By-Laws §101(3).

Granting of preliminary relief, it is true, will temporarily prevent the appointment from eligible lists of those who have, after the arduous process of taking the existing type of examinations, successfully passed and been placed on the eligible list. However, they will not be denied an equal opportunity in the future to qualify under such examination procedures as are found to be constitutionally permissible and, pending trial of the case, they would be eligible for appointment as Acting Principals or Acting Assistant Principals. Thus the balance of hardships tips decidedly in favor of plaintiffs, and, pending final determination of the merits, the effect of preliminary relief would be to preserve the status quo until the issues are resolved. Checker Motors Corp. v. Chrysler Corp., 405 F.2d 319, 323 (2d Cir. 1969).

Having in mind that the existing examination system is not believed by the Chancellor of the New York City District to be a workable one, we do not envisage any great harm to the public as a result of preliminary relief. On the contrary, such relief may possibly lead the Board of Examiners, after taking another hard look at its examination procedures, to consider an overhaul that will not only eliminate racial discrimination, but lead to procedures that will be more adaptable to the Community School Board type of administration.

Lastly we cannot overlook the fact that various persons having the duty of selecting supervisory personnel, such as members of community school boards, have stated in affidavits filed with the court that they have often found that holders of licenses from the Board of Examiners do not possess the ability to perform the duties of a supervisory position for which a candidate is sought, with the result that in order to select qualified personnel it has been necessary to appoint unlicensed candidates on an acting basis. (See Affs. of Dr. Edwin J. Haas, Edythe J. Gaines, and Peter J. Strauss).

For the foregoing reasons a preliminary injunction will issue restraining defendants from (1) conducting further examinations of the type found to be unconstitutionally discriminatory against Blacks and Puerto Ricans, and from (2) promulgating eligible lists on the basis of such examination procedures.

The foregoing shall constitute our findings of fact and conclusions of law as required by Rule 52(a), F.R. Civ. P.

We take this opportunity to express appreciation to the parties for their thorough papers, and to the amici for their briefs, which were of assistance in resolving the difficult and complex issues.

Settle order on notice.

Notes and Problems

1. Without much elaboration, the court in Chance v. Board of Examiners used the chi-square test to examine the likelihood of random low-pass rates among black and Puerto Rican supervisor candidates. Because there are three relevant categories in the race variable, the chi-square procedure has three E_j's and O_j's with which to work. The same procedure is used as with two categories, but remember to adjust the degrees of freedom to the increased number of categories. Theoretically, there is no limit to the number of categories you may use, subject to the warning that the expected number for each category must be five or greater or a corrective measure taken. As an exercise, use a chi-square test to analyze the likelihood that the pass rates for Assistant Principal, Day Elementary Schools, 1965 Examination, occurred by chance. Use the data from the table on page 188.

2. If you add to the information presented in the table on page 190, you can determine which of the nation's five largest school systems is most and least likely to have chosen its principals on a random basis, i.e., without regard to race. To do this by means of a chi-square test, what additional information do you need? The court considers this problem on page 191.

3. As an example of chi-square testing with more than two categories, consider the problem confronting an enforcement officer from the New Jersey Gambling Commission. The officer's task is to determine whether a particular die at the Boardwalk Regency Casino is fair. He rolls the die sixty times, coming up with the following observed values:

Value on die	1	2	3	4	5	6
Frequency	12	9	17	4	8	10

As is often the case, the difficult task is to determine the expected values. Here, how often would you expect to see a "1" come up in 60 tries? About 10 times on average. This is true of every number if the die is fair. The null hypothesis is that there is no relationship between your choice of die and the outcome of a roll. After setting up the table of observed and expected values, test this null hypothesis by proceeding through steps 1-4 (see pages 173-178) and checking the chi-square table for the appropriate number of degrees of freedom.

4. A client complains to you of racial discrimination in jury selection. She presents you with the following data taken from the community voter rolls and jury lists of ten juries, each composed of ten jurors:

EXHIBIT V-5.
Proportion of Each Race in the Community

Source	Anglo	Black	Chicano	Oriental	Other
From jury list:	57%	26%	9%	7%	1%
From voter list:	42%	38%	12%	5%	3%

Establish the null hypothesis and test it using a χ^2 test. Note that, because the E_j for one category is less than five, you will need to reduce the number of categories by combining two of them and doing the test with only four categories.

5. The relatively small proportion of black principals in New York City can be dramatically demonstrated by a bar graph. As an exercise, do so. For each of the five largest school systems in the country, calculate the disparity in terms of the number of standard deviations between expected and observed percentages of black principals. Use the data on page 190 to find the observed percentages. Use the following percentages of qualified applicants who are black to calculate expected values:

EXHIBIT V-6.
Expected Values of Applicants for Principal

City	Percentage Black of All Qualified Applicants	Percentage Puerto Rican of All Qualified Applicants
New York	13%	1%
Detroit	26%	3%
Los Angeles	12%	5%
Chicago	22%	2%
Philadelphia	26%	1%

Using this data it is also possible to describe graphically the disparities in Puerto Rican pass rates. This data also allows you to rank these cities by the randomness of their selection procedure by using a chi-square test.

Problem: In re Kroger

A statistical expert in the case of In re Kroger Co., Federal Trade Commission Slip Opinion No. 9102, June 11, 1979, used a chi-square test to evaluate comparative price claims made by Kroger's supermarket chain. Kroger established a sample list of grocery basket items called the "Price Patrol" and advertised that the prices of those items rose less at Kroger's supermarkets than at competing markets. The complaint focused on price increases and decreases that occurred before particular items appeared on the sample list. There are two ways this list might be deceptive. One is that the sample list was chosen not at random, but rather to favor Kroger in some way. The second is for Kroger to raise prices on other goods in the store but to allow fewer increases on Price Patrol items. The expert's testimony focused on this latter method of deception.

> In analyzing the raw data. . . , Professor Klayle, an expert in statistical techniques, performed a Chi-square analysis of the data. The value of the Chi-square indicates whether there is an association between the observed and the expected values. In the case of the Price Patrol, there were fewer Price Patrol items showing price increases than one would expect. The number of items on the Price Patrol showing price decreases was larger than expected. And the high Chi-square results indicate that the likelihood of these changes being due to chance alone is extremely small. Dr. Klayle concluded that, because of the high Chi-square values, it is extremely unlikely that these variables of classification, that is whether an item was on a current Price Patrol list and whether the price of the item increased or decreased from the previous week, are not associated.

What data would you need to perform such a test and, once gathered, how would you set up the chi-square test to analyze it?

D. THE Z TEST

Once the standard deviation is understood, the next statistical test is easy. You may have wondered why the court opted for a test of "two or three standard deviations from the mean" to determine whether an observed value was too far from the expected value to be considered a chance occurrence. The formula for the *Z test* is merely a mathematical trick to enable us to calculate the probability of a chance occurrence once we know how many standard deviations the observed value is from the expected value. For any observation, we can calculate what is called a *Z score*. By looking up the *Z* score on a *Z* table, we can determine the probability that the observation occurred by chance. The *Z* table is sometimes called the *Normal table* because it reveals the frequency with which observations from a normally distributed sample occur within any number of standard deviations from the mean.

After reading Chinese for Affirmative Action v. FCC (*CAA*), we compared the results using a standard deviation test and chi-square test. To use the *Castaneda* test, we calculated the number of standard deviations from the mean for a particular observation. The Z test performs an identical exercise. The formula for the Z score is:

$$Z = \frac{X_i - M}{SD}$$

Note that the Z score is nothing more than the number of standard deviations our observed value is from the mean. The X_i is the particular observation the probability of which we wish to know, M is the mean or expected number, and SD is the standard deviation. If the Z score is negative, it is because the observed value is less than the expected value, e.g., fewer Asian-Americans hired than would be expected on the basis of the percent of the population that is Asian-American. If the Z score is positive, then more Asian-Americans are hired than one would expect. Using the *CAA* data for Asian-American employees:

$$Z = \frac{(9 - 17)}{4} = \frac{-8}{4} = -2.00$$

Statistically, the question we ask is: what is the probability of getting a Z score of -2.00 if the null hypothesis is true, i.e., if race has no relationship to employment at the radio station? A more straightforward approach is: what is the chance that only nine Asian-Americans would be hired if there were no discrimination? The answer is given by the Z table appearing in Exhibit V-7 on the following pages.

Down the left side of the table are Z scores with one decimal place, i.e., one number after the decimal point, e.g., 2.0, 2.6, 3.4. Across the top is a second number after the decimal point, e.g., 0.02, 0.03, 0.07. To find the number associated with Z score 1.96, for instance, go down to the 1.9 row and over to the 0.06 column to the number .4750.

For our example, the number associated with a Z score of -2.00 is .4773, obtained by tracing down to the 2.0 row and over to the 0.00 column. In percentage terms, .4773 is 47.73%, and it can be interpreted as the percentage of random observations that will be *less* than two standard deviations below the mean. Since 50% of all observations will be somewhere below the mean or expected value, we can calculate that 2.27% of them (50%–47.73%) will be two standard deviations *or more* below the mean. If our question is, "What is the probability of having an observed number of Asian-American employees that is 2.00 standard deviations below the expected number?" then the answer is 2.27%. If the question is, "What is the chance that

EXHIBIT V-7.
Z Values

Z	0.00	0.01	0.02	0.03	0.04	0.05	0.06	0.07	0.08	0.09
0.0	.0000	.0040	.0080	.0120	.0160	.0199	.0239	.0279	.0319	.0359
0.1	.0398	.0438	.0478	.0517	.0557	.0596	.0636	.0675	.0714	.0753
0.2	.0793	.0832	.0871	.0910	.0948	.0987	.1026	.1064	.1103	.1141
0.3	.1179	.1217	.1255	.1293	.1331	.1368	.1406	.1443	.1480	.1517
0.4	.1554	.1591	.1628	.1664	.1700	.1736	.1772	.1808	.1844	.1879
0.5	.1915	.1950	.1985	.2019	.2054	.2088	.2123	.2157	.2190	.2224
0.6	.2257	.2291	.2324	.2357	.2389	.2422	.2454	.2486	.2517	.2549
0.7	.2580	.2611	.2642	.2673	.2704	.2734	.2764	.2794	.2823	.2852
0.8	.2881	.2910	.2939	.2967	.2995	.3023	.3051	.3078	.3106	.3133
0.9	.3159	.3186	.3212	.3238	.3264	.3289	.3315	.3340	.3365	.3389
1.0	.3413	.3438	.3461	.3485	.3508	.3531	.3554	.3577	.3599	.3621
1.1	.3643	.3665	.3686	.3708	.3729	.3749	.3770	.3790	.3810	.3830
1.2	.3849	.3869	.3888	.3907	.3925	.3944	.3962	.3980	.3997	.4015
1.3	.4032	.4049	.4066	.4082	.4099	.4115	.4131	.4147	.4162	.4177
1.4	.4192	.4207	.4222	.4236	.4251	.4265	.4279	.4292	.4306	.4319
1.5	.4332	.4345	.4357	.4370	.4382	.4394	.4406	.4418	.4429	.4441
1.6	.4452	.4463	.4474	.4484	.4495	.4505	.4515	.4525	.4535	.4545
1.7	.4554	.4564	.4573	.4582	.4591	.4599	.4608	.4616	.4625	.4633
1.8	.4641	.4649	.4656	.4664	.4671	.4678	.4686	.4693	.4699	.4706
1.9	.4713	.4719	.4726	.4732	.4738	.4744	.4750	.4756	.4761	.4767

z	0	1	2	3	4	5	6	7	8	9
2.0	.4773	.4778	.4783	.4788	.4793	.4798	.4803	.4808	.4812	.4817
2.1	.4821	.4826	.4830	.4834	.4838	.4842	.4846	.4850	.4854	.4857
2.2	.4861	.4864	.4868	.4871	.4875	.4878	.4881	.4884	.4887	.4890
2.3	.4893	.4896	.4898	.4901	.4904	.4906	.4909	.4911	.4913	.4916
2.4	.4918	.4920	.4922	.4925	.4927	.4929	.4931	.4932	.4934	.4936
2.5	.4938	.4940	.4941	.4943	.4945	.4946	.4948	.4949	.4951	.4952
2.6	.4953	.4955	.4956	.4957	.4959	.4960	.4961	.4962	.4963	.4964
2.7	.4965	.4966	.4967	.4968	.4969	.4970	.4971	.4972	.4973	.4974
2.8	.4974	.4975	.4976	.4977	.4977	.4978	.4979	.4979	.4980	.4981
2.9	.4981	.4982	.4983	.4983	.4984	.4984	.4985	.4985	.4986	.4986
3.0	.4987	.4987	.4987	.4988	.4988	.4989	.4989	.4989	.4989	.4990
3.1	.4990	.4991	.4991	.4991	.4992	.4992	.4992	.4992	.4993	.4993
3.2	.4993	.4993	.4994	.4994	.4994	.4994	.4994	.4995	.4995	.4995
3.3	.4995	.4995	.4996	.4996	.4996	.4996	.4996	.4996	.4996	.4997
3.4	.4997	.4997	.4997	.4997	.4997	.4997	.4997	.4997	.4997	.4998
3.5	.4998	.4998	.4998	.4998	.4998	.4998	.4998	.4998	.4998	.4998
3.6	.4998	.4998	.4999	.4999	.4999	.4999	.4999	.4999	.4999	.4999
3.7	.4999	.4999	.4999	.4999	.4999	.4999	.4999	.4999	.4999	.4999
3.8	.4999	.4999	.4999	.4999	.4999	.4999	.4999	.4999	.4999	.4999
3.9	.5000	.5000	.5000	.5000	.5000	.5000	.5000	.5000	.5000	.5000

Source: R. Fisher and F. Yates, Statistical Tables for Biological, Agricultural and Medical Research 45 (Table II) (6th ed. 1963), and E. Lovejoy, Statistics for Math Haters 239 (1975). Reprinted with permission of the authors and publishers.

only nine Asian-Americans would be hired, when we expect 17 if there were no discrimination?" then the answer is again 2.27%. This statistical test is called a *one-tailed test,* because it is considering only the possible observations that are on one side of the mean, in this case below the mean.

Typically, however, the Z test is used as a *two-tailed test.* The Supreme Court in *Castaneda* suggests that if the observed value is two or three standard deviations from the mean, then the null hypothesis that there is no relationship between race and selection for the jury would be suspect. There are two ways in which an observed value may be two or three standard deviations from the mean—it may be two or three standard deviations above *or* below the mean. Either event will cause us to reject the null hypothesis, and the practice in courts has been to add together the probabilities of getting a particular Z score above *and* below the mean. Thus, if our Z score is 2.00, the probability of being 2.00 or more standard deviations below the mean is 2.27%, as shown above, and the probability of being 2.00 or more standard deviations above the mean is 2.27%, by the same logic. The probability of a non-discriminatory observation 2.00 or more standard deviations from the mean (on either side of the mean) is 2.27% plus 2.27%, or 4.54%. This is the result of a two-tailed test.

To remember the distinction between one- and two-tailed tests, look at the underlying question you wish to answer. If you are asking either for the probability of being below the mean by a given amount or for the probability of being above the mean by a given amount, then use a one-tailed test, because you are only looking in one direction. If you are asking for the probability of being a given number of standard deviations away from the mean, then use a two-tailed test, because you must consider both the possibility of being above the mean and the possibility of being below.

In *CAA,* the null hypothesis was that race had no relationship to employment decisions. Two possibilities would cause us to reject this null hypothesis. One is that Asian-Americans are favored in the hiring process. The other is that they are disfavored in the hiring process. The percentage 2.27% is the probability of a nondiscriminatory outcome that is 2.00 or more standard deviations below the expected number. The probability 2.27% is also the chance of a non-discriminatory outcome that is 2.00 or more standard deviations above the expected number. So the chance of a random result that is 2.00 or more standard deviations from the mean in either direction is 4.54%. If we conclude that there was discrimination here, we have a 4.54% chance of being wrong. Thus, the probability of a false rejection of the null hypothesis is 4.54%, and the probability of a type 1 error is 4.54%.

Compare this 4.54% with the results of a chi-square test on these same numbers. The chi-square test indicated that the chance of improperly rejecting the null hypothesis was between 5% and 3% (and probably closer to 5%). The Z test indicates more precisely that the chance of a type 1 error is 4.54%. When such precision is important, the Z test is more useful. The chi-square test has the advantage, however, of being useful when there are more than two categories for which observed and expected values are to be analyzed.

In Gay v. Waiters' and Dairy Lunchmen's Union, Local 30, note the two-tailed Z test applied to employment figures. It will be useful for you to try to replicate the calculations presented. Also, pay attention to the care with which the court determines the expected number of minority employees.

GAY v. WAITERS' AND DAIRY LUNCHMEN'S UNION, LOCAL 30
489 F. Supp. 282 (N.D. Cal. 1980)

SCHWARZER, District Judge.

I. INTRODUCTORY STATEMENT

This is an action for alleged discrimination by defendants in hiring, promoting and transferring black males into waiter positions. Plaintiffs sue on behalf of themselves and a class consisting of black males who were denied employment as waiters. The only defendants remaining in the case are The St. Francis Hotel Corporation ("St. Francis") and Hilton Hotels Corporation ("Hilton"). Plaintiffs' principal claim is based on 42 U.S.C. §1981, but they also charge that defendants conspired with the labor union representing waiters in violation of 42 U.S.C. §1985(3) and breached the collective bargaining agreement in violation of 29 U.S.C. §185. . . .

III. FINDINGS OF FACT. . .

FINDINGS OF FACT RESPECTING THE UNION

Both the St. Francis and Hilton at all relevant times were members of the Hotel Employers Association, an employers' association which represented these and other hotels in collective bargaining. Through their membership in the association, the St. Francis and Hilton were at all relevant times parties to collective bargaining agreements covering waiters and others engaged in food service. . . .

At all relevant times, the union operated a hiring hall in San Francisco for waiters, bus persons and captains, among others. The collective bargaining agreement required employers to apply to the union for referrals to fill any vacancy, and the union to refer applicants on a non-discriminatory basis. Section 3(g) provided for an adjustment board to hear any appeals based on a failure to comply with the referral provisions of the agreement. No appeals to the adjustment board were ever made.

In the early 1970's the union maintained a classification board which interviewed members and determined whether to rate them as experienced waiters. When requests for experienced a la carte waiters were received at the hiring hall, the dispatcher would select from among available members those whom he considered qualified for the position. This practice, as well as the classification board, was discontinued sometime after the filing of this action. There was in any event a shortage of experienced a la carte waiters and few were in the hiring hall available for referral. The hotels, unable to meet their needs for experienced waiters through the hiring hall, generally hired directly, subject only to the requirement that the new waiter be or become a union member.

For the most part, referrals were limited to temporary extra banquet waiters for which the hotels depended on the hiring hall. Men were dispatched to these jobs on rotation designed to give each of them as nearly the same number of jobs as feasible.

In the early 1970's, about 150 members would generally be in the hiring hall available for some type of waiter work. In later years, the number increased to about 300. In 1977, Local 2 reorganized the operation of the hiring hall and undertook to persuade the hotels to make regular use of it.

When the union referred a waiter to a hotel in response to a request for help, it issued a so-called work slip. The slip was taken by the waiter to the hotel and evidenced his good standing as a union member. When a hotel requested applicants for permanent positions, the union on occasion issued an interview slip. More frequently, the hotel would select the person it wished to hire for a particular permanent position from among applicants from various sources and, before hiring, send him to the union to obtain a work slip evidencing his good standing. When this procedure was followed, the notation "by request" was made on the slip.

The union maintained copies of these slips in its records. However, there is only an informal and incomplete record of requests for referrals received by the union and none which reflects requests to which the union was unable to respond.

Until 1972, the union had fewer than 100 black male members. In 1973 it had 127. In 1976, when membership information was computerized, 8,000 of its 18,000 members were coded as waiters (which included waitresses and bus persons). Of these 8,000 names, 5,343 (66%) were race coded. Blacks totaled 4.5%. The union did not encourage membership by those who did not hold jobs, generally urging a person to find a job before applying.

Based on the available records, plaintiffs computed the rate at which black waiters were referred to jobs by the hiring hall as follows:

Union Referrals to St. Francis Hotel, 1970-1977

	Black	Total	% Blacks
By Request	10	187	5.3
Others	9	106	8.5
All Referrals	19	293	6.5

Union Referrals to All Employers, 1970-1977

	Black	Total	% Blacks
By Request	85	3,020	2.8
Others	555	7,177	7.7
All Referrals	640	10,197	6.3

Union Referrals to Hilton Hotel, 1974-1977

	Black	Total	% Blacks
By Request	4	112	3.6
Others	5	53	9.4
All Referrals	9	165	5.4

Union Referrals to All Employers, 1974-1977

	Black	Total	% Blacks
By Request	47	1396	3.4
Others	453	5418	8.4
All Referrals	500	6814	7.3

The union's Equal Employment Opportunity (EEO 3) reports show referrals of black males for all jobs in proportion to total referrals for the respective periods of August and September as follows:

August/September	1970	1971	1972
Total referrals	2551	2467	3550
Black males	263	252	320
	10.3%	10.2%	9% . . .

D. Liability of Defendant Hotels on the Class Claim

1. THE STANDARD OF PROOF UNDER SECTION 1981

At the threshold lies the question of the quantum of proof required of plaintiffs. This case is brought under Section 1981, not Title VII of the Civil Rights of 1964 (42 U.S.C. §2000e et seq.). Plaintiffs chose to present it as a case of intentional racial discrimination, contending that their proof of disparate impact on blacks is sufficient to support an inference of purposeful disparate treatment. If this were so, this Court would not reach the question whether Section 1981 requires that the defendants' intent to discriminate be proved. However, because plaintiffs' proof may not support such an inference, the Court must address this issue.

Until the decision in Washington v. Davis, 426 U.S. 229, 96 S. Ct. 2040, 48 L. Ed. 2d 597 (1976), it was generally assumed that the same standard of proof applied under Section 1981 and Title VII. Employment practices having a disparate racial impact were held unlawful, regardless of intent, unless justified as a business necessity.

In Washington v. Davis, plaintiffs alleged, among others, a claim under Section 1981 that the District of Columbia's examination for police applicants violated the Fifth Amendment. The examination was alleged to have resulted in the exclusion of disproportionately high numbers of black applicants. There was no allegation of intentional discrimination. The District Court granted summary judgment for defendants on this claim; the Court of Appeals reversed; and the Supreme Court, in turn, reversed the judgment of the Court of Appeals. The Court held, in part, that, in the absence of a racially discriminatory purpose, official action does not violate the equal protection guarantees of the Fifth and Fourteenth Amendments simply because it has a racially disproportionate impact.

Washington v. Davis did not decide whether in actions under Section 1981 plaintiffs must prove discriminatory purpose. The Court's discussion focused on proof requirements under the Fifth and Fourteenth Amendments; Section 1981 was not mentioned in this context. The Court's order that summary judgment be granted for defendants has nevertheless been taken to imply that Section 1981 requires proof of intent to discriminate. However, such a reading of the opinion is not required; summary judgment on the Section 1981 claim is adequately supported by the Court's additional holding that the defendant had proved the examination to be job related. Moreover, the focus on the constitutional flaw in the decision under review may have been dictated by the Court's action in raising the intent issue sua sponte.

The conclusion that Washington v. Davis left open the issue under Section 1981 was also reached by the entire panel in Davis v. County of Los Angeles, 566 F.2d 1334, 1340, 1350 n. 10 (9th Cir. 1977), *vacated as moot,* 440 U.S. 625, 99 S. Ct. 1379, 59 L. Ed. 2d 642 (1979). Indeed, Justice Powell observed, in dissenting from the holding that Davis v. County of Los Angeles had become moot, that the case presented "the important question — heretofore unresolved by this Court — whether cases brought under 42 U.S.C. §1981, like those brought directly under the Fourteenth Amendment, require proof of racially discriminatory intent or purpose." 440 U.S. at 637, 99 S.Ct. at 1386.

In Davis v. County of Los Angeles, supra, the majority of the Court of Appeals panel held that proof of liability under Section 1981 should track the requirements of Title VII — that is, that no proof of intent is required. 566 F.2d at 1340. In vacating that case as moot, however, the Supreme Court expressly noted that its action deprived the opinion of the Court of Appeals of precedential effect. 440 U.S. at 634 n. 6, 99 S. Ct. at 1384 n. 6. Thus, the question must be considered as open in the Ninth Circuit. Its disposition turns on several factors discussed in the opinions in Davis v. County of Los Angeles and in other recent lower court decisions.

The legislative history of Section 1981 shows that although it rests in part upon the Thirteenth Amendment, its origin was closely tied to the Fourteenth Amendment. A number of courts have relied on this connection in finding the same intent requirement as under the Fourteenth Amendment. The legislative history also shows that Congress in enacting the first precurser of Section 1981 meant it "to deal with the most egregious forms of reactionary white conduct; . . . not . . . to fashion an evidentiary presumption designed to root out more subtle forms of discrimination." Note, Racially Disproportionate Impact of Facially Neutral Practices — What Approach under 42 U.S.C. Section 1981 and 1982?, 1977 Duke L.J. 1267, 1279-80.

Legislative history is not conclusive; the present day interpretation of Section 1981 is not rigidly bound by "the sentiments of the Reconstruction Congress." Runyon v. McCrary, 427 U.S. 160, 191, 96 S. Ct. 2586, 2604, 49 L. Ed. 2d 415 (1976) (Stevens, J., concurring). However, the policy concerns which led the Supreme Court in Washington v. Davis to hold that the Fourteenth Amendment requires proof of intent support a similar interpretation of Section 1981. The Court there thought it appropriate to "await legislative prescription" before extending to new areas the rule that facially neutral action is illegal if in practice it benefits or burdens one race more than another. Washington v. Davis, 426 U.S. at 248, 96 S. Ct. at 2051.

Title VII of the Civil Rights Act of 1964 was such a legislative prescription in the field of employment. Title VII proscribes conduct based on proof of its racially discriminatory impact alone, Griggs v. Duke Power Co., 401 U.S. 424, 432, 91 S. Ct. 849, 854, 28 L. Ed. 2d 158 (1971), but it does so in the context of a procedural scheme which subjects claims to a prior administrative screening investigation and a uniform short statute of limitations. Section 1981 cases may be filed without prior screening and within the varying time limits provided by state laws. The absence of mechanisms to control the filing of what may be baseless claims, or claims which have been rendered more difficult to defend by the passage of time warrants, as a matter of policy, judicial adherence to a higher standard of proof under Section 1981, at least until further legislative action.

Moreover, Section 1981 applies to a whole range of public and private contractual relationships other than employment. As discussed above, the legislative history of Section 1981 lacks any indication of intent to outlaw conduct on the basis of racially disparate impact alone. Section 1981 therefore does not supply the "legislative prescription" that the Supreme Court regarded as necessary before expanding the rule into new areas.

These considerations lead the Court to agree with the majority of courts that have decided the issue that proof of discriminatory intent is required in actions brought under Section 1981. This conclusion has important implications, discussed below, regarding the requisites of a prima facie case based, as in this action, primarily on statistics.

2. ELEMENTS OF THE PRIMA FACIE CASE

Plaintiffs contend that the defendant hotels engaged in a pattern or practice of purposeful discrimination against black male applicants for positions as waiters. The record is devoid of any direct evidence of intent to discriminate. It does, however, reflect a series of employment practices which may have tended to exclude blacks.

The record shows that both defendants preferred to promote or transfer from within rather than to hire from the outside. Both had experience requirements which, however, were vague, unwritten, and left entirely to the discretion of the various room or service managers to administer. Aside from experience, hiring criteria used were subjective, turning on such matters as appearance, demeanor and job stability. Until recently, the hotels hired permanent waiters largely without resort to the union hiring hall. Thus there was no established procedure for giving public notice of all job openings. The defendants also had in effect rules and practices which caused some prospective applicants to encounter difficulties in their attempts to make application; their employment offices were open to the public only for limited periods, applications were not always accepted, and there was no regular procedure for interviewing all applicants. Finally, some apparently qualified black applicants were rejected without explanation.

The aggregate impact of these practices might well have been felt more severely by blacks seeking waiter positions than by whites. Whether it is sufficient to constitute a prima facie case of purposeful discrimination turns on the analysis of defendants' hiring and work force statistics in the following sections.

Teamsters v. United States, 431 U.S. 324, 339, 97 S. Ct. 1843, 1856, 52 L. Ed. 2d 396 (1977), established that purposeful employment discrimination may be proved by statistical evidence. The Court said:

> Statistics showing racial or ethnic imbalance are probative in a case such as this one only because such imbalance is often a telltale sign of purposeful discrimination; absent explanation, it is ordinarily to be expected that nondiscriminatory hiring practices will in time result in a work force more or less representative of the racial and ethnic composition of the population in the community from which employees are hired. Evidence of longlasting and gross disparity between the composition of a work force and that of the general population thus may be significant. . . .

431 U.S. at 340 n. 20, 97 S. Ct. at 1857 n. 20. The Court agreed that the government had made out a prima facie case under Title VII by proving, in addition to individual instances of discrimination, that in communities where 10 to 50 percent of the population was black, defendant employed not a single black line driver before the suit was

commenced. In Hazelwood School District v. United States, 433 U.S. 299, 307, 97 S. Ct. 2736, 2741, 53 L. Ed. 2d 768 (1977), also a Title VII case, the Court again stated: "Where gross statistical disparities can be shown, they alone may in a proper case constitute prima facie proof of a pattern or practice of discrimination."

Statistical evidence of racial disparity may therefore be probative of purposeful discrimination. See Washington v. Davis, 426 U.S. at 242, 96 S. Ct. at 2048; Village of Arlington Heights v. Metropolitan Housing Development Corp., 429 U.S. 252, 266, 97 S. Ct. 555, 563, 50 L. Ed. 2d 450 (1977); Personnel Administrator of Massachusetts v. Feeney, 442 U.S. 256, 99 S. Ct. 2282, 2295 n. 23, 60 L. Ed. 2d 870 (1979). Justice Stevens, concurring in Washington v. Davis, explained the evidentiary connection as follows: "Frequently the most probative evidence of intent will be objective evidence of what actually happened rather than evidence describing the subjective state of mind of the actor. For normally the actor is presumed to have intended the natural consequences of his deeds." Id., 426 U.S. at 253, 96 S. Ct. at 2054. He added that "the line between discriminatory purpose and discriminatory impact is not nearly as bright, and perhaps not quite as critical, as the reader of the Court's opinion might assume." Id., 426 U.S. at 254, 96 S. Ct. at 2054.

While perhaps not bright, the line between impact and intent is critical in this case. If the intent requirement is to have meaning in a statistics-based case, the evidence must show more than that the defendant could or should have known that its actions had a disparate impact on blacks. As the Court said in Personnel Administrator of Massachusetts v. Feeney, supra, 99 S. Ct. at 2296 "'[d]iscriminatory purpose,' however, implies more than intent as volition or intent as awareness of consequences. [citation omitted] It implies that the decisionmaker . . . selected . . . a particular course of action at least in part 'because of,' not merely 'in spite of,' its adverse effects upon an identifiable group." (footnotes omitted). *Feeney* upheld a veterans' preference law against a claim of unconstitutional sex discrimination. The Court found that where adverse impact was an unavoidable consequence of a legitimate legislative policy untainted by evidence of a prohibited purpose, no inference of discriminatory intent arose. Id., 99 S. Ct. at 2296 n. 25. Whether such an inference may be drawn from evidence of disparate impact turns on a sensitive and practical evaluation of the full factual context. Id., 99 S. Ct. at 2296 n. 24. In cases of employment discrimination, where proof of intent rests on statistical evidence of impact, that impact must be "longlasting and gross" to sustain the inference that the defendant acted for the purpose of creating or maintaining it. *Teamsters,* supra, 431 U.S. at 340, n. 20, 97 S. Ct. at 1856 n. 20.

Another feature of statistical evidence that must be borne in mind is that, contrary to the illusion of certainty which it may create, it does not afford a mathematically precise basis for decision. *Hazelwood* recognizes this fact, directing the district court to evaluate not simply a single set of employment statistics but the entire range of available data, including work force, applicant flow and hiring rates. 433 U.S. at 308 n. 13, 97 S. Ct. at 2742 n. 13. Inferences drawn from the statistical proof must, moreover, be tempered by the other evidence which brings "the cold numbers convincingly to life." *Teamsters,* 431 U.S. at 339, 97 S. Ct. at 1856. As the Supreme Court noted in *Hazelwood,* quoting from *Teamsters,* "[S]tatistics . . . come in infinite variety . . . [T]heir usefulness depends on all of the surrounding facts and circumstances." 433 U.S. at 312, 97 S. Ct. at 2744.

a. The Relevant Labor Market

The statistical analysis must begin with a determination of the pool of individuals available and qualified to become waiters for the defendant hotels and the percentage of black males in that pool. Whether or not an inference of discriminatory impact and purpose arises from the percentage of blacks hired or promoted by the defendants depends on the choice of the proper labor market for statistical comparison. See *Hazelwood,* supra, 433 U.S. at 311 n. 17, 97 S. Ct. at 2743 n. 17. The parties are in fundamental disagreement on this issue, and have presented largely contradictory expert opinions.

In the opinion of plaintiffs' expert, Dr. John H. Pencavel, the relative availability of black waiters should be measured by reference to the percentage of black males, 21 to 64 years of age, living in the San Francisco/Oakland Standard Metropolitan Statistical Area (SMSA), weighted in accordance with the frequency of applications from within and without San Francisco, and earning less than the 1975 median earnings of permanent banquet waiters at defendant hotels, adjusted to 1970 dollars. Applying these factors, Dr. Pencavel arrived at a black availability range of 15.1% to 17.9%, varying with the median earnings at the two hotels. Each of these factors will be discussed in the following sections.

i. The relevant geographic area

Because the percentage of black residents varies among the counties in the SMSA, the percentage of blacks in the relevant labor market will vary depending on the geographic definition of the market. Dr. Pencavel chose to use data from the San Francisco/Oakland SMSA, weighted in accordance with the ratio of applications for all waiter positions received from within and without San Francisco, respectively. Of all waiter applications which were retained by the St. Francis and the Hilton, approximately 85 percent came from residents of the City and County of San Francisco, 13% came from other counties in the SMSA, and 2% came from outside the SMSA. Thus, Dr. Pencavel weighted the San Francisco data by a factor of 85/98 and the data from other counties in the SMSA by 13/98.

Dr. Pencavel's approach is sound. It is reasonable to estimate the geographic composition of the available labor pool by reference to the distribution of residence of the actual applicants. Plaintiffs concede that the applications Dr. Pencavel relied on do not document the residence of all actual applicants. Not all written applications were retained, nor were all applications in writing. However, there is no reason to suppose that the residential distribution of the group whose applications were retained does not approximate that of the actual applicant pool as a whole.

Defendants contend that weighting results in double-counting. That would be true if San Francisco figures were averaged with SMSA figures. However, the procedure used by Dr. Pencavel was to back out the San Francisco figures from the SMSA aggregates before applying the weighting factors. This method avoids double counting.

Defendants' expert, Dr. Seymour L. Wolfbein, argued that the judgment of the Bureau of the Census to aggregate the counties of the SMSA as an economically integrated unit should be accepted without modification. This judgment, however, does not preclude an attempt to arrive at a more precise measure of the percentage of blacks in the work force available for the particular jobs at issue here. Refinement of SMSA data by weighting is therefore not inconsistent with their design. Inasmuch as 85% of waiter applications submitted to defendants came from San Francisco residents, it is reasonable to conclude (in the absence of contrary evidence) that approximately the same proportion of all those available to become waiters for defendants resided in San Francisco.

ii. The relevant age bracket

Waiters employed in establishments serving alcoholic beverages are required by state law to be 21 years of age. No waiter over 64 years old appears to have been hired by defendants. Accordingly, Dr. Pencavel used an age bracket of 21 to 64 years, inclusive. The choice is not seriously disputed by defendants and is reasonable.

iii. The relevant population segment

Dr. Pencavel favored the use of general population, rather than the more restricted civilian labor force data, although he acknowledged that both should be considered. Labor force data include all persons who, at the time of the census, are either employed or actively seeking employment.

Census data have been criticized for tending to undercount certain segments of the population such as the poor or disadvantaged, and hence proportionately undercounting blacks more than whites. Dr. Pencavel argued that this phenomenon is aggravated in the labor force statistics by the phenomenon of discouraged workers, i.e., those who have given up seeking work actively. Pencavel's opinion was that discouraged workers are more prevalent among black males than white males. Dr. Wolfbein countered by citing findings that the rate of participation in the labor force is the same among black and white males.

The criticisms of the census data appear to the Court to be too speculative and imprecise to warrant rejection of the work force data for present purposes. That conclusion is supported by the arguments against the use of general population data in place of work force data. The former include children, students, inmates of institutions, military personnel, the disabled and those not actively seeking work. The percentage of blacks in each of these categories differs from the percentage of nonblacks.[23] The birthrate for blacks, for example, is higher than for whites and may therefore explain the higher percent-

---

23. The differences are shown in the following table, computed from 1970 Census data for San Francisco/Oakland SMSA, weighted in the same manner as plaintiffs' data:

	Weighted SMSA	
	Nonblacks	Blacks
Males, 16 years and over	100 %	100 %
Civilian labor force	70.0	71.0
Armed Forces	3.9	2.6
Not in labor force	26.1	26.4
Inmates of institutions	1.0	1.6
Enrolled in school	6.2	6.7
Other — Under 65 years old	7.6	12.5
65 years and over	11.3	5.5

Source of data: 1970 Census of Population, Vol. 1, Characteristics of the Population, Part 6: California at 499 (Table 85), 611 (Table 92) (April 1973).

age of blacks among students not in the labor force and among those under 65 years old. Illness and disability may have a higher incidence among blacks than others. Similarly, racial discrimination may have led to a relatively higher percentage of black persons not actively seeking work. The exclusion of those population segments not likely to supply applicants for waiter positions makes the civilian work force data a more reliable indication of the available labor pool.

iv. The relevant time

The last available complete census data come from the 1970 Census. Plaintiffs contend that the percentage of blacks in the population has significantly increased since that time. Data from the 1975 Current Population Survey (CPS), offered by plaintiffs, tend to refute this assertion. Comparison of these data with the 1970 San Francisco SMSA Census shows that the percentage of blacks in the population, as well as their percentage in the civilian labor force, rose by only a fraction of one percent in the higher earnings bracket and declined in the lower earnings bracket.[24]

Plaintiffs claimed that projections prepared by the California Employment Development Department for current years for which no census data are available show an increase in the percentage of blacks in the population and work force in San Francisco. The projections appear to be prepared from various public health data and are conceded by plaintiffs to be considerably less reliable than census or CPS data; they certainly afford no basis for comparison with earlier census or CPS data and must therefore be rejected. The Court accepts

---

24. The figures presented by Dr. Pencavel reflect the percentage of black males, aged 21 to 65, in the San Francisco/Oakland SMSA within different earnings brackets. Using the 1975 median earnings of banquet waiters employed by the St. Francis ($16,267) and the Hilton ($11,970), respectively, and adjusting for inflation to compute the median earnings in 1970, the following comparison may reasonably be made, based on the earnings brackets available from census data:

1970		1975	
Earning $11,000 or less		Earning $17,000 or less	
Population	13.4%	13.5%	
Labor Force	12.4%	12.9%	
Earning $8,500 or less		Earning $12,000 or less	
Population	16.4%	15.8%	
Labor Force	15.7%	15.2%	

Any comparison of black representation within particular earnings brackets over time must, however, be viewed with caution. It is likely to be affected by the relative rise in the income of black workers and the relative improvement in their status, as shown by the increase in the black percentage of white collar and craft workers and the decrease in the black percentage of laborers.

the 1970 census and the 1975 CPS data as the most recent reliable data.

v. *The relevant earnings brackets*

Dr. Pencavel contended that the availability of workers for particular jobs depends in the terms of employment offered, the most obvious aspect of which is pay. Normally persons will not be interested in a job paying less than they are currently earning. Dr. Pencavel therefore narrowed the available population further to that segment "whose earnings are less than those they can expect to earn if employed as waiters by the defendant hotels." Plaintiffs' Exhibit 66 at page 10. The black percentage of that segment is claimed to be the closest estimate of actual black availability for the jobs at issue.

Dr. Pencavel's analysis was based on what defendants' records showed to be the median earnings of permanent banquet waiters in 1975. There are obvious difficulties involved in reliance on those figures.

The premise for the use of earnings brackets is that prospective applicants make a choice based on information concerning the earnings in the job. There is no evidence to show what prospective applicants knew, believed, or were told (by published notices, interviewers or other means) about the earnings of defendants' waiters. There is certainly no reason to believe that any applicants would be influenced by the figures computed by plaintiffs for purposes of the trial, i.e., the 1975 median earnings of permanent banquet waiters at the St. Francis and the Hilton, respectively. A median earnings figure is a construct that would not be known by anyone who did not know the whole range and distribution of waiters' actual incomes. Banquet waiter earnings are used because other waiters probably under-report their tips; however, for this reason the Court can only speculate about the range of earnings of other waiters. In any event, there is no evidence to show what potential applicants expected banquet or other waiters at the hotels to earn, other than vague references to the St. Francis being the best paying hotel.

Even if potential applicants had more complete information than is before the Court, their expectations about waiter earnings would inevitably be uncertain. Lack of predictability of the number of hours he will work and the amount of tips he will receive preclude an applicant from accurately forecasting his future earnings. Even among banquet waiters, who receive a fixed percentage in lieu of tips, the spread in earnings is wide. An expected earnings cutoff, although it may be useful in estimating the available labor pool for certain other occupations, is of doubtful validity for hotel waiters.

216   Chapter V. Significance Testing

The record discloses other reasons militating against use of the earnings cutoff. The position of waiter has strong non-monetary incentives and disincentives. For example, the irregular working times may, regardless of the compensation, make the job attractive to certain persons but unattractive to those seeking normal hours. Personal service positions appeal to some but repel others. Whether the impact of considerations such as these is random without regard to race is not known. Dr. Wolfbein also questioned whether, in light of recent evidence concerning the aspirations of young blacks, it can be assumed that interest in becoming a waiter is equal among racial groups. Cf. United States v. Commonwealth of Virginia, 454 F. Supp. 1077, 1097 (E.D. Va. 1978). These factors, although not subject to being quantified, inject further uncertainty into the use of an earnings cutoff; if nothing else, they illustrate the significant role of subjective elements in determining availability.

For these reasons, the Court rejects as too speculative the use of an earnings cutoff to define the availability of black males.

[The Court has, in any event, considerable doubt concerning the validity of the earnings cutoff method. Census data show no correlation between median earnings and black participation in particular job categories, as shown by the following summary for the San Francisco SMSA:

Percentage of Black Workers by Occupational Group
(In Order of Median Earnings)

Industry	Annual Median Earnings	Total # Workers	% Blacks
Other Craftsmen and Kindred Workers	$9697.00	23,098	11.5%
Other Mechanics and Repairmen	9679.00	15,429	5.7%
Transport Equipment Operatives	9482.00	43,579	12.5%
Cabinet Makers	9406.00	1,127	4.9%
Bakers	8672.00	1,784	8.9%
Operatives Except Transport	8644.00	72,492	13.1%
Clerical and Kindred Workers	8629.00	77,933	9.7%
Salesclerk — Retail Trade	8096.00	14,400	5.0%
Laborers Exc. Farm	8073.00	49,991	21.3%
Apparel Craftsmen and Upholsterers	7994.00	1,808	6.5%
Service Workers Exc. Private Household	7657.00	80,540	14.5%
Farmers and Farm Managers	7214.00	1,706	6.1%
Religious Workers	6845.00	2,447	8.7%
Farm Laborers and Farm Foreman	5429.00	3,241	12.7%
Private Household Workers	4409.00	978	24.4%

Source: 1970 Census of Population, Vol. 1, General Social and Economic Characteristics, Part 6: California at 1823-26 (Table 175) (April 1973).   (fn. 28)]

D. The Z Test   217

vi. The relevant employment qualifications

In *Hazelwood,* the Supreme Court stated that "a proper comparison [is] between the racial composition of Hazelwood's teaching staff and the racial composition of the qualified public school teacher population in the relevant labor market." 433 U.S. at 308, 97 S. Ct. at 2742. In a footnote following this statement, the Court distinguished *Teamsters,* saying that "the comparison between the percentage of Negroes on the employer's work force and the percentage in the general areawide population was highly probative, because the job skill there involved — the ability to drive a truck — is one that many persons possess or can fairly readily acquire." 433 U.S. at 308 n. 13, 97 S. Ct. at 2742 n. 13.

Plaintiffs contend that the job skill involved in the position of waiter is readily attainable by members of the general population with normal intelligence and physical capabilities. That contention is amply supported by the evidence. Plaintiffs' witnesses as well as various employees of defendants testified that the necessary skills can be acquired on the job in a matter of days or weeks. The requisite skills vary, of course, between restaurants offering a complex menu with French service and those offering simple fare directly served out of the kitchen. Moreover a waiter's competence is undoubtedly enhanced by experience. Nevertheless, the skills required to work as a waiter in these hotels are those that many persons possess or can readily acquire. Accordingly, general population and civilian work force data are the proper frame of reference.

Defendants respond that as a matter of policy they hire only experienced waiters in order to maintain the standards they have set for themselves as luxury hotels. This argument is properly considered as a part of their rebuttal case, discussed below. Prior work experience is not so manifestly a job-related qualification for waiters that plaintiffs should bear the burden of proving, as part of their prima facie case, the percentage of blacks among experienced waiters.

vii. Finding and conclusion respecting black male availability

As the Court observed in *Hazelwood,* "what the hiring figures prove obviously depends upon the figures to which they are compared." 433 U.S. at 310, 97 S. Ct. at 2743. The "determination of the appropriate comparative figures" was held by the Court to require evaluation and findings by the district court respecting the relevant factors and considerations.

Based on the record and the considerations discussed and evaluated in the foregoing paragraphs, the Court finds the appropriate comparison is with a labor market consisting of all males, aged 21 to

64, in the civilian labor force in the weighted San Francisco/Oakland SMSA as shown in the 1970 census.

The statistical data offered by plaintiffs produce a black availability figure of 11.1% in that market, or availability pool. Exhibit 66, Table 3. The choice of a particular figure should, however, not be permitted to create the illusion of certainty or precision. As Dr. Pencavel observed, no single figure provides a reliable estimate of availability. The entire process is perhaps more calculated to exclude gross error than to lead to precise answers. The Court concludes that the analysis producing an availability figure of 11.1% is the one which, in the circumstances of this case, appears to be the most rational and valid.[31]

b. Comparison of Defendants' Hiring Data

i. Hiring, transfer and promotion data

Inasmuch as the question to be decided is whether defendants engaged in purposeful discrimination during the years at issue, comparisons must be based on defendants' employment practices — hiring, transfer and promotion of black males into waiter positions — during that period. Because of the nature of the statistical tests applied, the question arises whether to use aggregate figures for the whole period or to divide it into parts. This choice necessarily has an effect on the outcome of the analysis. The same percentage of blacks actually hired could be statistically significant or not, depending on the number of hires; the more years grouped together, the larger the number, the more significant statistically is a given percentage variation from the number of black hires expected.

Use only of data for the entire period for either defendant, however, would be questionable. Statistical testing of the type used here assumes that the employment practices being evaluated remained the

---

[31]. The Court notes that although at trial plaintiffs' expert advocated a market having a range of 15.1% to 17.9% black males, in his pretrial deposition, taken before he had refined his analysis for trial, he estimated the range of availability of black males to be between 11% and 20%.

Additional support is provided by occupational census data. Although the Court concludes below that the occupational census data reflecting the percentage of black waiters in the San Francisco SMSA are unacceptable as a measure of availability, the range between the San Francisco (6.98%) and the national figure (15.7%) nevertheless provides a useful reference. Exhibit 77. The national figure corresponds more closely to the overall percentage of blacks in the population, suggesting the absence of pervasive racial bars to access to waiter jobs. It is therefore reasonable to expect that the true San Francisco availability figure would fall somewhere above the actual San Francisco figure of 6.98% and below a maximum of 15.7%. For the reasons discussed above in connection with the definition of the relevant labor market, however, the 15.7% figure cannot simply be adopted as the measure of local availability.

same over the whole period. Material differences in the way selection practices within a single period affected black applicants would lead to invalid or misleading test results for that period. Plaintiffs argued that the hotels changed their practices so as to increase their rate of hiring and promoting blacks during the course of this lawsuit. Plaintiffs did not attempt to support this argument with statistical tests comparing the rates before and after any alleged change in defendants' practices. The test statistics below suggest that the hiring practices of the St. Francis may have changed after 1975 and those of the Hilton after 1977. These years end the periods of the statistically most significant divergence between actual and expected hiring of blacks by the respective hotels. But in the absence of direct evidence of a change in hiring practices, the Court cannot rely on these periods to the exclusion of data for later years.

Reliance on one test of a single time period would be especially inappropriate, moreover, where the issue is intent and the test is whether the disparities were so "longlasting and gross" that an inference of purposeful discrimination arises. A more satisfactory indication of intent is the pattern of cumulative results. The racial makeup of progressively increasing numbers of hires, year by year, is relevant to what the defendants might have known about the racial impact of their hiring practices over time. The tables below organize the annual data into consecutive periods, each starting with the earliest year for which figures are in evidence.

The data [appear in tables on the following page.]

ii. *Statistical analysis — prima facie case*

Having determined the pool of available workers, the percentage of blacks in that pool, and the rate at which blacks were hired by defendants, it is possible to draw certain statistical conclusions. The statistical test employed estimates the likelihood that the variation between the observed data (i.e., the actual number of blacks hired) and the expected (based on the percentage of blacks in the availability pool) is consistent with random selection, given the total number hired.

[The Z statistic is computed as a part of a statistical test, the so-called Z test. It is based on an assumed normal distribution of observed values around any expected value. The variation between observed values and the expected value (here the percentage of blacks in the availability pool) is measured in terms of standard deviation (or standard error). The Z statistic reflects the number of standard deviations by which the observed value differs from the expected value. By reference to tables based on the normal distribution curve, the Z sta-

St. Francis Hotel Hire, Promotion and Transfer Data

[1] Period	[2] Positions Filled	[3] Blacks	[4] % Black	[5] Expected No. of Blacks at 11.1%	[6] Z Statistic
1970	12	0	0%	1	−1.22
1970-71	35	1	2.9%	4	−1.55
1970-72	111	6	5.4%	12	−1.91
1970-73	153	9	5.9%	17	−2.05
1970-74	193	11	5.7%	21	−2.39
1970-75	256	13	5.1%	28	−3.07
1970-76	293	17	5.8%	33	−2.89
1970-77	329	21	6.4%	37	−2.72
1970-78	371	25	6.7%	41	−2.67
1970-79	390	38	7.2%	43	−2.46

Hilton Hotel Hire, Promotion and Transfer Data

Period	Positions Filled	Blacks	% Black	Expected No. of Blacks at 11.1%	Z Statistic
1974	42	3	7.1%	5	−0.82
1974-75	59	3	5.1%	7	−1.47
1974-76	77	3	3.9%	9	−2.01
1974-77	125	6	4.8%	14	−2.24
1974-78	179	11	6.1%	20	−2.11
1974-79	206	17	8.3%	23	−1.30

tistic can be used to determine the probability that the observed value is the product of random selection or occurrence.

The formula for calculating the standard deviation is derived from an equation which mathematically describes the normal distribution. This formula was set out by the Supreme Court in *Hazelwood,* supra, 433 U.S. at 311 n. 17, 97 S. Ct. at 2743 n. 17 and in another context in Castaneda v. Partida, 430 U.S. 482, 496-97 n. 17, 97 S. Ct. 1272, 1281, 51 L. Ed. 2d 498 (1977): other courts have applied the formula in employment discrimination cases. See note 39 infra. The standard deviation is equal, in this case, to the square root of the product of the probability of drawing a nonblack (.889) and the probability of drawing a black (.111), multiplied by the square root of the sample size. Given a sample size of 100, for example, the formula in this case would be:

$$\sqrt{.111 \times .889} \times \sqrt{100} = 3.14$$

The standard deviation (or standard error) for this sample would therefore be approximately 3. Statistical theory tells us that if 100 samples were drawn independently and at random from the same population group, in 95 of the draws the observed number of blacks would be within two standard deviations of the expected number of blacks, i.e., approximately between 17 and 5. To put it differently, if there were fewer than 5 or more than 17 blacks in a sample of 100, the probability of obtaining that number as a result of random selection is less than five percent.

The Z statistic is equal to the difference between the expected number (11.1% of the sample) and the actual number of blacks hired divided by the standard deviation:

$$Z = \frac{(\text{Actual Number}) - (\text{Expected Number})}{(\text{Standard Deviation})}$$

The Z statistic thus reflects the number of standard deviations between expected and actual hires. Under the formulae stated above, the standard deviation tends to decrease as the sample size increases; as a necessary corollary, the Z statistic tends to increase as the standard deviation decreases.

The probability that a particular observed value would occur by chance can be estimated by consulting a table that shows the probability, in a normal distribution, of a random variation equal to the Z statistic. A Z statistic of 2.00, for example, indicates a less than 5% likelihood that the observed value would occur by chance. Such a result is described as being significant at the .05 level. See F. Mosteller, R. Rourke, & G. Thomas, Probability with Statistical Applications 130-46, 270-91, 292-98, 304-307 (2d ed. 1970).

Plaintiffs used the same statistical method to calculate Z statistics, based on an expected rate of black hires of 15.1%. (fn. 33)]

The Supreme Court first discussed this kind of statistical analysis in Castaneda v. Partida, 430 U.S. 482, 97 S. Ct. 1272, 51 L. Ed. 2d 498 (1977), in the context of alleged discrimination in the selection of grand jurors. In that case it held proof that over an 11 year period only 39% of the persons summoned were Mexican-Americans where Mexican-Americans constituted 79.1% of the county's population to be sufficient to establish a prima facie case of purposeful discrimination. In a footnote, the Court explained the statistical analysis and concluded:

> As a general rule for such large samples, if the difference between the expected value and the observed number is greater than two or three standard deviations, then the hypothesis that the jury drawing was random would be suspect to a social scientist. The 11-year data here reflect a difference between the expected and observed number of Mexican-Americans of approximately 29 standard deviations. A detailed calculation reveals that the likelihood that such a substantial de-

parture from the expected value would occur by chance is less than 1 in 10^{140}.

The data for the 2½-year period during which the State District Judge supervised the selection process similarly support the inference that the exclusion of Mexican-Americans did not occur by chance. Of 220 persons called to serve as grand jurors, only 100 were Mexican-Americans. The expected Mexican-American representation is approximately 174 and the standard deviation, as calculated from the binomial model, is approximately six. The discrepancy between the expected and observed values is more than 12 standard deviations. Again, a detailed calculation shows that the likelihood of drawing not more than 100 Mexican-Americans by chance is negligible, being less than 1 in $10.^{25}$

430 U.S. at 497 n. 17, 97 S. Ct. at 1281 n. 17. The Court cited the foregoing statement in *Hazelwood*, where it found the observed and the expected values in two time periods to differ by more than five and six standard deviations, respectively. 433 U.S. 309 n. 14, 97 S. Ct. at 2742 n. 14. It went on to observe:

> Because a fluctuation of more than two or three standard deviations would undercut the hypothesis that decisions were being made randomly with respect to race, 430 U.S. at 497 n. 17, 97 S. Ct. at 1281 n. 17, each of these statistical comparisons would reinforce rather than rebut the Government's other proof.

433 U.S. at 311 n. 17, 97 S. Ct. at 2743 n. 17.

Having previously concluded that plaintiffs are required to prove purposeful discrimination in this case, the Court must determine whether the statistical disparities here are sufficient to allow such an inference. Although, as the Supreme Court has observed, fluctuations of more than two or three standard deviations "would undercut the hypothesis that decisions were being made randomly," it does not follow that an inference of purposeful discrimination must arise. Where as here, the plaintiffs' proof rests essentially on comparisons with an availability percentage constructed for trial purposes, the disparity must be so gross as not to be dependent on defendants' contemporary knowledge of that rate. The weight to be given the Z statistic must, moreover, take into account the strength of the underlying data. In this case, the availability percentage, as previously discussed, is the product of numerous choices by the Court among alternatives and is at best a reasonable estimate, not a precise calculation derived from scientific observation.

To evaluate the statistical evidence in an intent case, one must also, insofar as possible, consider the data in the light in which they appeared to the defendant at the time. In the case of the St. Francis, after hiring no blacks in 1970 and only one in 1971, the rate of hiring increased to over five percent on a cumulative basis. Only for the entire 1970-75 period does the disparity exceed three standard deviations. In the case of the Hilton, the cumulative rate of hiring ranged between 3.9% and 7.1% during the precomplaint period; the Z statistic did not exceed 2.24.

Thus, even though a Z statistic of three indicates a probability of random occurrence of only .22%,[37] viewing the data as a whole the Court cannot conclude that they raise an inference of purposeful discrimination. "The mathematical conclusion that the disparity between . . . two figures is 'statistically significant' does not, however, require an *a priori* finding that these deviations are 'legally significant.'" United States v. Test, 550 F.2d 577, 584 (10th Cir. 1976) (jury selection case). The Supreme Court, while noting that disparities "greater than two or three standard deviations" would be suspect to a social scientist, has never accepted that level as sufficient to raise an inference of intent.[38] In the cases in which it has applied this analy-

37. The probability that the other observed values were the result of random selection is approximately as follows:

	Period	Z	% Probability of Chance Occurrence
St. Francis:	1970	1.22	22.24%
	1970-71	1.55	12.12%
	1970-72	1.91	5.62%
	1970-73	2.05	4.04%
	1970-74	2.39	1.68%
	1970-75	3.07	.22%
	1970-76	2.89	.38%
	1970-77	2.72	.66%
	1970-78	2.67	.76%
	1970-79	2.46	1.38%
Hilton:	1974	0.82	41.22%
	1974-75	1.47	14.16%
	1974-76	2.01	4.44%
	1974-77	2.24	2.50%
	1974-78	2.11	3.48%
	1974-79	1.30	19.36%

Source of probabilities: P. Hoel, Introduction to Mathematical Statistics 391 (Table II) (4th ed. 1971). Probabilities given are for a two-sided test. Mosteller, *supra,* at 302-307.

38. It must be borne in mind that the probability of random occurrence of a value is not the obverse of the probability that it is the result of deliberate action. The mathematical theory underlying the Z test is that values of considerable disparity from the expected value will occur, although based on the normal bell-shaped distribution curve the chance of their occurrence lessens sharply the greater the disparity. But the fact that a particular value having a Z statistic of 2 will occur at random only once in twenty does not mean that there is a 95% chance that it was the product of deliberate action. See Mosteller, Rourke & Thomas, supra, at 308-11.

sis to determine the presence of purposeful discrimination, it has relied on disparities ranging from five to 29 standard deviations. See *Castaneda,* supra, 430 U.S. at 497, 97 S. Ct. at 1281; *Hazelwood,* supra, 433 U.S. at 311, 97 S. Ct. at 2743.[39] Statistical disparities considerably more gross and long-lasting than those found here being required to support an inference of purposeful discrimination, the Court finds and concludes that plaintiffs have failed to establish a prima facie case as to their class claims under Section 1981.

3. DEFENDANTS' REBUTTAL EVIDENCE

Although the foregoing findings make it unnecessary to consider defendants' rebuttal evidence, the interest in a complete and final disposition of the case leads the Court nevertheless to deal with it.

a. Applicant Flow Data

In rebuttal of plaintiffs' case, defendants offered a variety of applicant flow statistics. They purported to show that the percentage of successful black applicants was not significantly different from that of white applicants.

In cases where general population or labor force data may not accurately reflect the segment of the work force available for the job, applicant data may be a more accurate frame of reference. *Hazelwood,* 433 U.S. at 308 n. 13, 313, 97 S. Ct. at 2742 n. 13, 2744. Such data may itself be skewed, however, by discriminatory practices which discourage or deflect potential applicants. Dothard v. Rowlinson, 433 U.S. 321, 330, 97 S. Ct. 2720, 2727, 53 L. Ed. 2d 786 (1977).

In this case, the record shows the applicant data of each defendant to be seriously flawed. In the case of the St. Francis, no record purporting to be an accurate and complete list of applicants and their race was maintained in the ordinary course of business. The St. Francis's data were derived from a walk-in log which contains only sporadic and fragmentary entries, from records kept only in recent years by interviewers and from an analysis of 600 applications by persons not on the walk-in logs or interviewer records.

The deficiencies of these records include the following:

39. See, also, Younger v. Glamorgan Pipe & Foundry Co., 20 Fair Empl. Prac. Cas. 776, 782 n. 8 (W.D. Va. 1979) (differences of 2.3 and 1.47 standard deviations found not to provide basis for "inferring that such . . . decisions were made or influenced because of race"). Cf. Thomas v. Parker, 19 Fair Empl. Prac. Cas. 49, 51 (D.D.C. 1979) (prima facie case of racially motivated adverse treatment of an individual was established by nonstatistical evidence plus statistics showing less than one chance in 10,000 that the observed value occurred by chance [a comparable Z statistic, in the terms discussed above, would be at least 3.90]).

(1) The race coding of the walk-in log is speculative and unreliable and may understate black applicants. Persons counted as black were only those whose names appeared on a list of black waiters prepared by plaintiffs for purposes of this lawsuit; black applicants not on plaintiffs' list were assumed to be non-black.
(2) The walk-in logs were kept only intermittently and the information on them was incomplete.
(3) The application procedures used may have discouraged black applicants at a stage prior to sign-in.
(4) The records did not include oral applications made to managers by temporary banquet waiters or employees seeking transfer or promotion.
(5) The impact of the internal promotion policy on the applicant data cannot be ascertained.

These deficiencies in the applicant flow data render any comparison between the rate at which black and non-black applicants were hired meaningless.

In the case of the Hilton, data were derived only from applications specifically for waiter positions during the period 1976-1978. The data were based not on records kept in the ordinary course of business but on a review of those applications for the period 1976-1978 which were retained. Their deficiencies include the following:

(1) The count of blacks was inaccurate and unreliable; persons were counted as blacks only if their names appeared on plaintiffs' list of black waiters. The Hilton itself noted two persons wrongly race coded by this method. Note to Hilton's Ex. A-1.
(2) Oral applications were not recorded.
(3) Application procedures may have deterred black applicants.
(4) Applications for internal promotions and transfers are not recorded.
(5) Only applications specifying waiter as the position sought and those not specifying a position but listing prior waiter experience were counted; applicants who specified captain or bus person but were qualified as waiters were not counted.

These data also are insufficient to permit one to make a meaningful comparison between the rates at which blacks and non-blacks were hired.

b. The Experience Requirement

Had the plaintiffs established a prima facie case, the issue would arise whether the inference of discriminatory intent could be rebutted by proof that defendants consistently imposed an experience requirement that accounted for the rate at which they hired and promoted blacks to be waiters. Both defendants contended and offered evidence to show that they did not hire as waiters persons without prior waiter experience. Plaintiffs disputed the evidence that any waiter experience, much less any minimum length, was consistently required. Because, as the following discussion shows, defendants failed to prove that the proportion of blacks among experienced waiters was smaller than among the labor force as a whole, the evidence that an experience requirement may have been imposed, consistently or not, is irrelevant. And because the low rate of hiring blacks cannot be attributed to the experience requirement, the Court need not reach the legal issue whether experience was a legitimate qualification.

Defendants' contention was addressed to the plaintiffs' statistical case. Arguing that waiter experience is a requisite element of the appropriate labor pool definition, defendants attempted to show that there was no disparity between the percentage of blacks they hired and the percentage in the pool of experienced waiters. Dr. Wolfbein, defendants' labor economist, urged the use of census occupational data for the SMSA to establish this percentage. Those data report as waiters persons who were employed as waiters at the time of the taking of the census or, if then unemployed, whose last job had been as waiter. The percentage of blacks among those shown as waiter in this census was 6.98% in the San Francisco/Oakland SMSA in 1970; the corresponding national figure was 15.7%.

The occupational census does not purport to reflect the total number of persons who are experienced or qualified in a particular occupation. Each person who has worked since 1960 is counted as experienced in only one job: his current job or, if unemployed at the time of the census, his immediately preceding job. The occupational census "presents a picture of occupational background rather than qualification." Smith v. Union Oil Co. of California, 17 Fair Empl. Prac. Cas. 960, 967-968, 17 Empl. Prac. Dec. ¶8411 (N.D. Cal. 1977). Thus an experienced waiter who is working at another job or is unemployed after having had another job does not show up as a waiter. Conversely, an inexperienced waiter newly hired at the time of the census would be counted as a waiter.

The significance of the occupational data varies with the degree of mobility in and out of the particular occupation. Persons engaged in professional or other relatively highly skilled occupations, such as engineers or scientists, are less likely to be found in jobs outside their occupations than those who, like waiters, are engaged in lower paid and more easily learned occupations. The occupational census is therefore a more accurate index of availability of the former than the latter. For purposes of showing the percentage of black males in the pool of experienced waiters, it is of limited utility. The method used does not produce an accurate count of the number of experienced waiters in the population.

If these shortcomings were racially neutral — if there were reason to believe that approximately the same percentage of whites as blacks were experienced waiters employed in another job at the time of the census — the census occupational data would have some persuasive force. The evidence, however, is to the contrary. The evidence previously discussed suggests race-based barriers to entry into the occupation in the San Francisco SMSA. In 1970, black males represented only 6.98% of all male waiters in the San Francisco SMSA, while representing 15.7% nationally. The local waiters union which had 4100 members in 1970 had only 78 black members; as late as 1976, no more than 4.5% of its members were blacks. It may be inferred that an experienced black waiter trying to change jobs within this area, or having moved here from elsewhere, had more difficulty obtaining a waiter job than a comparably experienced white. As a result it is likely that experienced black waiters had to take other jobs more often than did whites. Thus, blacks may well have comprised a disproportionately large fraction of the experienced waiters omitted by the method used in compiling occupational census figures. The possibility of such a racial bias is strong enough to warrant rejecting, as a measure of availability, the percentage of blacks among those counted in the occupational census as waiters.

The Court therefore finds the 1970 census occupational figure for experienced black waiters in this area too unreliable an indication of actual prior experience to define the available labor pool, even if experience were a valid qualification. That is not to say that census occupational data may not in the proper case serve as a useful guideline for availability. In this case, however, conditions prevailing in the labor market cause them, and the statistical calculations they underlie, to be entitled to little weight.

The nonstatistical evidence gives no more support to the contention that the experience requirement explains why fewer blacks were hired than otherwise. Defendants did not claim that any of the named plaintiffs or class members who testified were unqualified. There is no evidence that any particular black was rejected by the defendants for lack of experience. Counsel for the St. Francis argued that experienced black waiters were scarce, citing union membership data for the early 1970's. However, union membership was a consequence of getting a waiter job, not of seeking one. The testimony of the individual and class plaintiffs shows, moreover, that many experienced black waiters sought work at the St. Francis and Hilton without success.

The Court finds that on this record, the experience requirement cannot be regarded as having placed a significant limitation on the number of blacks available for employment as waiters. Hence, it could not be accepted as an explanation for any disparity in the rate at which blacks were hired.

V. CONCLUSION

For the reasons stated, the complaint must be dismissed and judgment entered for defendants, the parties to bear their own costs.

It is so ordered.

Notes and Problems

1. The court in Gay v. Waiters' and Dairy Lunchmen's Union, Local 30 used the Z test to calculate the percentage chance of random disparity between black and white referrals by the union. The Z test is the most convenient way to translate standard deviations into probabilities. As an exercise, confirm the Z scores calculated by the court on page 220. Beware of two problems with that table. The first is that there is an error in column (3), labelled *Blacks*. What is it? Check those numbers by using column (2), *Positions Filled,* and column (4), *% Black*. The second problem is that if you use the court's numbers in column 5, *Expected No. of Blacks at 11.1%*, you will not match their Z statistics in column 6. What error does the court make in column 5? To find out, ignore the court's calculations in that column and make your own. Using the Z tables, check the associated probabilities given by the court in footnote 37.

2. If you are working with percentages rather than raw numbers (e.g., 12% of employees are minorities, rather than 126 employees are minorities), you can use the Z formula:

$$Z = \frac{P_1^* - P_1}{\sqrt{\frac{P_1 \times P_2}{N}}}$$

where P_1^* is the observed proportion of minorities, P_1 is the expected proportion, P_2 is the expected proportion of nonminorities, and N is the total number of employees in the group being examined (the size of the group of suspect composition). Using column 4, % *Black,* column 2, *Positions Filled,* and a relevant black population of 11.1%, apply the proportional formula for the Z test, given above, to calculate the probability that the black proportion of total employees hired at the Hilton Hotel during the years 1974-1979 could occur by chance.

3. Your choice of legal remedies to address your client's discrimination complaint will affect the utility of your statistical evidence. As the law stands in 1982, a Z score of two or three might constitute a prima facie case of employment discrimination under Title VII, but it is less likely to do so under either a §1981 claim or an equal protection claim. Note the court's difficulty with this issue. It has not been established what level of statistical certainty is enough to prove "intent" to discriminate. Title VII requires only "disparate impact." This court suggests on page 224 that "purposeful discrimination" might be inferred from statistical evidence, though it would look for disparities "more gross and long-lasting than those found here." The Congress is currently considering an amendment to Title VII that would replace the disparate-impact test as currently applied with a standard closer to that adopted for §1981 and equal protection claims.

4. A major part of each party's statistical presentation is proof of what is the expected number or proportion of people in each category. Plaintiffs will argue for as high a number as possible and defendants will argue for a low figure. Frequently the highest arguable proportion of a minority group will be the percentage of minority members in the community as a whole. But, *Castaneda* aside, courts rarely accept this for the expected number, preferring instead to determine the percentage in the *relevant* population. The consideration by the court in Gay v. Waiters' and Dairy Lunchmen's Union, Local 30 of relevant geographic area, age bracket, population segment, time, earnings, and employment qualifications is a noteworthy example of the type of descriptive statistical exercise through which advocates and finders of fact must go before drawing inferences from the available data.

The fact that an observation is more than two or three — or 1000 — standard deviations from the mean does not prove what *caused* the disparity between expected and observed. It only indicates the role that chance played in the outcome. The court's close attention to the "relevant" population is an attempt to eliminate as many other innocent causal explanations for the disparity as possible.

Problem: Alabama Nursing Home Association v. Califano

In Alabama Nursing Home Association v. Califano, 465 F. Supp. 1183 (N.D. Ala. 1979), plaintiffs unsuccessfully challenged the Medicaid method of reimbursement to nursing-home facilities in Alabama on the grounds that ceilings on payment rates were not reasonably related to cost. HEW regulations provided that states must reimburse nursing homes at a rate high enough to cover relevant costs of an "efficiently and economically operated facility." The Alabama plan set a rate equal to the full cost for the *sixtieth* percentile of facilities. To calculate this rate, the state made a list of rates for each nursing home from highest to lowest. Setting the rate ceiling equal to the full cost for the sixtieth percentile means that the reimbursement would cover all costs for 60% of the nursing homes on the list, starting from the one with the lowest rates and moving up to those with higher rates. Plaintiffs alleged this was unreasonable and that "the lowest ceiling which would meet the requirement of reasonable cost-related reimbursement would be the mean cost plus one-half standard deviation." Test your understanding of the Z test and standard deviations by calculating what percentage of nursing homes would be fully reimbursed under the plaintiff's plan if nursing homes were normally distributed around the mean in terms of their costs.

CHAPTER VI

INFERRING FROM A SAMPLE

A. POPULATIONS AND SAMPLES

When testing statistical significance and calculating means and standard deviations, we have, up to this point, had the advantage of relatively complete information concerning the population about which we were drawing statistical conclusions. Whenever we used a Z test or a chi-square test, the relevant information, such as expected and observed numbers, was given to us or could be calculated from other data or surveys that were already available. At times, however, all of this information is not available. To appreciate the consequences of this lack of information, we must first distinguish between the notions of a *population* and a *sample* of that population. If, for instance, we want to know with 100% certainty how many lawsuits the average taxpayer in New York City had been involved in, we would have to make inquiries of each taxpayer, add up the total number of experiences reported, and divide by the total number of taxpayers. This would give us the *population mean*, the arithmetic average number of lawsuits per taxpayer. Such a survey might be prohibitively expensive, however, and if we could get a reasonable estimate of the average number of lawsuits by asking a random selection of New Yorkers, the estimate might be acceptable for our purposes. The estimate of the population mean obtained from surveying less than the entire population is called a *sample mean*. The CBS/New York Times poll cited occasionally by Dan Rather is such a sampling. In order to draw conclusions about the opinions of the country as a whole, with a margin of error of plus or minus 2%, the pollsters usually survey between 800 and 1200 randomly selected people.

Once the sampling of people has been surveyed, the sample mean, sample variance, and sample standard deviation (which is also referred to as the *standard error of the sample*) can be calculated. The statistical difference between the population measures and the sample means, variances, and standard errors is how certain we are that the measure is accurate. If we carefully counted all the members of the population, we would be 100% certain that the calculated mean or variance was accurate. If we survey only a sample of the total population, we cannot be so sure. It is this difference in accuracy that leads us to distinguish between population means (M), variances (SD^2), and standard deviations (SD), on the one hand, and sample means (m), variances ($s.e.^2$), and standard error ($s.e.$) on the other, even though they mean the same thing for both populations and samples. The process of drawing conclusions about the population as a whole from evidence gleaned from a random sample is thus an important part of statistical inference.

Drawing conclusions from a random sample involves two steps. The first is making a *point estimate* of the population statistic, such as the mean, variance, or standard deviation, relevant to the legal issue at hand. The fact that a sample may not perfectly reflect the characteristics of the population from which it is drawn suggests that point estimates from different samples of the same population may differ. In order for the point estimate to be useful, the second step is to determine how different the point estimate is likely to be from the true, but unknown, population statistic.

B. SAMPLE MEANS AND STANDARD ERRORS

The limits of likely deviation of the estimate from the true (population) value form a *confidence interval* around the estimate. For ininstance, Dan Rather's news report, "47% of voters favor a tax cut, subject to a margin of error of 2%," means that, based on its sample, CBS is fairly certain that between 45% and 49% of voters favor a tax cut. Instead of an estimated point (i.e., 47%), we can develop an interval (i.e., 45% to 49%) in which we can be fairly certain that the true value lies.

The *sample mean* (m) is the best point estimate of the population mean (M). It is calculated in the same way:

$$m = \frac{\Sigma(x_i)}{n}$$

where the x_i's are the values of all the items in the sample and n is the number of items in the sample. For the population, recall that the mean is:

$$M = \frac{\Sigma(X_i)}{N}$$

where the X_i's are the values of all the items in the population and N is the size of the population.

This is not particularly complicated. You can imagine a lawsuit where the issue involved is how much pollution the ABC factory emitted last month. The relevant regulation prohibits emission in excess of 60 tons per month. It is too expensive for the Environmental Protection Agency to have inspectors spend a whole month at the factory measuring the entire population, in this case, measuring the emissions for each day of the month. Instead, the inspectors choose six days as a sample and measure the pollution. The total monthly pollution will be 30 times the daily mean tons of pollution. The results are reported in Exhibit VI-1. The first column shows the measurements actually made, i.e., the results of the sample. The second column shows the actual emissions for every day of the month, which we know simply because we are omniscient. The inspectors do not have this additional information.

In this example, 1.9 tons is the sample mean, while 2.1 tons is the population or true mean daily pollution. The estimated monthly pollution is 30 days times 1.9 tons per day, which equals 57 tons. The true monthly pollution is 63 tons. The sample does not show a violation of the law. But is the estimate acceptable for purposes of proof? Recall that the inspectors do not know the true monthly pollution.

To answer this question, we have to know the standard error of the sample. This should be no surprise. The standard deviation has shown up in all of the statistical tests done so far, and *standard error* is just the name given to the standard deviation of a sample. But, so far, we have been able to calculate the true standard deviations, because we knew the true means, expected values, and observed values of the populations. Now we only know those in our sample. If we knew the population mean, M, we would calculate the variance as:

$$SD^2 = \frac{\Sigma(X_i - M)^2}{N}$$

Because we don't know M, we use m, the sample mean, which is our best estimate of M. Because we don't know all of the X_i's, we use the x_i's that are in our sample. Because the number of observations in the samples is less than the total number of possible observations, we use

EXHIBIT VI-1.
Pollution from ABC Factory

Date in November	(1) Tons Measured (x_i)	(2) Tons Actually Emitted (X_i)
1	1.3	1.3
2	—	2.2
3	—	2.5
4	—	1.5
5	1.5	1.5
6	—	3.3
7	—	3.0
8	—	2.3
9	2.1	2.1
10	—	2.1
11	—	0.7
12	—	0.9
13	1.2	1.2
14	—	1.0
15	—	2.1
16	—	1.9
17	—	3.8
18	—	3.2
19	—	3.5
20	—	1.2
21	—	3.2
22	2.3	2.3
23	—	3.3
24	—	0.0
25	—	2.4
26	—	2.3
27	—	0.8
28	—	2.3
29	3.0	3.0
30	—	2.2

$$m = \frac{\Sigma(x_i)}{n} = \frac{11.4}{6} = 1.9 \qquad M = \frac{\Sigma(X_i)}{N} = \frac{63.0}{30} = 2.1$$

n instead of N. This suggests that the formula for sample variance would be:

$$s.e.^2 = \frac{\Sigma(x_i - m)^2}{n} \quad (Wrong!)$$

But, for an obscure mathematical reason that has something to do with degrees of freedom,[1] we divide, not by n, but by $n - 1$ to calculate the *sample variance*. The correct formula for sample variance is:

$$s.e.^2 = \frac{\Sigma(x_i - m)^2}{n - 1} \quad (Correct!)$$

Thus, the variance for our sample days of pollution is:

$$
\begin{aligned}
s.e.^2 = &\ (x_1 - m)^2 = (1.3 - 1.9)^2 = (-0.6)^2 = .36 \\
+ &\ (x_2 - m)^2 = (1.5 - 1.9)^2 = (-0.4)^2 = .16 \\
+ &\ (x_3 - m)^2 = (2.1 - 1.9)^2 = (+0.2)^2 = .04 \\
+ &\ (x_4 - m)^2 = (1.2 - 1.9)^2 = (-0.7)^2 = .49 \\
+ &\ (x_5 - m)^2 = (2.3 - 1.9)^2 = (+0.4)^2 = .16 \\
+ &\ \underline{(x_6 - m)^2 = (3.0 - 1.9)^2 = (+1.1)^2 = 1.21} \\
= &\ \Sigma(x_1 - m)^2 = \ 2.42
\end{aligned}
$$

$$= \frac{\Sigma(x_1 - m)^2}{(n - 1)} = \frac{2.42}{(6 - 1)} = \frac{2.42}{5} = 0.48 = s.e.^2$$

and the standard error, $s.e. = \sqrt{s.e.^2} = \sqrt{0.48} = 0.70$.

C. CONFIDENCE INTERVALS

Remember our goal, which is to determine whether our monthly pollution estimate of 57 tons is an acceptable one for the purposes of proof. We arrived at 57 tons by multiplying the sample mean ($m = 1.9$ tons per day) by the number of days in the month (30 days), i.e., 1.9 times 30 equals 57 tons per month. We must see how confident we are that 1.9 is an accurate measurement of the mean. What we are going to do is establish a *confidence interval* around that sample mean value.

Two pieces of information accompany a confidence interval. The first is the interval itself, the range (e.g., between 45% and 49% for the CBS poll) within which the true mean is likely to fall. The second is how likely the true mean is to fall within that range, i.e., how sure we are that the true mean is in the stated interval. For large samples, we determine the range of the interval and the degree of certainty

---

1. See D. Suits, Statistics: An Introduction to Quantitative Economic Research 62-63 (1963).

from the standard error of the sample (*s.e.*) and the *Z* table, which we have already discussed. The "degree of certainty" is referred to as the *confidence level*.

Remember that the *Z* table presents the probability that a given observation will be a certain number of standard deviations away from the expected (or mean) value. In establishing a confidence interval around a sample mean, we are employing the *Z* table to calculate the probability that the population mean will be within a specific range around the sample mean.

1. The Standard Error of the Sample Mean

Just as observations differ from one another, means from different samples will vary. To determine how much a particular sample mean is likely to vary from the actual population mean, we use the standard error of the sample, *s.e.*, to calculate the *standard error of the sample mean, s.e.$_m$*, also referred to as the *mean standard error*. The formula for the standard error of the sample mean is:

$$s.e._m = \sqrt{\frac{SD^2}{n}} \cong \sqrt{\frac{s.e.^2}{n}}$$

where SD^2 is the variance of the population, and n is the sample size. Since we often do not know the population variance, we estimate it as $s.e.^2$, which is the variance within the sample.

Consider our population example. Our estimate of the sample mean is 1.9 tons per day. One standard error is 0.70 tons per day, so the standard error of the mean is $\sqrt{0.70^2/6}$, or 0.3. If we believe the population mean is within one standard error, *s.e.$_m$*, of the sample mean, then the confidence interval is 1.9 tons plus or minus 0.3 tons, or from 1.6 to 2.2 tons per day. How certain can we be that the population mean is within this range, i.e., what is our confidence level? Refer to the table of *Z* values (on page 200) and look at $Z = 1.00$ (corresponding to the probability of an observation within one mean standard error of the mean). The table gives the figure .3413. That is the probability, 34.13%, that the true mean is within one standard deviation *above* the sample mean. Twice the percentage, 34.13% times 2 equals 68.26%, is the probability that the population mean is within one standard error above or below the sample mean. That is to say, we can be about 68% certain that the true mean lies in this interval.

If the actual mean is within this range of 1.6 to 2.2 tons per day, is the regulation violated? This rate of pollution on a monthly basis is from 1.6 times 30 to 2.2 times 30 tons per month, or from 48 to 66 tons per month. The lower boundary of our confidence interval, 48 tons, does not violate the regulatory maximum of 60 tons per month, but the upper boundary does. All that we can say is that we are 68% sure that the true amount of pollution lies in that range. We cannot conclude that the regulation has been violated, nor can we say, with very great certainty, that it has not.

Consider a number of other possible results of a sample. What would happen if we had calculated a sample standard error of 0.70, a mean standard error of 0.3, and a sample mean of 2.8? Then, establishing a one-standard-deviation confidence interval around the mean would indicate a 68% chance that the population mean lay between 2.5 and 3.1. The monthly pollution estimate is between 2.5 times 30 and 3.1 times 30 tons, or between 75 and 93 tons per month. Since both ends of the range exceed the regulatory maximum we can say with a 68% confidence level that the regulation has been violated. Is 68% confidence enough to impose civil and possibly criminal penalties on the owners or operators of the factory?

We have confronted the question of degree of certainty before. A general rule of "two or three standard deviations" from *Castaneda* is the clearest indication from the Supreme Court in discrimination cases. Other courts have suggested that 95% certainty is sufficient for a prima facie case of discrimination. How certain do we need to be in this environmental protection context?

In our example involving a standard error of 0.70 tons, a corresponding $s.e._m$ of 0.3, and a sample mean of 2.8 tons per day, a three-standard-error confidence interval would indicate a monthly pollution emission of from 57 to 111 tons.[2] Referring again to the Z table, we note that we can be 99.74% certain that the true mean lies in this three-standard-deviation range (.9974 = 2 × 0.4987, the figure from the table). But the lower boundary of this range, 57 tons, does not violate the regulation, so we cannot say that we are 99.74% sure that the regulation has been violated. If we had information for the whole month, we would be 100% sure that the regulation had been violated, but we do not and therefore cannot, on the basis of the available statistics, make such a conclusion with great confidence.

In common practice (statistical practice, that is, not legal practice), there are three confidence levels that are usually used, 99%, 95%, and 90%. These confidence levels allow for a 1%, 5%, and 10%

---

2. The calculations are:

57 tons per month = [2.8 tons − (3 std. dev. × 0.3 tons)] × 30 days

111 tons per month = [2.8 tons + (3 std. dev. × 0.3 tons)] × 30 days

chance of a type 1 error, respectively. The probability of a type 1 error is called the *significance level*. A 1% or .01 significance level allows a 1% chance of error. A 5% or .05 significance level allows for 5% chance of a type 1 error. A 10% or .10 significance level allows for a 10% chance of error. The Z scores corresponding to these significance levels are called the *critical values* for these significance levels and are as follows:

EXHIBIT VI-2.
Critical Values of Z Corresponding to Three Common Significance Levels

Significance Level	.10	.05	.01
Critical Value	1.645	1.960	2.575

These critical values of Z are commonly used in calculating confidence intervals.

Cases at law have revealed two "common tests" for a prima facie case in the discrimination context, the two-to-three-standard-deviations test and the 95% confidence level test. The latter test is analogous to the .05 significance level test for which the critical Z value is 1.96. The two-to-three-standard-deviations test is passed by a Z score greater than 2.00 to 3.00. The chance of finding guilty an innocent party who just meets .05 significance level is 5%. The chance of finding guilty an innocent party who just meets the two-to-three-standard-deviations test is 4.54% to 0.26%.

2. Choosing a Sample Size

In discussing confidence intervals, we ignored the problem of deciding how large a sample we need to take in order to get an acceptable estimate of the population values. While actually taking the sample is more likely to be a task for a statistical expert, the lawyer should be aware of the necessary size of the sample so that he or she can estimate the cost to the client of using this method of proof. The key to choosing a sample size is deciding how accurate a measure of the population value is needed.

Suppose we want to estimate the mean daily discharge of an air pollutant from the ABC factory within one-half ton either way at a 95% level of confidence. How many days must we sample to make this estimate? The level of accuracy chosen means that 95% of the observations must be within one-half ton of the mean value. The Z score corresponding to a confidence level of 95% is 1.96 mean standard errors. So one-half ton must be equal to 1.96 mean standard errors. The calculations proceed as follows:

Step 1. Determine the desired size of one standard error of the mean. If one-half ton equals 1.96 mean standard errors, then one mean standard error equals .5 divided by 1.96, or .25 tons.

To calculate the sample size, n, we must use the statistical relationship:

$$s.e._m = \sqrt{\frac{SD^2}{N}} \cong \sqrt{\frac{s.e.^2}{n}}$$

where $s.e._m$ is the standard error of the mean we have just calculated, where $s.e.^2$ is the sample variance (which is an estimate of the population variance SD^2 and which must still be calculated), and where n is the sample size, which we don't know. To find n, we have to know SD^2, the population variance.

Step 2. Estimate the population variance. This is a difficult step because we may have no good idea of what this value is. But there are several methods available to approximate this value, and the first is to estimate it by using data from another factory with a similar technology. A second is to use data from the last test we made of this factory's emissions. If no other information is available, an estimate might be made using other facts casually available, such as daily variation in output of whatever product the ABC factory produces, combined with engineers' reports relating to the capabilities of whatever pollution reduction equipment might have been installed. Assume that one of these methods reveals the population variance, calculated from the monthly data on page 234, of 0.81 tons.

Step 3. Calculate n. The formula given above can be rearranged as:

$$n = \frac{SD^2}{s.e._m^2} = \frac{0.81 \text{ tons per day}}{(0.26 \text{ tons})^2} = \frac{0.81}{0.07} \text{ days} = 11.6 \text{ days}$$

A sample of 12 days should give us the desired level of accuracy.

If we are willing to accept less accuracy we can do with a smaller sample. There are two ways in which we can decrease the accuracy.

The first is to allow for a one-ton margin of error rather than plus or minus one-half ton. This changes the calculation of n as follows:

Step 1. Determine the size of one mean standard error. One mean standard error would be one ton divided by 1.96 (the critical value for a .05 significance level), which equals 0.51 tons.
Step 2. Estimate the population variance. Use the same estimate, $SD^2 = 0.81$ tons per day.
Step 3. Calculate n.

$$n = \frac{SD^2}{s.e._m^2} = \frac{0.81 \text{ tons per day}}{(0.51 \text{ tons})^2} = \frac{0.81}{0.26} \text{ days} = 3.1 \text{ days}$$

The second way to lower the accuracy is to decrease the confidence level, e.g., to 90%:

Step 1. One mean standard error equals one-half ton divided by 1.645 (the critical value for a 10% significance level), or 0.30 tons.
Step 2. Use the same estimate, $SD^2 = 0.81$ tons per day.
Step 3. Calculate n.

$$n = \frac{SD^2}{s.e._m^2} = \frac{0.81 \text{ tons per day}}{(0.30 \text{ tons})^2} = \frac{0.81}{0.09} \text{ days} = 9 \text{ days}$$

As an exercise, calculate the sample size needed for an estimate of the mean within one ton at a 90% confidence level. What sample size would be needed for an estimate of the mean within 500 pounds at a 99% confidence level? When the estimated sample size is larger than the population, it is not possible to get the desired accuracy without surveying the entire population.

The next several cases use confidence intervals to infer information about populations from samples. In reading them, try to determine what the population is and how it is sampled. What hypotheses are these statistics being used to test? Are they rejected or not?

RESERVE MINING CO. v. EPA
514 F.2d 492 (8th Cir. 1975)

BRIGHT, Circuit Judge.
The United States, the States of Michigan, Wisconsin, and Minnesota, and several environmental groups seek an injunction ordering Reserve Mining Company to cease discharging wastes from its

iron ore processing plant in Silver Bay, Minnesota, into the ambient air of Silver Bay and the waters of Lake Superior. On April 20, 1974, the district court granted the requested relief and ordered that the discharges immediately cease, thus effectively closing the plant. United States v. Reserve Mining Co., 380 F. Supp. 11 (D.Minn. 1974). Reserve Mining Company appealed that order and we stayed the injunction pending resolution of the merits of the appeal. Reserve Mining Co. v. United States, 498 F.2d 1073 (8th Cir. 1974). We affirm the injunction but direct modification of its terms. As to other issues brought before us by appeals during the course of this complex litigation, we affirm in part and reverse in part.

I. INTRODUCTION

A. SUMMARY OF CONTROVERSY

In 1947, Reserve Mining Company (Reserve), then contemplating a venture in which it would mine low-grade iron ore ("taconite") present in Minnesota's Mesabi Iron Range and process the ore into iron-rich pellets at facilities bordering on Lake Superior, received a permit from the State of Minnesota to discharge the wastes (called "tailings") from its processing operations into the lake.

Reserve commenced the processing of taconite ore in Silver Bay, Minnesota, in 1955, and that operation continues today. Taconite mined near Babbitt, Minnesota, is shipped by rail some 47 miles to the Silver Bay "beneficiating" plant where it is concentrated into pellets containing some 65 percent iron ore. The process involves crushing the taconite into fine granules, separating out the metallic iron with huge magnets, and flushing the residual tailings into Lake Superior. . . .

II. HEALTH ISSUE

The initial, crucial question for our evaluation and resolution focuses upon the alleged hazard to public health attributable to Reserve's discharges into the air and water.

We first considered this issue on Reserve's application for a stay of the district court's injunction pending a determination of the merits of its appeal. We noted the usual formulation of the applicable standards to be met by the party seeking a stay. One of those standards addresses the likelihood of success by the moving party on the merits of the appeal. In applying this standard we made a preliminary assessment of the merits of Reserve's appeal from the trial court's injunction order. We noted that the "rather drastic remedy ordered by the district court . . . was a response to the finding of a substantial danger to the public health," and that our preliminary assessment of whether such a substantial danger was presented "should control our action as to whether to grant or deny a stay." 498 F.2d at 1076-1077.

In this preliminary review, we did not view the evidence as supporting a finding of substantial danger. We noted numerous uncertainties in plaintiffs' theory of harm which controlled our assessment, particularly the uncertainty as to present levels of exposure and the difficulty in attempting to quantify those uncertain levels in terms of a demonstrable health hazard. As we stated then, ". . . it is not known what the level of fiber exposure is, other than that it is relatively low, and it is not known what level of exposure is safe or unsafe." 498 F.2d at 1082. In confirmation of our view, we noted the opinion of Dr. Arnold Brown, the principal court-appointed expert, that no adverse health consequences could be scientifically predicted on the basis of existing medical knowledge. Additionally, we noted the district court's conclusion that there is "'. . . insufficient knowledge upon which to base an opinion as to the magnitude of the risks associated with this exposure.'" 498 F.2d at 1083. We thought one proposition evident:

> [A]lthough Reserve's discharges represent a possible medical danger, they have not in this case been proven to amount to a health hazard. The discharges may or may not result in detrimental health effects, but, for the present, that is simply unknown. [Id.]

On the basis of the foregoing we forecast that Reserve would likely prevail on the merits of the health issue. We limited this forecast to the single issue before us whether Reserve's plant should be closed immediately because of a "substantial danger" to health:

> While not called upon at this stage to reach any final conclusion, our review suggests that this evidence does not support a finding of *substantial danger* and that, indeed, the testimony indicates that such a finding should not be made. *In this regard,* we conclude that Reserve appears likely to succeed on the merits of its appeal on the health issue.

498 F.2d at 1077-1078. (Emphasis added) We reached no preliminary decision on whether the facts justified a less stringent abatement order.

As will be evident from the discussion that follows, we adhere to our preliminary assessment that the evidence is insufficient to support the kind of demonstrable danger to the public health that would justify the immediate closing of Reserve's operations. We now address the basic question of whether the discharges pose any risk to public health and, if so, whether the risk is one which is legally cognizable. This inquiry demands separate attention to the discharge into the air of Silver Bay and the discharge into Lake Superior.

A. THE DISCHARGE INTO AIR

As we noted in our stay opinion, much of the scientific knowledge regarding asbestos disease pathology derives from epidemiological studies of asbestos workers occupationally exposed to and inhaling high levels of asbestos dust. Studies of workers naturally exposed to asbestos dust have shown "excess" cancer deaths and a significant incidence of asbestosis. The principal excess cancers are cancer of the lung, the pleura (mesothelioma) and gastrointestinal tract ("gi" cancer).

Studies conducted by Dr. Irving Selikoff, plaintiffs' principal medical witness, illustrated these disease effects. Dr. Selikoff investigated the disease experience of asbestos insulation workers in the New York-New Jersey area, asbestos insulation workers nationwide, and workers in a New Jersey plant manufacturing amosite asbestos. Generally, all three groups showed excess cancer deaths among the exposed populations, as well as a significant incidence of asbestosis. With respect to cancer generally, three to four times the expected number of deaths occurred; with respect to lung cancer in particular, five to eight times the expected number; and with respect to gastrointestinal cancer, two to three times that expected. Dr. Selikoff described the increase of gastrointestinal cancer as "modest."

Several principles of asbestos-related disease pathology emerge from these occupational studies. One principle relates to the so-called 20-year rule, meaning that there is a latent period of cancer development of at least 20 years. Another basic principle is the importance of initial exposure, demonstrated by significant increases in the incidence of cancer even among asbestos manufacturing workers employed for less than three months (although the incidence of disease does increase upon longer exposure). Finally, these studies indicate that threshold values and dose response relationships, although probably operative with respect to asbestos-induced cancer, are not quantifiable on the basis of existing data.

Additionally, some studies implicate asbestos as a possible pathogenic agent in circumstances of exposure less severe than occupational levels. For example, several studies indicate that mesothelioma, a rare but particularly lethal cancer frequently associated with asbestos exposure, has been found in persons experiencing a low level of asbestos exposure. Although Dr. Selikoff acknowledged that these studies of lower-level exposure involve certain methodological difficulties and rest "on much less firm ground" than the occupational studies, he expressed the opinion that they should be considered in the assessment of risks posed by an asbestos discharge.

At issue in the present case is the similarity of the circumstances of Reserve's discharge into the air to those circumstances known to result in asbestos-related disease. This inquiry may be divided into two stages: first, circumstances relating to the nature of the discharge and, second, circumstances relating to the level of the discharge (and resulting level of exposure). . . .

The Level of Exposure

The second major step in the inquiry of the health aspects of Reserve's air emissions is an assessment of the amount of the discharge and the resulting level of exposure. Two principal issues are raised: first, what in fact is the level of exposure; second, does that level present a cognizable risk to health? The district court found the level "significant" and comparable to the levels associated with disease in nonoccupational contexts. 380 F. Supp. at 48.

The first issue was addressed at length in our stay opinion. We noted there the great difficulties in attempted fiber counts and the uncertainties in measurement which necessarily resulted. 498 F.2d at 1079-1080. Commenting on these difficulties, Dr. Brown stated that the fiber counts of the air and water samples could establish only the presence of fibers and not any particular amount, i.e., such a count establishes only a qualitative, and not a quantitative, proposition. The district court recognized these difficulties in counting fibers and observed that "[t]he most that can be gained from the Court [ordered] air study is the very roughest approximation of fiber levels." 380 F. Supp. at 49.

A court-appointed witness, Dr. William F. Taylor, made the most sophisticated attempt to use the fiber counts in a quantitative manner. By taking the average fiber count of five testing sites in Silver Bay, Dr. Taylor concluded that the burden of fibers in the air of Silver Bay exceeded that present in St. Paul, Minnesota, (used as a control) by a margin which could not be attributed to chance.

[The fiber concentration found was 0.0626 fibers per cc, with a 95 percent confidence interval of from 0.0350 to 0.900 fibers per cc. (Although we indicated in the stay opinion that this count, like the other fiber counts, is subject to a nine-fold margin of error, 498 F.2d at 1078 n. 7, Dr. Taylor's testimony indicates that this particular calculation, embodying as it does the average of several readings, is subject to the lesser margin of error indicated above). It is significant that this concentration, even at its upper range, is far below the legally permissible level for occupational settings, and, thus, obviously below those levels typically associated with occupational exposure to asbestos.

Dr. Taylor warned that his Silver Bay computation, based on only several days of sampling during a particular time of the year, could not be extrapolated to represent the average annual burden of fibers in the air of Silver Bay. (fn. 34)]

The experts indicated that the counting of fibers represents a scientifically perilous undertaking, and that any particular count can only suggest the actual fiber concentration which may be present. Nevertheless, Dr. Taylor's computation indicating some excess of asbestiform fibers in the air of Silver Bay over that of the control city of St. Paul appears statistically significant and cannot be disregarded. Thus, as we indicated in the stay opinion and as the district court concluded, while the actual level of fibers in the air of Silver Bay is essentially unknown, it may be said that fibers are present at levels significantly higher than levels found in another Minnesota community removed from this air contamination.

Given the presence of excess fibers, we must now assess the effects of this exposure on the public. We note first, as we did in the stay opinion, that the exposure here cannot be equated with the factory exposures which have been clearly linked to excess cancers and asbestosis. Our inquiry, however, does not end there. Asbestos-related disease, as noted earlier, has been associated with exposure levels considerably less than normal occupational exposure. The studies indicating that mesothelioma is associated with the lower levels of exposure typical of residence near an asbestos mine or mill or in the household of an asbestos worker are of significance. Although these studies do not possess the methodological strengths of the occupational studies, they must be considered in the medical evaluation of Reserve's discharge into the air.

Of course, it is still not possible to directly equate the exposure in Silver Bay with the exposure patterns in these nonoccupational studies. The studies typically do not attempt to quantify the level of exposure and, as noted above, it is not possible to assess with any precision the exposure level in Silver Bay; thus, exposure levels may be compared only on the most general basis. Furthermore, it is questionable whether Reserve's operations may be equated with those of an asbestos mine or mill; for, while we concur in the trial court's finding that Reserve discharges fibers similar, and in some cases, identical to amosite asbestos, it is also true, as testified by plaintiffs' own witnesses, that only a portion of Reserve's discharge may be so characterized. Additionally, it is also true that at least some of the fiber counts reported to the court reflect *all* amphibole fibers present, thereby including fibers inconsistent with amosite asbestos. Even if all the amphibole fibers inconsistent with amosite could still be attributed to Reserve's discharge, it remains uncertain whether the disease effects attributable to amosite may be extended to these other fibers, or whether the varying forms of asbestos possess differing pathogenic properties.

Conclusion

Plaintiffs' hypothesis that Reserve's air emissions represent a significant threat to the public health touches numerous scientific disciplines, and an overall evaluation demands broad scientific understanding. We think it significant that Dr. Brown, an impartial witness whose court-appointed task was to address the health issue in its entirety, joined with plaintiffs' witnesses in viewing as reasonable the hypothesis that Reserve's discharges present a threat to public health. Although, as we noted in our stay opinion, Dr. Brown found the evidence insufficient to make a scientific probability statement as to whether adverse health consequences would in fact ensue, he expressed a public health concern over the continued long-term emission of fibers into the air. . . . The existence of this risk to the public justifies an injunctive decree requiring abatement of the health hazard on reasonable terms as a precautionary and preventive measure to protect the public health. . . .

Notes and Problems

1. The relevant factual difficulty in Reserve Mining Co. v. EPA was determining the concentration of asbestiform in the air near the Silver Bay taconite ore processing plant. The court reports the findings of Dr. Taylor as a concentration of 0.0626 fibers per cubic centimeter, with a 95% confidence interval from 0.0350 to 0.900 fibers per cubic centimeter. This must be a typographic error. The court (or Dr. Taylor) may have meant an interval from 0.0350 to 0.0902. A 95% confidence interval means that each endpoint of the interval must be 1.96 standard errors from the mean. Dr. Taylor reports the mean as 0.0626 fibers per cc., so if 0.0350 is the lower endpoint of the interval, then 1.96 mean standard errors is equal to the difference between 0.0626 and 0.0350, or (0.0626 − 0.0350 =) 0.0276 fibers per cc. If the lower endpoint of the interval is 0.0626 minus 0.0276, which equals 0.0350, then the upper endpoint must be 0.0626 plus 0.0276, which equals 0.0902 fibers per cc.

C. Confidence Intervals   247

2. To practice the confidence interval process, repeat the type of calculation Dr. Taylor must have performed. Since the court doesn't report the individual results of his samples, we need to invent some fictitious data. Proceed as follows to make the report:

Step 1. Decide what level of confidence we want to have in our findings. A 95% confidence interval is customary, so adopt that one.

Step 2. Do a sampling of the population. If we sample three days, our results might be:

EXHIBIT VI-3.
Asbestiform Population Sample

Day	Fiber Concentration (per c.c.)
1	0.03835
2	0.08685
3	0.06270

Step 3. Calculate the sample mean:

$$m = \frac{\Sigma(x_i)}{n} = \frac{0.03835 + 0.08685 + 0.06270}{3}$$

$$= \frac{0.1879}{3} = 0.0626$$

Step 4. Calculate the mean standard error from the variance of the sample, where:

$$s.e.^2 = \frac{\Sigma(x_i - m)^2}{(n - 1)}$$

Given our sample and a calculated mean value of 0.0626, the calculations proceed like this:

$$
\begin{aligned}
s.e.^2 = & (x_1 - m)^2 = (0.03835 - 0.0626)^2 = (0.02425)^2 = 0.000588 \\
+ & (x_2 - m)^2 = (0.08685 - 0.0626)^2 = (-0.02425)^2 = 0.000588 \\
+ & (x_3 - m)^2 = (0.06270 - 0.0626)^2 = (0.0001)^2 = 0.00000 \\
= & \Sigma(x_i - m)^2 = \overline{0.001176}
\end{aligned}
$$

$$= \frac{\Sigma(x_i - m)^2}{(n - 1)} = \frac{0.001176}{(3 - 1)} = \frac{0.001176}{2} = 0.000588 = s.e.^2$$

$$s.e._m = \sqrt{\frac{s.e.^2}{n}} = \sqrt{\frac{0.000588}{3}} = 0.014$$

Step 5. From the Z table, find the critical value for the confidence level. The critical value for 95% is 1.96 standard errors.

Step 6. Determine the endpoints of the confidence interval. The lower endpoint is the mean minus 1.96 standard errors. The upper endpoint is the mean plus 1.96 standard errors. One mean standard error is 0.014 (from step 4), so 1.96 mean standard errors is 0.0274 (=0.014 × 1.96). The mean is 0.0626, so the endpoints are 0.0626 ± 0.0274.

Step 7. State the confidence interval. Here, we can be 95% certain that the population mean is between 0.0352 and 0.0900 fibers per cubic centimeter.

3. To test your understanding of how the Z table works in this process, figure out a 70.16% confidence interval for the data in Note 2. To find the Z score corresponding to 70.16%, look at the table. This Z score means that 70.16% of the observations will be within "Z" standard errors of the mean. Remember that to use the Z table we double the percentage after each Z score. For a Z score of 1.96, for instance, we doubled the percentage .4750, found on the table, to get the 95% confidence level. Our task in this problem is to find the percentage in the table which, when doubled, gives .7016. We find it associated with a Z score of 1.04. The critical value for 70.16% is 1.04 mean standard errors. Use this value in steps 6 and 7 to find the confidence interval. You should get an interval of 0.0626 ± 1.04 times 0.014, or 0.0626 ± 0.01456 fibers per cc. The general formula for the interval is $m \pm z \times s.e._m$. What would the interval be for a 98.02% level of confidence? What about for a 0.10 significance level?

Problem: Marathon Oil Co. v. EPA

One issue in Marathon Oil Co. v. EPA, 564 F.2d 1253, at 1266 (9th Cir. 1977), was the appropriateness of the method by which the EPA set its effluent discharge standards for the oil, mud, grease, and soaps that are washed from offshore oil drilling platforms. The following excerpt from that opinion details the EPA's method.

The EPA's Mode of Analysis

Before turning to petitioner's specific objections, it is useful to outline briefly the statistical approach taken by the EPA in setting the instant discharge standards. The approach closely paralleled the EPA's typical method of setting effluent limitations. Broadly stated, the EPA sets effluent standards by looking to the average performance of the best existing pollution technology in the industry. More particularly, the EPA first determines the "best practicable control technology currently available." (BPCTCA) If possible, the EPA then collects effluent discharge data from those plants in the industry that utilize BPCTCA in an exemplary fashion. The EPA's final task is to utilize the data to compute an effluent discharge standard average of this data, which typically will be a number that exemplary BPCTCA plants can meet. This is done by using so-called "confidence intervals." Since the effluents being treated typically vary in quality and make-up from the time of one data collection to the time of another, the quality of the effluent discharged will also vary considerably. For example, in the instant case, one of the offshore platforms viewed by the EPA to be using BPCTCA in the treatment of deck drainage showed effluent qualities ranging from 1.1 mg/1 on June 3 and June 24 to 86.5 mg/1 on June 27, with considerable variance between these two figures. In computing a standard based on a "confidence interval" of, say, 99 percent, the EPA determines a mg/1 standard that the effluent discharge data from an exemplary plant can be expected to fall inside of 99 percent of the time. The EPA used "confidence intervals" of 99 percent and 97.5 percent, respectively, in setting the produced water and deck drainage limitations under challenge here.

This standard requires that the daily effluent of produced water be less than 9.25 pounds per day (564 F.2d at 1259). This figure was obtained by testing the average daily effluent of several plants with the best practicable control technology currently available, computing the mean and standard errors, and then computing the upper endpoint of the confidence interval.

Assuming that one standard error of the mean is 0.70 pounds per day, state the entire confidence interval for the produced water limitations.

Section 301(b)(1)(A) of the Federal Water Pollution Control Act (33 U.S.C.A. §§1251 et seq.) requires that the EPA limit discharges to those achievable by use of BPCTCA technology. In *Marathon Oil*, plaintiffs urged that the standard set by the use of 99% confidence intervals was stricter than that allowed by §301. Evaluate this argument by considering whether a plant conscientiously using the best practicable control technology currently available would be able to comply with the standard 100% of the time.

UNITED STATES v. GENERAL MOTORS CORP.
377 F. Supp. 242 (D.D.C. 1974)

GASCH, District Judge.

This matter is before the Court on plaintiff's motion for summary judgment and defendant's opposition thereto. Plaintiff's complaint, filed on November 6, 1970, is the first action of its kind brought pursuant to the National Traffic and Motor Vehicle Safety Act of 1966, as amended (Act of September 9, 1966, 80 Stat. 718 et seq., 15 U.S.C. §1381 et seq.) (hereinafter referred to as the Act). Specifically, plaintiff seeks a judgment declaring that defendant has violated Section 108(a)(4) (15 U.S.C. §1397(a)(4)) of Title I of the Act by failing to issue safety defect notifications in regard to 15 × 5.50 Kelsey-Hayes disc wheels (hereinafter referred to as WHEELS).

It was the determination of the Director of the National Highway Safety Bureau (hereinafter referred to as the Director), acting pursuant to Section 113(e)(2) (15 U.S.C. §1402(e)(2)), that the WHEELS contain a "defect which relates to motor vehicle safety." Accordingly, by letter dated November 4, 1970, the Director notified the defendant of his determination and directed General Motors to furnish the notification specified in Section 113(c) of the Act (15 U.S.C. §1402(c)).

Defendant counterclaims for judicial review of this determination by the Director and asks this Court to declare the November 4, 1970, determination and direction unlawful and void. Further, defendant seeks an injunction restraining the enforcement of the Director's order.

I. Factual Background

Defendant is a corporation organized under the laws of Delaware and a "manufacturer" within the meaning of Section 102 of the Act (15 U.S.C. §1391). During the years 1960 through 1965, defendant manufactured approximately 321,743 three-quarter ton GMC and Chevrolet pickup trucks (hereinafter referred to as TRUCKS). It is estimated that approximately 200,000 of these TRUCKS were equipped with the WHEEL which was optional equipment.[6]

This action has its origin in 1968 when the National Highway Safety Bureau (NHSB) (now the National Highway Traffic Safety Administration), following up on reports of failures of these WHEELS, requested information from the defendant in connection with vehicles equipped with the WHEEL. As a result of this preliminary investigation, NHSB issued an Investigation Report (Part I) on April 2, 1969. At this time, the report noted that failures had been reported to both NHSB and General Motors and attention was focused on alleged "flaws" or "cracks" in the gutter that might be a cause of the failure of

---

6. The reason that such a high percentage of purchasers chose this optional equipment, it has been submitted, is that the tires could be installed, changed, or repaired in the field without resort to a service station. This factor appealed to the camper enthusiast. Consequently, it has been estimated by defendant that 810,000 of these WHEELS were installed on the TRUCKS.

the WHEEL. The defendant, by letter of April 19, 1969, argued that there was no manufacturing defect and further that the number of failures, admitted by General Motors to be "significant" could be solely attributed to owner abuse through overloading.[7]

A. INITIAL ADMINISTRATIVE ACTIONS

1. The Letters and the "Settlement"

Notwithstanding defendant's contention then, and its contention now, that owner abuse through overloading was the cause of the WHEEL failures, General Motors saw fit to inform owners of the TRUCKS, in a letter of May 28, 1969, that there existed "a serious safety risk in certain 15 × 5.50 three-piece disc wheels." Furthermore, in this letter defendant noted that

> [T]he problem is serious under conditions of heavy loading, such as when the truck is equipped with an over-the-cab type camper. Specifically, if one of these trucks is overloaded and the tires are overinflated, these wheels are subject to sudden fracturing and breaking apart during use. Such occurrences result in immediate loss of air from the tire, which can come off the wheel, possible loss of control of the vehicle, and substantial risks of serious harm to persons and property in the vicinity. . . . In view of the potential safety hazard arising from the continued use of three-piece wheels if you have an over-cab camper or other heavy special body on your truck, or may attach one in the future, we urge that you replace the wheels as soon as possible. Also, if your truck was purchased as a used vehicle and you are unable to determine that it was equipped with a camper or other special body or not overloaded by owners prior to your purchase, a prudent course of action would be to replace the wheels. . . .

This May 28 letter was mailed to 280,000 TRUCK owners. Upon further investigation, however, the Acting Federal Highway Administrator, by letter of August 25, 1969, notified defendant that a preliminary determination had been made that a defect relating to motor vehicle safety existed with respect to the TRUCKS equipped with the WHEELS.

Defendant was offered an opportunity to respond to the allegations as provided by Section 113(e) (15 U.S.C. §1402(e)). Taking advantage of this opportunity, defendant, by letter of September 11, 1969, while denying the existence of a defect and again asserting

---

7. It is worthy of note that General Motors, upon consideration of the "significant" number of WHEEL failures, cancelled the use of the WHEEL at the end of the 1965 model year and substituted a 16-inch wheel and tire which increased the rear-wheel load-carrying ability by 20 percent within the same gross vehicle weight.

owner abuse as the cause of the failures, made the candid admission that

> [A] brief period of excess loading can cause a crack to occur in a wheel. Thereafter, such a crack may develop to the point of wheel failure with further use of the wheel even under proper load.

Prior to the final determination by the government, the defendant made a settlement offer which would include the replacement, at defendant's expense, of the WHEELS on all TRUCKS *equipped with campers or other special bodies.* This settlement was agreed upon and the Section 113 proceeding was properly closed. However, the agency reserved its right to take further action "if it becomes necessary in the interests of safety."

Commencing October 7, 1969, the defendant mailed a second letter to TRUCK owners offering to replace the WHEELS in accordance with the terms of the settlement. In carrying out this agreement, defendant gave the owners the option of selecting as a replacement a 15 × 5.50 two-piece wheel which General Motors described as "having an increased safety factor over the three-piece 15 inch wheel," or a 16 × 6.00 two-piece wheel. It is noteworthy that the 15-inch replacement wheel had exactly the same tire-wheel combination load carrying capacity as the Kelsey-Hayes three-piece disc wheel.

The termination of the Section 113 proceeding was announced on October 9, 1969. The defendant, pursuant to government request, stated that as of February 23, 1970, it had replaced 22,452 WHEELS in accordance with the October 7, 1969, offer. As of the present, it has been reported by General Motors that 62,229 WHEELS have been replaced. . . .

II. SUMMARY JUDGMENT

On the basis of the record before this Court, plaintiff has urged the granting of summary judgment. Prior to discussing at length the facts which plaintiff alleges establish that there is no genuine issue of material fact, it is necessary to examine the statutory context and statutory grounds upon which plaintiff bottoms its motion. This examination is undertaken to define precisely what plaintiff must establish to meet its burden of proving the existence of a defect. In view of the fact that this is the first case to raise this question, it is of utmost importance to examine the Act carefully to determine Congressional intent. . . .

The conclusion of the Court, therefore, upon a reading of the entire statute and the plain language of the Act is that Congress meant to have notifications issued by the manufacturer if there existed a "large number of failures." The language "defect in performance" forcefully demands such a result. . . .

[I]t appears to this Court that one very clear way that a "defect in performance" can be indicated is by looking to the number of failures. Inasmuch as the overriding purpose of the Act is uniform and prompt notification of defects, it is beyond peradventure that plaintiff's "large number of failures" theory supports that Congressional purpose. . . .

Accordingly, the examination of facts upon which plaintiff bases its summary judgment motion, infra, will be directed at discovering whether plaintiff has shown that no genuine issue of material fact exists as to the occurrence of a "large number of failures" and if so, whether defendant has a ground of defense which raises a genuine issue of material fact which must be tried.

C. THE FACTS SUPPORTING PLAINTIFF'S MOTION FOR SUMMARY JUDGMENT AND GENERAL MOTORS' DEFENSE

After a careful review of this voluminous record, it is the Court's conclusion that plaintiff has made a prima facie case that a "large number of failures" do exist thus establishing, under this Court's test, a safety-related defect.

The primary undisputed facts that lead the Court to this conclusion are the following. First, it is undisputed that General Motors received 2,361 reports of failures. Perhaps more importantly, the defendant states in its pleadings that "General Motors has never denied that a significant number of wheels failed."

Additionally, the government has conducted a stratified random sampling of a population constructed from the 1,319 owners of TRUCKS identified by General Motors as having reported WHEEL failures. That study conducted by Dr. G. Koch had as its goal the collection of affidavits from sample owners detailing whether or not WHEEL failures had actually occurred on any TRUCK owned by a person in the sample population. Further, once such affidavits were obtained that information was utilized to draw inferences as to the overall population.

Dr. Koch and the government did succeed in gathering 154 affidavits which revealed 393[26] actual WHEEL failures. Moreover, based

---

26. Altogether, with the assistance of the *amici curiae*, 160 affidavits are on record revealing 436 actual WHEEL failures. These affidavits are attached to the Plaintiff's Motion for Summary Judgment.

on the random sampling and the use of 95 per cent confidence intervals, Dr. Koch concluded that: (1) there would be at least 670 owners taken from the list of those who reported WHEEL failures to General Motors who would be willing to furnish affidavits regarding the failure of WHEELS on TRUCKS if they had been contacted for that purpose; this group would have reported 1,400 failures; and (2) based upon the list of owners who had reported WHEEL failures to the agency combined with the list furnished by General Motors, 707 owners on the combined list, if asked, would have furnished affidavits detailing some 1,503 WHEEL failures.

The 160 affidavits that are attached to the government's motion report 436 WHEEL failures. The most crucial factor in determining that summary judgment should be granted to plaintiff is that defendant has *not disputed* in any significant respect the authenticity of these affidavits. In fact, Ralph Morrison, Manager of General Motors Field Campaigns Service Department, has admitted through deposition that based upon the over 2,300 reported failures, it is more likely than not that 700 TRUCK owners would have had failures.

The Court concludes that the facts outlined above, and a review of the entire record herein, establishes a prima facie case that a "large number of failures" were in fact suffered, thus establishing a safety-related defect.[28]

The next question that the Court must address is whether the defendant has raised a genuine issue of material fact as to the aforementioned considerations.

The defendant did file an affidavit by a Dr. Joskow, a statistician, who concluded, based on information and belief, that the owner affidavits, attached to the government's motion, might be inaccurate because the owners might not recall what it was they experienced and, additionally, because the government personnel assisting in the accumulation of affidavits might be biased thus tainting the information. The Court is unable to conclude that Dr. Joskow's affidavit raises any genuine issue of material fact. Dr. Joskow does not question the methods used to make Dr. Koch's projections, only the *possible* inaccuracies that *might* have arisen due to the methods of accumulating the affidavits. Nowhere in the record is there a hint of any *factual* information indicating that the government affidavits were incorrect or otherwise contained false statements. . . .

---

28. The question as to what constitutes a "large number of failures" must necessarily be left to the determination of the Court on the merits of each individual case. The terminology "large" is, of course, relative and on the facts and surrounding circumstances of this action, the Court is convinced that the number of failures shown must be found to be "large" as a matter of law.

III. CONCLUSION

It is the opinion of the Court, therefore, that the uncontradicted evidence establishes that the WHEEL fails unpredictably and catastrophically in large numbers. The affidavits demonstrate the existence of this danger and this danger exists regardless of whether the cause of the failures has been found. It is admitted that the WHEELS fail even following a short period of overloading and it is unnecessary, under this Court's view of the statute, for the government to detail the loading history of each TRUCK and WHEEL.

Therefore, the Court holds that plaintiff has established de novo as a matter of law that the WHEELS contain a defect which relates to motor vehicle safety, within the meaning of the Act, in that the WHEELS are subject to sudden and catastrophic failure resulting in accidents and injuries to persons using the highways.

Defendant has presented no evidence negating this proof or sufficient to raise a genuine issue as to there being a large number of failures.

Accordingly, the Secretary of Transportation was warranted in finding, and did not abuse his discretion in so doing, that the WHEELS contain a defect which relates to motor vehicle safety. The Secretary was likewise correct in holding that purchasers of the TRUCKS equipped with the WHEELS are entitled to the notification provided by the Act pursuant to the November 4, 1970, letter.

It appearing to the Court that defendant was directed by the Secretary to furnish the statutory notification, and defendant has failed in compliance, this Court must conclude that plaintiff is entitled, as a matter of law, to summary judgment in this case.

Notes and Problems

1. A clear opportunity for statistical proof arose in United States v. General Motors Corp. when the court decided that proof of "a *large number* of wheel failures" is enough to require issuance of a safety defect notification. Whether in fact a large number had occurred was determined by use of a "stratified" sample of owners of the three-quarter ton pick-up trucks. A stratified sample is one of three broad categories of samples: simple random samples, clustered samples, and stratified samples. A *simple random sample* of a population is one in which every item in the population has an equal probability of being selected and each item is selected independently of whether any other particular item is selected. A *clustered sample* differs from a simple random sample in that items are drawn from the population

into the sample in groups, rather than individually. For instance, we might sample the truck owners by interviewing 1319 of the truck owners in Detroit, which would be less costly than interviewing 1319 truck owners scattered all over the country. A clustered sample will be an accurate reflection of the population as a whole only if the group chosen to be sampled is similar to the population as a whole in its relevant characteristics. A *stratified sample* differs from a simple random sample in that "subsamples" are taken from preselected portions of the population. The stratified sample is made up of simple random samples of the subpopulations. These subgroups are selected to give maximum accuracy to the sample by ensuring that all important groups are represented. Thus a stratified sample of truck owners would ensure that some city dwellers, some rural drivers, some Northerners and some Southerners were selected, while a simple random sample might, just by chance, ignore one of these groups.

The choice of sample type is up to the researcher. The choice between simple random and clustered samples is a tradeoff between cost and accuracy; the more homogeneous the cluster is and the more unlike the population it is, the more inaccurate the sample. The stratified sample is more accurate than a simple random sample because it incorporates some facts already known about the population. But it may be particularly costly to obtain useful information about which subgroups of the population are important and where they are located.

2. The stratified sample was made in order to establish a confidence interval for the total number of failures. The feature of this sample that distinguishes it from the air-quality data in *Reserve Mining Co.* is that one of the variables, whether an affidavit was returned or not, is a *categorical variable*. The measurement of a categorical variable indicates whether an item, a truck owner surveyed, is in the "affidavit returned" category or not. We give the categorical variable a measurement of one if an item is in the category and zero if it is not. The data from a hypothetical sample of ten owners might look like Exhibit VI-4. Six of the ten owners surveyed furnished affidavits revealing a total of nine failures.

Because we have a categorical variable (affidavit returned) and we want to project the expected percent of the population that would furnish an affidavit, we must use a slightly different approach to calculating the standard error of the mean. The formula is:

$$s.e._{proportional\ mean} = \sqrt{\frac{P_1 \times P_2}{n-1}}$$

where P_1 is the percentage reporting, P_2 is the percentage not reporting, and n is the number in the sample. Using this formula for the

EXHIBIT VI-4.
Affidavit Returned

Owner No.	Yes = 1 No = 0	Failures Revealed
1	1	2
2	1	0
3	0	0
4	1	3
5	1	1
6	0	0
7	0	0
8	1	2
9	0	0
10	1	1

standard error of the proportional mean, construct a 95% confidence interval around the mean for the ten-owner sample above. You should get an interval of 0.6 ± 1.96 × 0.16, or 0.6 ± 0.31, which means that between 29% and 91% would return affidavits. Given this interval, how many of the 2,361 owners reporting failures to General Motors would submit affidavits?

Also estimate the number of failures that this group would report. For this estimate we do not have a categorical variable. The number of failures reported can take on any positive value. So we must use the more complicated formula for a sample variance:

$$s.e.^2 = \frac{\Sigma(x_i - m)^2}{n - 1}$$

and then calculate the mean standard error:

$$s.e._m = \sqrt{\frac{s.e.^2}{n}}$$

Note: The Use of Student's *t* for Small Samples

1. As is true for chi-square tests, samples of small size create inaccuracies in calculations of confidence intervals. The reason is that means of different small samples of a population are likely to deviate more from one another than are means of large samples from the same population. Because the standard deviation (error) of the sample mean is used in computing the confidence interval, a substitute for the Z score that takes sample size into account must be used whenever the sample is small. The substitute is found on Student's *t* table, Exhibit VI-5. Because the *t* table has values for large and small

EXHIBIT VI-5.
Critical Values for Student's t

Degrees of Freedom	Significance Level One-Tailed Test											
	.45	.40	.35	.30	.25	.20	.15	.10	.05	.025	.01	.005	.0005
	Two-Tailed Test											
	.90	.80	.70	.60	.50	.40	.30	.20	.10	.05	.02	.01	.001
1	.158	.325	.510	.727	1.000	1.376	1.963	3.078	6.314	12.706	31.821	63.657	636.619
2	.142	.289	.445	.617	.816	1.061	1.386	1.886	2.920	4.303	6.965	9.925	31.598
3	.137	.277	.424	.584	.765	.978	1.250	1.638	2.353	3.182	4.541	5.841	12.924
4	.134	.271	.414	.569	.741	.941	1.190	1.533	2.132	2.776	3.747	4.604	8.610
5	.132	.267	.408	.559	.727	.920	1.156	1.476	2.015	2.571	3.365	4.032	6.869
6	.131	.265	.404	.553	.718	.906	1.134	1.440	1.943	2.447	3.143	3.707	5.959
7	.130	.263	.402	.549	.711	.896	1.119	1.415	1.895	2.365	2.998	3.499	5.408
8	.130	.262	.399	.546	.706	.889	1.108	1.397	1.860	2.306	2.896	3.355	5.041
9	.129	.261	.398	.543	.703	.883	1.100	1.383	1.833	2.262	2.821	3.250	4.781
10	.129	.260	.397	.542	.700	.879	1.093	1.372	1.812	2.228	2.764	3.169	4.587
11	.129	.260	.396	.540	.697	.876	1.088	1.363	1.796	2.201	2.718	3.106	4.437
12	.128	.259	.395	.539	.695	.873	1.083	1.356	1.782	2.179	2.681	3.055	4.318
13	.128	.259	.394	.538	.694	.870	1.079	1.350	1.771	2.160	2.650	3.012	4.221
14	.128	.258	.393	.537	.692	.868	1.076	1.345	1.761	2.145	2.624	2.977	4.140
15	.128	.258	.393	.536	.691	.866	1.074	1.341	1.753	2.131	2.602	2.947	4.073

16	.128	.258	.392	.535	.690	.865	1.071	1.337	1.746	2.120	2.583	2.921	4.015
17	.128	.257	.392	.534	.689	.863	1.069	1.333	1.740	2.110	2.567	2.898	3.965
18	.127	.257	.392	.534	.688	.862	1.067	1.330	1.734	2.101	2.552	2.878	3.922
19	.127	.257	.391	.533	.688	.861	1.066	1.328	1.729	2.093	2.539	2.861	3.883
20	.127	.257	.391	.533	.687	.860	1.064	1.325	1.725	2.086	2.528	2.845	3.850
21	.127	.257	.391	.532	.686	.859	1.063	1.323	1.721	2.080	2.518	2.831	3.819
22	.127	.256	.390	.532	.686	.858	1.061	1.321	1.717	2.074	2.508	2.819	3.792
23	.127	.256	.390	.532	.685	.858	1.060	1.319	1.714	2.069	2.500	2.807	3.767
24	.127	.256	.390	.531	.685	.857	1.059	1.318	1.711	2.064	2.492	2.797	3.745
25	.127	.256	.390	.531	.684	.856	1.058	1.316	1.708	2.060	2.485	2.787	3.725
26	.127	.256	.390	.531	.684	.856	1.058	1.315	1.706	2.056	2.479	2.779	3.707
27	.127	.256	.389	.531	.684	.855	1.057	1.314	1.703	2.052	2.473	2.771	3.690
28	.127	.256	.389	.530	.683	.855	1.056	1.313	1.701	2.048	2.467	2.763	3.674
29	.127	.256	.389	.530	.683	.854	1.055	1.311	1.699	2.045	2.462	2.756	3.659
30	.127	.256	.389	.530	.683	.854	1.055	1.310	1.697	2.042	2.457	2.750	3.646
40	.126	.255	.388	.529	.681	.851	1.050	1.303	1.684	2.021	2.423	2.704	3.551
60	.126	.254	.387	.527	.679	.848	1.046	1.296	1.671	2.000	2.390	2.660	3.460
120	.126	.254	.386	.526	.677	.845	1.041	1.289	1.658	1.980	2.358	2.617	3.373
∞	.126	.253	.385	.524	.674	.842	1.036	1.282	1.645	1.960	2.326	2.576	3.291

Source: R. Fisher and F. Yates, Statistical Tables for Biological, Agricultural and Medical Research 46 (Table III) (6th ed. 1963). Reprinted with permission.

numbers, you can always use this t value instead of the critical Z score. To find the appropriate critical t score, you must determine (1) what level of confidence you require, (2) what the appropriate number of degrees of freedom is, and (3) whether you want to have a one-tailed or two-tailed confidence interval.

The degrees of freedom for the t statistics are the same in the confidence interval application as for the χ^2 statistic: degrees of freedom equals n minus one, where n is the sample size.

So far we have calculated only two-tailed confidence intervals. As with the Z test, it is possible to make one- or two-direction comparisons with confidence intervals. Whenever we are searching for a range within which the population measure falls, we estimate an interval on either side of that figure and we calculate values both above and below the sample mean. Because we are searching for a 95% interval on either side of the mean, the interval has "two tails," that is, we look in both directions. A two-tailed interval for the sample mean percentage of truck owners returning affidavits (using the t table instead of Z, because we have only 10 observations) would give an interval of $m \pm t_{critical} \times s.e._m$, where the critical t value is for a two-tailed test, the significance level is .05, and there are ten minus one equals nine degrees of freedom. The table, under .05 (two-tailed) significance level and nine degrees of freedom, shows 2.262 as the critical t value. So the interval is $0.6 \pm 2.262 \times 0.16 = 0.6 \pm .36$, or from 0.24 to 0.96. The mean and standard error of the mean do not change. The only difference is that the critical t is substituted for the Z.

Note that the interval is bigger using the t value. This is because small samples give less precise results. The larger the sample, the closer the t value gets to the Z value and the more identical the resulting interval. Many statisticians use $n = 30$ as the cutoff point below which one should use the t value, but in reality one can always use the t value instead of Z. Using the t value just doesn't make much difference when n is large.

2. Using the t table, recalculate the interval for the estimated number of failures that would be reported by the truck owners.

3. Using the general t formula for the two-tailed confidence interval:

$$M = m \pm t_{critical} \times s.e._m$$

recalculate the confidence intervals for the problems in Notes 2 and 3 following Reserve Mining Company v. EPA (pages 247-248), where the hypothetical sample size was only three.

C. Confidence Intervals   261

Problems: A.B.G. Instrument & Engineering, Inc. v. United States

1. A familiarity with the proportional approach to confidence intervals, as described in the previous problems, enables one to test whether a delivery from a seller meets specifications contractually required by a buyer. In A.B.G. Instrument & Engineering, Inc. v. United States, 593 F.2d 394 (Ct. Cl. 1979), the Court of Claims discussed the inspection standards for nose fuze adapters, an artillery shell component. Under the criterion specified in the standards, lots ranging in quantity from 10,001 to 35,000 pieces required a sample size of 315 pieces. The opinion does not say what proportion of the shipment must be defective for the shipment to be rejected, but let us assume that any more than 5% defective is a reasonable standard for rejection. One issue in that case was whether the government was entitled randomly to select and test only 20 units *and* reject the shipment of 20,000 units if all 20 in the sample were faulty. The court stated, inter alia, that "it is a fair bet that a larger sample would have revealed an even greater number of disqualifying defects"; the court found for the government.

Using confidence intervals, we can determine whether proof of 20 defects violates the more-than-5% rejection standard. If the remaining, unsampled portion of the 315 required units were perfect, can we still say with a great enough certainty that the shipment is rejectable? Construct a 99% confidence interval for a sample of 315, of which 20 (or 6.35%) are defective. Remember that you have a categorical variable and are testing a proportion. What if the standard called for a 10% significance level? A 34% significance level? As noted above, the opinion did not reveal either the confidence level or the proportion required for rejection. If the rejection standard were 1%, at 95% certainty could we reject the shipment, given our 20-unit sample?

2. The example above presents an opportunity to calculate a one-tailed confidence interval. The type of question answered by a one-tailed interval is, "Can we be 95% certain that the population mean is greater than X?" where X is the legal maximum (5% defective in our example). Notice that we are only looking at "greater than," so we are not trying to calculate a range on either side of the sample mean. We have a sample mean of 0.0635. The standard error of that mean is 0.014 using the proportional formula. The critical t for a one-tailed test at a .05 significance level is the same as the Z for such a large number, 1.645. The lower end of a one-tailed, 95% confidence

interval is the sample mean minus $t_{critical}$ times the standard error of the mean, or:

$$0.0635 - 1.645 \times 0.014 = 0.0635 - 0.0230 = 0.0405.$$

Since 0.0405 is less than the legal maximum of .05, we cannot say that, based on a one-tailed test, we can be 95% certain this shipment violates the standard. Could we be 80% certain?

3. This approach has broad applicability. Consider a pollution example like the one with which we started this chapter (see page 234). The mean for a sample of six days was 1.9 tons per day of emission. If the standard error of the mean was 0.3 tons, could we be 90% sure that the actual daily average was greater than a statutory maximum of 1.5 tons? If there were a legal *minimum* average amount of pollution of 2.4 tons per day, could we be 95% certain that the sampled pollution was *less* than the legal standard? This latter question involves finding the upper end of a one-tailed confidence interval.

PRESSEISEN v. SWARTHMORE COLLEGE
442 F. Supp. 593 (E.D. Pa. 1977)

BECHTLE, District Judge.

INTRODUCTION

Plaintiff is a former Assistant Professor in the Education Program at Swarthmore College ("Swarthmore"). On February 29, 1972, plaintiff received notification from Swarthmore that she would not be reappointed as an Assistant Professor for the 1972-73 academic year due to "logistical" considerations. After exhausting the appropriate administrative remedies, plaintiff commenced this class action alleging that this nonrenewal was based solely on account of her sex, in violation of, inter alia, the Equal Employment Opportunity Act of 1972, Pub. L. No. 92-261, 86 Stat. 103 (1972), 42 U.S.C. §2000e et seq. (Supp. V), amending, Civil Rights Act of 1964, 78 Stat. 253. . . .

PROMOTIONS

In their attempt to establish a prima facie case that Swarthmore has a policy of discriminating against women with respect to promotions, the plaintiffs rely upon three types of evidence—statistical evidence, evidence of the promotion procedures and individual examples which seek to bolster the statistical evidence. Before the Court addresses the latter two types of evidence, we believe it is appropriate to discuss initially not only plaintiffs' statistical evidence, but in addition, defendants' statistical evidence and plaintiffs' rebuttal evidence with respect to the statistics.

The ranks of the faculty at Swarthmore from junior to senior are Instructor, Assistant Professor, Associate Professor and Professor. Part-time faculty not appointed to regular rank are designated Lecturer. In their case in chief, the plaintiffs offered the following statistical evidence: (1) The differences between men and women in time from receipt of the highest degree to promotion or appointment to Instructor at Swarthmore, between the 1966-67 and 1975-76 academic years, are as follows: The median time for males is 2.5 years; the median time for females is 3 years. The mean time for males is 3.2 years; the mean time for females is 5.7 years. The standard deviation for males is 2.36 years; the standard deviation for females is 5.2 years.

(2) The differences between men and women in time from receipt of the highest degree to promotion or appointment to Assistant Professor at Swarthmore, between 1966-67 and 1975-76, are as follows: The median time for males is 3 years; the median time for females is 4 years. The mean time for males is 3.3 years; the mean time for females is 5.8 years. The standard deviation for males is 2.18 years; the standard deviation for females is 4.82 years.

(3) The differences between men and women in time from receipt of highest degree to promotion or appointment to Associate Professor at Swarthmore, between 1966-67 and 1975-76, are as follows: The median time for males is 7 years; the median time for females is 9 years. The mean time for males is 7.6 years; the mean time for females is 16.4 years. The standard deviation for males is 3.77 years; the standard deviation for females is 14.03 years.

(4) The differences between men and women in time from receipt of highest degree to promotion or appointment to Professor at Swarthmore, between 1966-67 and 1975-76, are as follows: The median time for males is 13 years; the median time for females is 15 years. The mean time for males is 13.9 years; the mean time for females is 14.7 years. The standard deviation for males is 5.54 years; the standard deviation for females is 2.52 years.

Dr. deCani was unable to make any statistical examination of the above data as far as statistical significance is concerned. He did comment, however, that, in his view, the above exhibits demonstrated that, for each rank, the average time to promotion or initial appointment from receipt of highest degree was longer on the average for women than for men on the faculty at Swarthmore, from 1966-67 to 1975-76.

Notes and Problems

1. In addition to estimating population means, confidence intervals can be used to determine whether there is a significant difference between two means. In Presseisen v. Swarthmore College, the plaintiff's expert, Dr. deCani, calculated the means and standard deviations for career variables for men and women in different faculty positions at Swarthmore College. The opinion states that "Dr. deCani was unable to make any statistical examination of the above data as far as statistical significance is concerned." With our knowledge of confidence intervals we are equipped to make that examination.

Using the mean time from receipt of the highest degree to promotion or appointment to Assistant Professor at Swarthmore and the associated standard deviations, construct a 95% confidence interval for males and another for females. Assume that 20 males and ten females have been promoted or appointed to this rank. Do the intervals overlap? Think about the meaning of confidence intervals. What would an overlap of the intervals indicate?

2. Calculate comparable confidence intervals for the other data presented by the plaintiff. If you were the plaintiff's attorney, would you have encouraged Dr. deCani to make a statistical examination of whether there was a significant statistical difference between the means for men and women? Assume that ten males and ten females were promoted to Instructor, 30 males and 12 females were promoted or appointed to Associate Professor, and 40 males and five females were promoted or appointed to Professor during the relevant time period.

3. Assuming that the average time for males to be promoted to the rank of Associate Professor was 8.0 years, can we be 95% certain that the length of time it took females to be promoted to that rank was longer? 99% certain? 90% certain?

D. CORRELATION COEFFICIENTS

Up to this point, we have tested only the significance of deviations of observed values of a particular variable from expected values of the same variable. But there are also statistical methods that allow us to examine the relationships between different variables. These methods are particularly useful when the value of one variable gives

D. Correlation Coefficients

us an indication of the value of the other variable. Two variables are said to be *correlated* if knowing the value of one helps us to predict the value of the other. Knowing the circumference of someone's head, for instance, helps one to infer hat size. Likewise, knowing someone's age helps one to predict remaining life expectancy.

When an increase in the value of one variable, e.g., head circumference, leads us to expect an increase in the value of the other, e.g., hat size, we say that there is a *positive correlation* between the two or that the variables are *directly related*. When an increase in the value of one variable, e.g., age, leads us to expect a decrease in the value of the other, e.g., remaining life expectancy, there is a *negative correlation* between the two variables, an *inverse relationship*.

If we were to graph the values of the two variables, the curve connecting the points on the graph would slope upwards (as you look from left to right) if the variables are positively correlated, as in Exhibit VI-6, and downwards if negatively correlated, as in Exhibit VI-7.

The correlation coefficient is a summary statistic that describes the relationship between two variables. If two variables are unrelated, the correlation coefficient, referred to as r, is zero. If the two variables are correlated perfectly positively, then the correlation coefficient, r, is equal to $+1$, and if the variables are perfectly negatively correlated, r equals -1.

Head circumference and hat size are perfectly positively correlated, because when head size is divided by three, that is your hat size. (I don't know why three, but that's the way it is.) Age and remaining life expectancy are negatively correlated, but not perfectly so, because there are other factors that determine life expectancy, such as health, family history, evil habits, and occupation.

We will expect r to vary between -1.00 and $+1.00$, thereby indicating the nature of the relationship. While there are several ways to calculate the correlation coefficient, the easiest to remember requires first calculating the Z statistic for each observation of each variable. Consider the following hypothetical case.

In the case of Kilroy v. Kilroy, 80 D.B.1 (1976), Kermit Kilroy ("Dad") was awarded custody of young Kelly Kilroy ("Kid"), while Karen Kruger Kilroy ("Mom") was awarded visitation rights of seven days per month. In 1976, Dad sought to have the award modified for Kid's benefit. He desired to minimize the disruptive influence of Mom's visits on Kid by cutting down her visitation rights to only two days per month.

Dad offered as proof of Mom's disruptive influence a record

266   Chapter VI.   Inferring from a Sample

EXHIBIT VI-6.
Perfect Positive Correlation

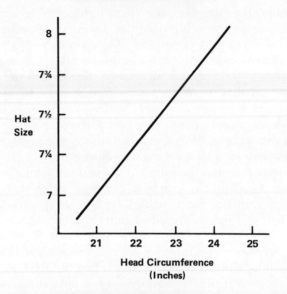

EXHIBIT VI-7.
Imperfect Negative Correlation

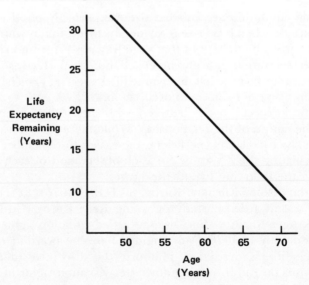

showing the length of Mom's visits and showing how many consecutive days Kid stayed home from school complaining of headaches after each of Mom's visits. The exhibit appears as follows:

EXHIBIT VI-8.
Plaintiff's Exhibit A

Year	Month	(X_i) Length of Visit (days)	(Y_i) Duration of School Absence (days)
1975	Sept.	7	15
	Oct.	3	12
	Nov.	2	6
	Dec.	5	12
1976	Jan.	4	9
	Feb.	2	3

The legal issue is whether there is a relationship between the duration of Mom's visits and Kid's school attendance. If there is a positive correlation, we might find it in Kid's interest to shorten Mom's visits. To calculate the correlation coefficient, we can use the following formula:

$$r = \frac{\Sigma(Z_{X_i} \times Z_{Y_i})}{N}$$

where r is the correlation coefficient, Z_{X_i} is the Z score for the ith observation of the variable labeled X (length of visit), Z_{Y_i} is the Z score for the ith observation of the variable labeled Y (duration of school absence), and N is the number of paired observations. Note that these observations are paired off. For instance, the length-of-visit observation for October 1975 ($X_2 = 3$) is paired with the duration-of-school-absence observation for October 1975 ($Y_2 = 12$). The Z score is the number of standard deviations a particular observation is from the mean and, you will recall, it can be positive or negative, depending on whether it is above or below the mean. The Z scores are:

$$Z_{X_i} = \frac{X_i - M_X}{SD_X} \quad \text{and} \quad Z_{Y_i} = \frac{Y_i - M_Y}{SD_Y}$$

To calculate r, first calculate the Z score for each X_i and Y_i. To do so, we must know the means and standard deviations for X and Y. The means are:

$$M_X = \frac{\Sigma X_i}{N} \quad \text{and} \quad M_Y = \frac{\Sigma Y_i}{N}$$

268   Chapter VI.   Inferring from a Sample

while the standard deviations are:

$$SD_X = \sqrt{\frac{\Sigma(X_i - M_X)^2}{N}} \quad \text{and} \quad SD_Y = \sqrt{\frac{\Sigma(Y_i - M_Y)^2}{N}}$$

For Plaintiff's Exhibit A, the means, standard deviations, and Z scores are shown in Plaintiff's Exhibit B:

EXHIBIT VI-9.
Plaintiff's Exhibit B

Month	X_i's	$X_i - M_X$	$(X_i - M_X)^2$	Z_{X_i}	Y_i's	$Y_i - M_Y$	$(Y_i - M_Y)^2$	Z_{Y_i}
Sept.	7	3.2	10.2	1.8	15	5.5	30.3	1.4
Oct.	3	−0.8	0.6	−0.4	12	2.5	6.3	0.6
Nov.	2	−1.8	3.2	−1.0	6	−3.5	12.3	−0.9
Dec.	5	1.2	1.4	0.7	12	2.5	6.3	0.6
Jan.	4	0.2	0.0	0.1	9	−0.5	0.3	−0.1
Feb.	2	−1.8	3.2	−1.0	3	−6.5	42.3	−1.6

$$\frac{\Sigma X_i}{N} = \frac{23}{6} = 3.8 \qquad \frac{\Sigma Y_i}{N} = \frac{57}{6} = 9.5$$

$$SD_X = \sqrt{\frac{\Sigma(X_i - M_X)^2}{N}} \qquad SD_Y = \sqrt{\frac{\Sigma(Y_i - M_Y)^2}{N}}$$

$$= \sqrt{\frac{18.6}{6}} \qquad\qquad = \sqrt{\frac{97.8}{6}}$$

$$= \sqrt{3.1} = 1.8 \qquad\qquad = \sqrt{16.3} = 4.0$$

Look at the list of Z scores by themselves to calculate the correlation coefficient, as in Plaintiff's Exhibit C:

EXHIBIT VI-10.
Plaintiff's Exhibit C

Observation	Z_{X_i}	Z_{Y_i}	$Z_{X_i} \times Z_{Y_i}$
1	1.8	1.4	2.5
2	−0.4	0.6	−0.2
3	−1.0	−0.9	0.9
4	0.7	0.6	0.4
5	0.1	−0.1	0.0
6	−1.0	−1.6	1.6

$$r = \frac{\Sigma(Z_{X_i} \times Z_{Y_i})}{N} = \frac{5.2}{6} = .87$$

This positive correlation of .87 indicates that Kid's school absences are directly related to the length of Mom's visit. Whether this correlation

D. **Correlation Coefficients**   269

is sufficiently high to influence the court is not a statistical question. It relates to the type of practical significance we have described as "importance." Nor does this positive correlation prove that the mother's visits *caused* Kid's absences. The correlation only demonstrates a relationship from which the court *may* infer causation.

Calculating the correlation coefficient looks like a complicated task. But even though the calculation is long, it presents no new computations other than multiplying the Z scores for the variables. There are nine steps to calculating r:

Step 1. Find the mean of the values of the first variable;
Step 2. Find the standard deviation for the values of the first variable;
Step 3. Find the mean of the values of the second variable;
Step 4. Find the standard deviation for the values of the second variable;
Step 5. Use the mean and standard deviation for the first variable to calculate the Z score for each value (observation) of the first variable;
Step 6. Use the mean and standard deviation for the second variable to calculate the Z score for each value (observation) of the second variable;
Step 7. Multiply the Z score for each value of the first variable by the Z score of the paired observation of the second variable;
Step 8. Add up the multiplied Z scores for each pair; and
Step 9. Divide the sum of the multiplied Z scores by the number of paired observations.

In statistical shorthand, these steps are:

Step 1. Calculate $M_X = \dfrac{\Sigma X_i}{N}$

Step 2. Calculate $SD_X = \sqrt{\dfrac{\Sigma(X_i - M_X)^2}{N}}$

Step 3. Calculate $M_Y = \dfrac{\Sigma Y_i}{N}$

Step 4. Calculate $SD_Y = \sqrt{\dfrac{\Sigma(Y_i - M_Y)^2}{N}}$

Step 5. For each X_i, calculate $Z_{X_i} = \dfrac{X_i - M_X}{SD_X}$

Step 6. For each Y_i, calculate $Z_{Y_i} = \dfrac{Y_i - M_Y}{SD_Y}$

Step 7. For each paired X_i and Y_i, calculate $Z_{X_i} \times Z_{Y_i}$
Step 8. Calculate $\Sigma(Z_{X_i} \times Z_{Y_i})$ and
Step 9. Calculate $r = \dfrac{\Sigma(Z_{X_i} \times Z_{Y_i})}{N}$

You will note from the statistical shorthand and the example that the nine steps are designed for calculating the correlation r coefficient for a population rather than for a sample. This should be evident from the use of M's, SD's, X's, Y's, and N's, instead of m's, $s.e.$'s, x's, y's, and n's. Substituting the sample notation, and introducing $n - 1$ for N where appropriate, the steps for the *sample* correlation coefficient, r_s, are:

Step 1. Calculate $m_x = \dfrac{\Sigma x_i}{n}$

Step 2. Calculate $s.e._x = \sqrt{\dfrac{\Sigma(x_i - m_x)^2}{n - 1}}$

Step 3. Calculate $m_y = \dfrac{\Sigma y_i}{n}$

Step 4. Calculate $s.e._y = \sqrt{\dfrac{\Sigma(y_i - m_y)^2}{n - 1}}$

Step 5. For each x_i, calculate $Z_{x_i} = \dfrac{x_i - m_x}{s.e._x}$

Step 6. For each y_i, calculate $Z_{y_i} = \dfrac{y_i - m_y}{s.e._y}$

Step 7. For each paired x_i and y_i, calculate $Z_{x_i} \times Z_{y_i}$
Step 8. Calculate $\Sigma(Z_{x_i} \times Z_{y_i})$ and
Step 9. Calculate $r_s = \dfrac{\Sigma(Z_{x_i} \times Z_{y_i})}{n - 1}$

Just as the mean for the sample is an estimate of the population mean, so also the sample correlation coefficient is an estimate of the population correlation coefficient. A confidence interval for the sample correlation coefficient is calculated in a manner similar to the confidence interval for the sample mean.[3] There is, however, a simpler method of testing the null hypothesis that no systematic relationship exists between the two variables. A null hypothesis tested by a correlation coefficient might also be stated, "The population correla-

---

3. This method is considered briefly in T. Wonnacott and R. Wonnacott, Introductory Statistics 291-295 (1969).

tion coefficient is not significantly different from zero," which would mean there is no relationship between the variables.

To test the significance of this correlation coefficient, we refer to an r table, such as appears in Exhibit VI-11. Down the side of the r table are the number of paired sample observations that underlie our

EXHIBIT VI-11.
Critical Values for the Correlation Coefficient r

Number of Pairs	Significance Level				
	.1	.05	.02	.01	.001
3	.98769	.99692	.999507	.999877	.9999988
4	.90000	.95000	.98000	.990000	.99900
5	.8054	.8783	.93433	.95873	.99116
6	.7293	.8114	.8822	.91720	.97406
7	.6694	.7545	.8329	.8745	.95074
8	.6215	.7067	.7887	.8343	.92493
9	.5822	.6664	.7498	.7977	.8982
10	.5494	.6319	.7155	.7646	.8721
11	.5214	.6021	.6851	.7348	.8471
12	.4973	.5760	.6581	.7079	.8233
13	.4762	.5529	.6339	.6835	.8010
14	.4575	.5324	.6120	.6614	.7800
15	.4409	.5139	.5923	.6411	.7603
16	.4259	.4973	.5742	.6226	.7420
17	.4124	.4821	.5577	.6055	.7246
18	.4000	.4683	.5425	.5897	.7084
19	.3887	.4555	.5285	.5751	.6932
20	.3783	.4438	.5155	.5614	.6787
21	.3687	.4329	.5034	.5487	.6652
22	.3598	.4227	.4921	.5368	.6524
27	.3233	.3809	.4451	.4869	.5974
32	.2960	.3494	.4093	.4487	.5541
37	.2746	.3246	.3810	.4182	.5189
42	.2573	.3044	.3578	.3932	.4896
47	.2428	.2875	.3384	.3721	.4648
52	.2306	.2732	.3218	.3541	.4433
62	.2108	.2500	.2948	.3248	.4078
72	.1954	.2319	.2737	.3017	.3799
82	.1829	.2172	.2565	.2830	.3568
92	.1726	.2050	.2422	.2673	.3375
102	.1638	.1946	.2301	.2540	.3211

Source: R. Fisher and F. Yates, Statistical Tables for Biological, Agricultural and Medical Research 63 (Table VII) (6th ed. 1963). Reprinted with permission.

test. Returning to our hypothetical case, we must assume that the data presented by Kermit Kilroy was a random sample of Mom's visits. Dad has presented six pairs of observations. By referring to the numbers to the right of six, we can determine the significance level for our rejection of the null hypothesis.

The correlation coefficient for the sample, r_s, turns out to be .85. To the right of six pairs, the table reveals that if the r score is as low as .7293, there is a 10% chance of a type 1 error. The higher the r score (e.g., 8114, .8822, and .9172), the lower the probability of a type 1 error, specifically, a 5%, 2%, and 1% chance, respectively. The sample correlation coefficient for the data offered by Dad is .85, indicating a confidence level between 95% and 98%.

Note that the larger the number of pairs of observations, the lower the correlation coefficient Dad would need in order to prove a significant statistical relationship. For instance, if, with 52 pairs of observations, the correlation coefficient was .38, he would be more than 99% certain that the population correlation coefficient was different from zero. The question of practical significance, however, arises again.

In the cases that follow, figure out how the correlation coefficient is being used. Determine what null hypothesis is being tested and note how the court evaluates the evidence.

BOSTON CHAPTER, NAACP, INC. v. BEECHER
371 F. Supp. 507 (D. Mass. 1974)

FREEDMAN, District Judge. . . .

[The NAACP, in an action against the state civil service department, and the United States, in an action against the City of Boston, alleged a violation of the Civil Rights Act of 1964 with respect to hiring of people for fire departments. The court held that the plaintiffs made out a prima facie case of discrimination in the use of the Fire Fighter Entrance Examination [FFEE], thereby shifting the burden to the defendants to justify use of the exam. — ED.]

VALIDATION

The burden has shifted to the defendants to prove a "manifest relationship" between the exam and the job. [Griggs v. Duke Power Co., 401 U.S. 424, 91 S. Ct. 849, 28 L. Ed. 2d 158 (1971).] In Castro v. Beecher [459 F.2d 725, at 732 (1st. Cir. 1972)], Court of Appeals stated that where the suspect classification (here the FFEE) is shown to have a racially discriminatory impact, the employer must show that the exam is in fact "substantially related" to job performance by coming forward with "convincing facts" establishing a fit between the qualifications and the job. Defendants' expert Costa testified that the employer cannot operate from some expert point of view that the tests are doing what they are supposed to do. It was his opinion that the courts and professional psychological standards always require that empirical evidence be obtained. The Court agrees.

Where it has been determined that an exam is suspect, the courts have required that a "fit" between the exam and job performance must be established by a study conducted according to professionally accepted standards. The Equal Employment Opportunity Commission Guidelines on Employee Selection Procedures, 29 C.F.R. 1607, et seq., suggest minimum standards for test validity and establish guides to be used in determining the validity of employment tests. "Courts confronted with challenges to public employment examinations predicated upon the equal protection clause of the Fourteenth Amendment have generally agreed that the Guidelines issued by the EEOC provide persuasive standards for evaluating claims of job-relatedness." [Vulcan Society v. Civil Service Commission, 360 F. Supp. 1265, 1273, n. 23 (S.D.N.Y., 1973).] And as was noted in Officers for Justice v. San Francisco, 371 F. Supp. 1328 (N.D. of Cal., 1973), at p. 1337, these Guidelines may indeed have the force of law in cases brought under Title VII of the Civil Rights Act of 1964. See *Griggs,* supra, at 433, of 401 U.S., at 854 of 91 S. Ct. where the Court stated:

> The Equal Employment Opportunity Commission, having enforcement responsibility, has issued guidelines interpreting §703(h) to permit only the use of job-related tests. The administrative interpretation of the Act by the enforcing agency is entitled to great deference. . . . Since the Act and its legislative history support the Commission's construction, this affords good reason to treat the guidelines as expressing the will of Congress.

The Civil Service defendants apparently anticipated their burden and during the pendency of these matters caused a validity study to be undertaken by Dr. Costa, who has been retained as the principal psychological testing consultant to the MDCS and was qualified by the Court to testify as an expert witness on employment testing practices. Some brief remarks about validity study procedures in general may be beneficial at this point.

There are two types of test validation studies recognized by the EEOC Guidelines. The Court will borrow from Judge Weinfeld's clear and concise descriptions of these studies in *Vulcan Society*, supra, 360 F. Supp. 1265, 1273 (1973).

> The preferred method of test validation is criterion-related or empirical validity, which includes what are referred to as the predictive and concurrent methods of validation. Predictive validation consists of a comparison between the examination scores and the subsequent job performance of those applicants who are hired. If there is a sufficient correlation between test scores and job performance, the examination is considered to be a valid or job-related one. Concurrent validation requires the administration of the examination to a group of current employees and a comparison between their relative scores and relative performance on the job.

Predictive validation is preferred where feasible over concurrent validation. A method less preferable than criterion related validity is known as "content validity." Again borrowing from Judge Weinfeld in *Vulcan*, supra, at p. 1274:

> Content validation is less preferable than the criterion-related methods of test validation but nevertheless a professionally accepted means of establishing job-relatedness where the more desirable empirical methods are impractical. [See EEOC Guidelines, 29 C.F.R. §1607.5(a).] An examination has content validity if the content of the examination matches the content of the job. For a test to be content valid, the aptitudes and skills required for successful examination performance must be those aptitudes and skills required for successful job performance. It is essential that the examination test these attributes both in proportion to their relative importance on the job and at the level of difficulty demanded by the job.

For either type of study, it is important that a careful job analysis be undertaken. The criteria used for an empirical validation study should be important criteria selected on the basis of a thorough job analysis. Likewise, a thorough knowledge of the job to be tested is necessary when constructing a content valid examination. "A job analysis is a thorough survey of the relative importance of the various skills involved in the job in question and the degree of competency required in regard to each skill. It is conducted by interviewing workers, supervisors and administrators; consulting training manuals; and closely observing the actual performance of the job." *Vulcan*, supra, at 1274.

Dr. Costa conducted what he described as a concurrent criteria-related validity study. A brief discussion of terms as described to the Court is in order. The mathematical relationship between the test scores and the measures of job performance is referred to as the correlation coefficient. A correlation coefficient of .00 means the study shows no relationship between the test and job performance, while a coefficient of 1.00 indicates a perfect relationship between the two. In order for a correlation coefficient to have significance, it must be both statistically and practically significant. Statistical significance means that the possibility of the results being reached by chance are minimal. Practical significance means that the coefficient shows a sufficiently high relationship between success on the test and successful job performance. Both Dr. Costa and plaintiffs' expert, Dr. Hunt, agreed that as a "rule of thumb" a coefficient of .3 would be the minimum level to indicate a satisfactory relationship. A lower coefficient would not be practically significant and would not justify use of the test.

Dr. Costa used two groups of subjects in his study. Group I consisted of 88 white Boston firefighters who had been recently appointed to the force from the list established in April 1972 as a result of the August 1971 FFEE. Group II consisted of 134 men who had been firefighters for approximately two years. The study implemented two "Performance Appraisal Systems." System I tested for job relatedness between the MDCS job predictors; i.e., exam score, training and experience, and the final score (general average mark), and the performance of 13 tasks considered by both experts to be important criteria of the job. The criteria consisted of such tasks as ladder extension, handling a pre-connected hose (both team-type performances consisting of nine sub-tasks with each man being tested at each sub-task), and such individual performances as air mask operation, extinguisher selection, securing lines and knots, and hose and hydrant operation. Only Group I was tested in Appraisal System I. The study was not done under actual firefighting conditions, but was conducted at the Boston Fire Department's training camp on Moon Island. Appraisal System II consisted of ratings of job performance obtained from company captains and lieutenants collected under actual firefighting conditions. The men were rated on twelve separate scales including understanding buildings, constructions, and fire behavior; mechanical ability; holding up under pressure and stress; carrying out orders; and overall job effectiveness. Both Groups I and II were rated in System II.

System II did not reveal any significant correlations between the two test criteria used (written exam score and general average mark) and successful job performance in either group tested. Dr. Costa admitted that if the results of System II were all that he had available upon which to base a conclusion, he would conclude that the exam was not a valid predictor of successful job performance.

Dr. Costa concluded that System I revealed seven significant correlations. In three of these, however, the predictor was the general average mark score. Since this score is a composite of the written exam score and the training and experience score, its utility in establishing the validity of the exam itself seems questionable. In addition, two of the significant correlations are based on a comparison of the number of errors made on the four best predicted criteria and the written test score. Frankly, the Court has had difficulty understanding the significance of these correlations. Plaintiff argues it is merely a restatement of the original thirteen criteria, only two of which have been shown to have a significant correlation with the written exam. This may be so. At any rate, the Court does observe that the Air Mask Operation (.346) and the Loop Man Position (.313) are the only two job tasks which show a significant relationship with the written exam score, and Dr. Hunt described these correlations as being barely significant.

Experts have interpreted the EEOC Guidelines to require that at least one relevant criterion be both statistically and practically significant. 29 C.F.R. §1607.5(c)(1). Dr. Hunt testified that if one significant correlation is to be accepted as evidence that an exam is job related, the one relevant criterion must represent an adequate measure of total job performance factors. If the one criterion only represents a small percentage of the important factors of the job, this should not be considered adequate to support a conclusion of test validity. Although she agreed that the 13 criteria implemented were important, it was Dr. Hunt's opinion that there were only two significant correlations and that those two did not represent an adequate measure of the total factors involved in the job. Based on Dr. Costa's study, she testified that she would not conclude the exam was job related. Dr. Costa is of the opinion that the exam has been shown to be valid according to EEOC Guideline minimum requirements and he has concluded that the exam is a valid predictor of job performance.

The Court has had no little difficulty in examining the testimony of both experts and in interpreting Dr. Costa's study and the results derived therefrom. Nor has the Court relished the thought of determining which expert opinion to accept on a subject about which the Court knows very little. It is easy for a member of the legal profession, however, to understand how two competent professionals can come to different conclusions given the same set of facts. And it appears, as with the law, that the area of psychological testing is not one where elements of preciseness and exactness usually obtain. The EEOC Guidelines, 29 C.F.R. §1607.5(c)(1), caution that a test should be closely scrutinized when it is valid against only one component of job performance. The Court feels close scrutiny should also be given where the test may be shown to be valid against a few components of job performance where those components represent only a small percentage of the total job. Close scrutiny seems particularly appropriate where the exam was not designed by experts. The evidence indicated that the FFEE was designed by persons with no training in psychological measurement or testing.

The Court has concluded that defendants have not met their burden of demonstrating that the exam is "in fact substantially related to job performance." See *Castro,* supra, 459 F.2d at 732. Arguably, the test has been validated according to the EEOC Guidelines' minimum standard. However, the Court cannot conclude that the study provides the *"convincing* facts establishing a fit between the qualification and the job," which *Castro,* supra, requires. The fact that only two, or perhaps four, significant correlations were found between the exam and components of job performance — components which represent only a fraction of those duties a firefighter encounters — and the fact that those correlations were only minimally significant does not constitute "convincing" evidence of job relatedness. There was also some criticism of the way in which the study was conducted, but the Court does not feel such criticism warrants discussion in light of this determination.

The Court does not wish to imply that it is establishing standards apart from those embodied in the EEOC Guidelines. However, these Guidelines must be read in the context of certain precedent binding upon this Court (*Griggs,* supra; *Castro,* supra) which clearly implies that facial compliance with the minimum standards may not be sufficient to meet the "heavy burden" which falls upon a public employer when required to prove the validity of a selection examination which has an adverse racial impact. . . .

UNITED STATES v. CITY OF CHICAGO
385 F. Supp. 543 (N.D. Ill. 1974)

MARSHALL, District Judge. . . .

THE EVIDENCE OF JOB RELATEDNESS

Plaintiffs having proved significant statistical disparities in both the 1971 patrolman's roster and the 1973 sergeants' roster, the burden rests with the defendants to persuade the court that their selection practices are job related. A job related selection practice is one which accurately appraises the ability and fitness of the candidate to perform the job he seeks or at least is designed to make that appraisal.

Therefore, if defendants' selection practices are job related, it follows that black and Hispanic candidates have less ability and are less fit to perform the jobs of patrolman sergeant than their white counterparts. Vulcan Society, etc. v. Commission, 490 F.2d 387, 392 (2d Cir. 1973).

Defendants do not, however, urge that conclusion. Quite the contrary: at every opportunity they have stated their conviction, based upon experience in the field, that blacks and Hispanics are as qualified as whites to perform the duties of patrolman and sergeant. If the conviction is firm, one would expect defendants themselves to question the job relatedness of their selection practices. And this is particularly so in light of their recruitment practices of the mid-1960's which did not produce the white-minority disparities they are presently experiencing.

However, rather than joining in a search for the cause of the disparities to the end that they might be remedied, defendants have chosen to lead the court "deep into the jargon of psychological testing." *Vulcan Society, etc.,* supra, 490 F.2d at 394. The result has been a virtual morass of competing theories advanced by professional testers of tests in which the debate has centered on predictive, concurrent, criterion and construct validation and the court has been left with the unwelcomed task of testing the testers. It is not amiss to observe that plaintiffs have not shunned the debate. . . .

THE 1973 SERGEANTS' ROSTER

As previously noted, this roster results from a formula in which the score on the 1973 written sergeants' exam comprises 60%, departmental efficiency ratings 30% and seniority 10%. While no challenge has been made against the seniority factor, the disproportionate results of the written exam and the efficiency ratings, separately and combined, are such as to require their validation by defendants.

The 1973 Sergeants' Exam

Defendant undertook to conduct a concurrent validity study of the 1973 sergeants exam under the direction of defendant, Dr. Charles A. Pounian. The sample used in the study was a matched control group of 176 incumbent sergeants, 88 of whom were black (out of 136 black sergeants) and 88 of whom were white (out of 1234 white sergeants). The 88 white sergeants were matched with the 88 black sergeants on the basis of age and efficiency ratings. The 1973 sergeants' exam was given to this sample group. The purpose of the study was to determine if there was a correlation between the participants' performances on the exam and their performances on the job.

D. Correlation Coefficients   279

The 88 white sergeants selected for the study were not shown to be representative of white sergeants or white patrolman-sergeant candidates. The 88 black sergeants, while representative of that group, were not representative of black patrolman-sergeant candidates. Since the vast majority of white and black patrolmen fail to obtain promotion to sergeant, incumbent sergeants cannot be deemed to be representative of patrolmen who aspire to the rank of sergeant.

The criteria used in the study were the various efficiency ratings received by the sample group of incumbent sergeants during the rating period of January to June, 1973, although defendants' own expert, Dr. Guion, testified to a "great deal of mistrust in the use of administrative rating systems for criterion measures." No reliability studies were performed on the performance ratings despite the fact that in a concurrent validity study the criteria used must represent a fair measure of performance on the job.

The range of test scores for the total sergeant group was 30 to 71, on a scale of 100. The range for the black sergeants was 30 to 71, and for the whites 32 to 70. The average test score for the blacks was 53.72 with a standard deviation of 7.63 and the average for the whites was 53.91 with a standard deviation of 7.44.[9]

Test scores of the black sergeants, the white sergeants and the total group were first plotted against each sergeant's overall efficiency rating. Dr. Pounian reported statistically significant correlations, which were found, however, to be erroneous. The initial erroneous correlation coefficient between the total group's test scores and their efficiency ratings was .247 (significant at the one per cent level). The correlation coefficient between the black sergeants' test scores and their efficiency ratings was .205 (significant at the 5% level). For the white sergeants the correlation coefficient was .319 (significant at the one per cent level). Dr. Pounian also correlated the test scores of the same group with their "quality of work" subrating taken from the overall efficiency ratings used above and reported a correlation coefficient of .507 (significant at the one per cent level). However, this correlation coefficient was also erroneous.

Dr. Pounian then re-analyzed the data and recomputed the correlation coefficients. The correlation coefficient reported between test

---

9. The "average" sergeant participant in the study would not have made the lowest cut-off (i.e. composite score of 70) on the 1973 sergeants' roster unless he had an efficiency rating of 92 or above for the period January-June, 1973. The "average" participant in the study would not make the realistic cut-off of 73.9 regardless of his efficiency ratings. I.e., .60(53.91) + .30(100) + 10 = 72.35. It follows that no incumbent sergeant participant who scored below the average would be promoted. To the layman, at least, these circumstances tend to show that either the test is not related to the sergeants' job or half of the incumbent sergeants are not related to their jobs.

scores and the overall efficiency ratings was .293 for the total group; and between test scores in the subrating "quality of work" was .514. However, these correlation coefficients, like their forerunners, were found to be erroneous.

A third set of corrected correlation coefficients was submitted by Dr. Pounian. These showed comparisons of test scores with the participants' overall efficiency ratings and, in addition, comparisons with all of the subratings "quality of work," "quantity of work," "dependability," "personal relationships" and "attendance and promptness." These correlation coefficients were substantially lower.[10] Thus, of the six correlations finally computed by defendants for the entire sample of 176 sergeants who took the exam, the highest two correlations were .25 and .26. None of the third set of correlations reaches a level of practical significance in the judgment of either of defendants' expert witnesses, Dr. Pounian or Dr. Ebel.

As might be expected, the testimony of plaintiffs' expert, Dr. Barrett, was that defendants' concurrent validity study had failed to show that the sergeants' examination was a valid or useful test.

The Supreme Court has recognized that the EEOC Guidelines deserve thoughtful consideration in cases of this type and they "demand that employers using tests have available 'data demonstrating that the test is predictive of or significantly correlated with important elements of work behavior. . . .'" Griggs v. Duke Power Co., 401 U.S. at 433, 91 S. Ct. at 855, (fn. 9). The data which defendants offered regarding the 1973 sergeants' exam was the above described concurrent validity study.

The concurrent validity study failed to satisfy the minimum requirements set forth in the EEOC guidelines in three key respects. First, the criteria used as measurements of job performance were the sergeants' efficiency ratings and there was no showing that the ratings had been developed through a proper job analysis or that they otherwise possessed validity. They were used simply because they were available.

Second, the sample of incumbent sergeants was not representative of patrolmen applicants. Incumbent sergeants had already demonstrated the ability to overcome the obstacles to promotion posed by a written test. In that respect, they were nonrepresentative

---

10. After the proofs were closed on the motions for preliminary relief and plaintiffs' briefs had been filed, defendants ascertained that two answers on the 1973 exam had been "miskeyed," i.e., the Commission's "correct" answers were incorrect. This provoked still a fourth set of correlations which were submitted by mail under date of August 23, 1974. That, in turn, provoked plaintiffs to demand additional discovery. To avoid delay, the court announced that it would not consider the "final" set of correlations.

of the applicant population and particularly the minority applicant population. See 29 C.F.R. §1607.5(b)(1).

Third, the study failed in its own objective. Because of reporting errors, levels of statistical and practical significance in some correlations between test scores and efficiency ratings, which the defendants' two experts initially observed, were later corrected and reduced to levels of marginal or no practical significance. See 29 C.F.R. §1607.5(b)(5).

Thus, the concurrent validity study offered by defendants constitutes an attempt, motivated by anticipated litigation and not in conformity with the EEOC Guidelines, to show a substantial relationship between the written sergeants' exam alone and later success on the job of sergeant. The study, in fact, did the opposite. Once the study's errors were corrected, an insubstantial relationship between success on the written test and success as a sergeant was revealed.

Defendants also attempted to validate the exam by eliciting the view of one expert, in the field primarily of educational testing, that the test was "content valid." Content validity is a form of "rational," rather than "empirical" validity. The analysis depends appreciably on opinions of psychologists. See, "Application of the EEOC Guidelines to Employment Test Validation: A Uniform Standard for Both Public and Private Employers," 41 George Washington L. Rev. 505, 518 (1973).

Here defendants' own concurrent validity study shows that incumbent sergeants, who were rated well by their supervisors, did not know enough of the content on the written test to score appreciably better than the patrolmen candidates. Thus, while patrolmen applicants knew 51% of the test items on the average, incumbent sergeants knew only 53% of the test items on the average.

For a test to be content valid, the knowledge, skills and aptitudes required for successful examination performance must be the knowledge, skills and aptitudes required for successful job performance. Vulcan Society, etc. v. Commission, 490 F.2d 387, 395 (2d Cir. 1973). Any test on which successfully performing incumbent employees do not know 47% of the answers cannot reasonably be considered content valid. The written test does not begin to satisfy the EEOC Guidelines' requirement that a content-valid test consists of "suitable samples of essential knowledge, skills and behavior composing the job in question." 29 C.F.R. §1607.5(a).

From all the evidence including defendants' own studies, the 1973 sergeants' exam has not been shown to have a demonstrable relationship to successful performance as a sergeant. Use of the 1973 sergeants' eligibility list derived in principal part from that discriminatory test must, therefore, be enjoined. . . .

Notes and Problems

1. The cases of Boston Chapter, NAACP, Inc. v. Beecher and United States v. City of Chicago illustrate the use of correlation coefficients in rather complex concurrent validation tests. These studies require not only an understanding of the basic statistical techniques but also of the arcane art of examination validation. Many of the most mathematically complex of employment discrimination cases occur in this context. The ability of EEOC administrative judges to deal with these statistics is astounding to those who thought the practice of law involved primarily verbal skills. Their decisions are often incomprehensible to those without a background in statistical methods.

2. As an exercise, compute the correlation coefficient for the sample data in Exhibit VI-12. The X variable measures the distance in miles from the pollution source at which we take our measurements of airborne asbestiform fiber concentration. The Y variable measures the concentration in parts per million. We have sampled the air at four testing sites.

EXHIBIT VI-12.
Airborne Asbestiform Data Samples

Test No.	Miles from Source	Fiber Concentration
1	5	60 p.p.m.
2	50	10 p.p.m.
3	100	8 p.p.m.
4	1000	8 p.p.m.

Is the correlation coefficient you calculated statistically significant? Looking at the data, can you tell why there is not a very high correlation? In addition to the actual results of our test, which should give you some clue, the small number of observations requires a relatively high r_s in order for the correlation to be statistically significant. If we had 20 observations and came up with the same r_s, would the result be statistically significant? Would it be practically significant if the issue is whether distance from the plant has a strong relationship to concentration of asbestiform fibers? Because there might be other factors affecting concentration of fibers at various distances from the pollution source, such as the presence of other factories, the evaluation of practical significance must be carefully done, and the possibility of alternative explanations for varying concentrations must be considered. One of the limitations of the correlation coefficient is that it only compares two variables at a time. It does not take into account other factors that may be relevant. Multiple regression, the subject of the next chapter, is a statistical method for summarizing the independent effects of numerous factors.

3. We have been using the correlation coefficient to perform a two-tailed test very much like the t and Z tests. If we have a sample with 21 pairs of observations, for instance, the table of critical values for r indicates that there is a 10% chance of getting an r_s greater than .3687 or less than $-.3687$ by chance, a 5% chance of getting an r_s greater than .4329 or less than $-.4329$ by chance, and so on. This corresponds to a t or Z test that indicates the probability of having an observed number that is a given number of standard deviations above or below the expected value.

Problem: In re National Commission on Egg Nutrition

The context in which practicing attorneys use correlation coefficients is often the evaluation of scientific studies relevant to the factual issue involved in a case. In In re National Commission on Egg Nutrition, 88 F.T.C. 89 (1976), the FTC considered whether this trade association had made false, misleading, deceptive, and unfair representations regarding the relationship between eating eggs and increased risk of heart attacks or heart disease. The correlation coefficient indicates the extent to which two variables fluctuate together, and the trade association would love to have been able to prove a zero or even negative correlation. The FTC reviewed the following studies:

(1) A study of 12,000 men in seven countries revealed a correlation coefficient, r_s, of 0.73 between the mean level of dietary serum cholesterol in each country and new deaths from coronary heart disease and definite nonfatal heart attacks.
(2) The study in (1) revealed an $r_s = 0.81$ for mean level and all new diagnoses of coronary heart disease clinical manifestations.
(3) The study in (1) revealed $r_s = 0.89$ between percent of calories from fat in the diet and mean level of dietary serum cholesterol.
(4) The study in (1) revealed $r_s = 0.84$ between percent of calories from saturated fat and subsequent incidence of coronary heart disease.
(5) Dr. Connor, using data from the World Health Organization, found $r_s = 0.83$ between average daily dietary cholesterol intake in 24 countries and their coronary heart disease mortality rates.

(6) Using another set of data from the World Health Organization and the Food and Agriculture Organization, Dr. Connor found $r_s = 0.762$ between coronary heart disease death rate in 30 countries and average daily intake of dietary cholesterol.
(7) Using the same data as in (6), Dr. Connor found $r_s = 0.666$ between the coronary heart death rate and average daily intake of eggs.
(8) Using slightly different data from the same sources as Dr. Connor, Dr. Stamler found $r_s = 0.617$ between 1964 coronary heart disease mortality for 20 countries and dietary level of cholesterol.
(9) Dr. Stamler also found $r_s = 0.546$ between 1964 mortality rate and percent of calories from saturated fat for the countries examined in (8).

Test the statistical significance of each of these results. Are there any that do not reach the .05 significance level? Which are most statistically significant? Which of these nine findings is of greatest practical significance in determining the relationship between eating eggs and the increased risk of heart attacks? Do you need more scientific information to evaluate the practical significance of some of these findings? What information would be useful?

E. SPEARMAN RANK CORRELATION COEFFICIENT

PENNSYLVANIA v. LOCAL 542, INTERNATIONAL
UNION OF OPERATING ENGINEERS
469 F. Supp. 329 (E.D. Pa. 1978)

A. Leon HIGGINBOTHAM, Jr., Circuit Judge. . . .
[Twelve blacks, on behalf of a class of minority workers, instituted this action against a local union and contractors' and trade associations, alleging employment discrimination in the union's membership practices, the operation of its referral system, and wages and hours of minority workers. — ED.]
Closely tied to plaintiffs' proof of a differential in minority hours and wages is the proof of an arbitrary system of referrals. Siskin's analysis on this subject was limited to District I, although in a separate analysis unionwide clustering of minorities with a limited number of employers was also shown to exist. First I will detail the District I study.

E. Spearman Rank Correlation Coefficient

Siskin's study of the District I referral system involved an analysis of seventeen out-of-work lists from among the various groups (five from Group I, one from Group II, five from Group 1-A, and six oilers and RA [Registered Apprentice] lists). The seventeen lists were the remainder after eliminating all lists during that period with less than forty names. This was to assure statistical significance. Each list had been used for one of any of the months between 1969 and 1971. A rank was then assigned to each person on the list according to his position. Referrals were counted based on the first referral date marked in a listee's work records. (A referral constituted any attempt to contact a worker including acceptances, refusals, or failures to achieve contact so long as noted in work records.) By computer, seventeen "selection" lists were created reflecting the actual order of referral.

By creating the selection list, Siskin was able to compare actual referral rankings to the work list ranking in order to determine the coefficient correlating the two lists positively, negatively or neutrally. The appropriate numerical correlation (the Spearman rank correlation coefficient ("r")) ranges from "-1" to "$+1$". A "$+1$" correlation would mean that the two lists are identical; a "-1" would mean that they bear a perfect reverse order correlation; a "0" correlation would mean that the relationship appears random. Based on Siskin's analysis the r correlation coefficients for the seventeen lists are as follows:

List	r
Group I	
#4	.20
#7	.55
#10	.52
#13	.46
#16	.62
Group II	
#2	.08
Group I-A	
#3	.22
#5	.40
#9	.37
#12	.43
#15	.54
RA & Oilers	
#1	.24
#6	.38
#8	.44
#11	.44
#14	.46
#17	.45

While all of the lists except list #2 from Group II were clearly on the positive correlation side, further analysis by Siskin revealed that virtually none of the lists reflecting actual referral rankings was *significantly* similar to the corresponding out-of-work list.

This conclusion was reached after calculating the "variance." In the present context the correlation coefficient (r) when squared (r^2) measures the variance in selection rank which is explained by the out-of-work list rank. The formula $1 - r^2$ measures the variance in selection not explained by the out-of-work list. Of the seventeen lists examined, on only one was more than one-third of the selection rank predictable or explainable based on out-of-work list rank. On another list the position on the out-of-work list explained only .6% of the selection list ranking.

List	Rank Correlation Coefficient (r)	Percent Variance Explained (r^2)	Percent Variance Not Explained $(1 - r^2)$
Grp. I-A			
# 3	.22	4.8	95.2
# 5	.40	16.0	84.0
# 9	.37	13.7	86.3
#12	.43	18.5	81.5
#15	.54	29.2	70.8
Grp. I			
# 4	.20	4.0	96.0
# 7	.55	30.3	69.7
#10	.52	27.0	73.0
#13	.46	21.2	78.8
#16	.62	38.4	61.6
Grp. II			
# 2	.08	0.6	99.4
Oilers & RA			
# 1	.24	5.8	94.2
# 6	.38	14.4	85.6
# 8	.44	19.4	80.6
#11	.44	19.4	90.6
#14	.46	21.2	78.8
#17	.45	20.3	79.7

The average for all lists indicates that 82.5% of variance is the result of factors *other than* order on the out-of-work list. Although there exists a possibility that selections based on skill could theoretically have created discrepancies in selection rank, plaintiffs point out, and I agree, that the low correlation on the single skill oilers and RA lists (a correlation very like that for the other lists) itself tends to disprove the theory that "skills" explains the variance.

E. Spearman Rank Correlation Coefficient

The next statistical test performed on the seventeen lists was to determine whether predictability of selections from out-of-work list ranking increased depending on the type of list. A table indicating the percent of explained variance within groups, with and without ranking those who were not referred at all, is below:

Group	Including Non-referred	Deleting Non-referred
I-A	15.7	14.2
I	20.6	21.9
II	0.6	2.0
Oilers & RA	15.8	14.1
	17.5	17.6

The percentage of explained variance is relatively small.

Siskin prepared yet another chart, an "expectancy chart," indicating the probability of selection for persons listed in respective quintiles of each out-of-work list:

Percentage Probability of Selection Order Compared to Work List Order

Quintile on Out-of-Work List	Percent Selected in Quintile of Selection List					
	1st	2nd	3rd	4th	5th	
1st	24.0%	24.0%	17.8%	17.5%	16.6%	100
2nd	29.0	28.3	15.5	15.1	12.1	100
3rd	32.2	28.9	17.2	12.6	9.3	100
4th	14.0	18.7	34.1	17.4	15.8	100
5th	2.1	1.1	16.7	36.4	43.8	100
	100	100	100	100	100	

As can readily be seen the probability of selection is not increased by being in the first or even second quintile on the out-of-work list, although someone in the fifth quintile of the out-of-work list is not at all likely to be selected among the first or second selection list quintiles. While this analysis in itself does not seek to identify race as the factor creating the lack of correlation, it confirms that the out-of-work list ranking is simply not the principal basis for selection. This corroborates plaintiffs' claims of discrimination in the sense that it proves there is much room for arbitrary and standardless selections. When combined with the other statistical disparities considering the race factor directly, this correlation study aids the inference of discrimination. . . .

Notes and Problems

1. In Pennsylvania v. Local 542, International Union of Operating Engineers, the court refers to Dr. Siskin's use of a variant of r called the *Spearman rank correlation coefficient*. The formula for the rank correlation coefficient, r_r, is:

$$r_r = 1 - \frac{6\Sigma(X_i - Y_i)^2}{N(N^2 - 1)}$$

The rank correlation coefficient is useful when the variables between which we wish to know the relationship cannot easily be measured. Some variables can only be ranked or placed in order, rather than measured. The reason for the ordering may be size, importance, time out of work, preference, or any number of other criteria. The measurements giving rise to the ordering, such as volume, degree of importance, hours out of work, strength of preference, are ignored once the ranking is made. Lists that include the measurements are called *cardinal* rankings, while lists that only indicate the resulting ordering are called *ordinal* rankings. The Spearman formula adapts the correlation coefficient to make it useful for ordinal rankings. Both the X and Y lists must be ordinal lists for the Spearman r to be calculated.

2. As a simple example, consider sample data in Exhibit VI-13. The days-out-of-work data come from the payroll office. The order-of-referral data come from the desk of the union officer in charge of referrals.

EXHIBIT VI-13.
Unranked Sample Data

Days Out of Work		Order of Referral	
Omar	25 days	1.	Ivan
Raoul	37 days	2.	Gabrielle
Ivan	92 days	3.	Omar
Gabrielle	9 days	4.	Raoul
Yvette	50 days	5.	Yvette
Olga	4 days	6.	Olga

The data in the order-of-referral list is already ranked. Ivan is first to be referred and Olga is last. The days-out-of-work list must be ranked as follows:

1. Ivan
2. Yvette
3. Raoul

E. Spearman Rank Correlation Coefficient

4. Omar
5. Gabrielle
6. Olga

The calculation of r_r ignores the absolute disparity in terms of number of days out of work. It only takes order or ranking into account. If days out of work is the X list and order of referral is the Y list, then the five (X_i, Y_i) observations appear as follows:

EXHIBIT VI-14.
Ordinal Ranking of Sample Data

Worker	Observation (X_i, Y_i)
Ivan	(1, 1)
Yvette	(2, 5)
Raoul	(3, 4)
Omar	(4, 3)
Gabrielle	(5, 2)
Olga	(6, 6)

The rank correlation coefficient, r_r, is .43. The computation is as follows:

$$r_r = 1 - \frac{6\Sigma(x_i - y_i)^2}{n(n^2 - 1)}$$

$$\Sigma(x_i - y_i)^2 = (1 - 1)^2 + (2 - 5)^2 + (3 - 4)^2 + (4 - 3)^2 + (5 - 2)^2 + (6 - 6)^2$$
$$= 0 + 9 + 1 + 1 + 9 + 0 = 20$$

$$\frac{6\Sigma(x_i - y_i)^2}{n(n^2 - 1)} = \frac{6(20)}{6(6^2 - 1)} = \frac{120}{210} = .57$$

$$1 - \frac{6\Sigma(x_i - y_i)^2}{n(n^2 - 1)} = 1 - .57 = .43 = r_r$$

To test the statistical significance of this r_r, we use a Spearman r table (Exhibit VI-15). For six pairs of observations, r_r must be greater than .829 to reach a .10 significance level or greater than .886 to reach a .05 significance level. The null hypothesis for this two-tailed test was, "There is no relationship between positions on the two lists." Do we reject this null hypothesis or not? Why? What is the factual implication of there being no relationship between positions on the two lists?

3. The data in Exhibit VI-16 indicate the ages of seven faculty members at Matchbook Law School and their average scores on student evaluations forms. On the forms filled out by students, a "5" is

EXHIBIT VI-15.
Critical Values for the Rank Correlation Coefficient r_r

Number of Pairs	Significance Level			
	.10	.05	.02	.01
5	0.900	——	——	——
6	0.829	0.886	0.943	——
7	0.714	0.786	0.893	——
8	0.643	0.738	0.833	0.881
9	0.600	0.683	0.783	0.833
10	0.564	0.648	0.745	0.794
11	0.523	0.623	0.736	0.818
12	0.497	0.591	0.703	0.780
13	0.475	0.566	0.673	0.745
14	0.457	0.545	0.646	0.716
15	0.441	0.525	0.623	0.689
16	0.425	0.507	0.601	0.666
17	0.412	0.490	0.582	0.645
18	0.399	0.476	0.564	0.625
19	0.388	0.462	0.549	0.608
20	0.377	0.450	0.534	0.591
21	0.368	0.438	0.521	0.576
22	0.359	0.428	0.508	0.562
23	0.351	0.418	0.496	0.549
24	0.343	0.409	0.485	0.537
25	0.336	0.400	0.475	0.526
26	0.329	0.392	0.465	0.515
27	0.323	0.385	0.456	0.505
28	0.317	0.377	0.448	0.496
29	0.311	0.370	0.440	0.487
30	0.305	0.364	0.432	0.478

Source: N. Johnson and F. Leone, 1 Statistics and Experimental Design in Engineering and the Physical Sciences 547 (2d ed. 1977). Reprinted with permission.

the best ranking, indicating high quality teaching, while "1" is the lowest ranking, indicating terrible teaching. Test the null hypothesis that there is no relationship between age and popularity of a teacher using the cardinal rankings supplied and the regular correlation coefficient.

Using the same idea, prepare two ordinal rankings of the seven teachers, calculate the Spearman rank correlation coefficient, and test the same null hypothesis. You will notice that the two correlation coefficients are not equal. The reason is that the Spearman correlation

EXHIBIT VI-16.
Student Evaluations of Law School Professors

Professor	Age	Average Evaluation Score
Bank	32	3.5
Belle	29	2.7
Braverman	45	3.4
Burnet	54	3.6
Ellision	41	2.9
Marhuney	60	4.8
Wechslaw	47	4.1

coefficient ignores quantitative differences between the scores and only considers the ranking of the scores.

Note: The Coefficient of Determination

In the family-law hypothetical of Kilroy v. Kilroy, with which we opened the discussion of correlation coefficients, we calculated an $r = .87$ between length of Kid's visit with Mom and Kid's absences from school. The significance of that population statistic was described as being "practical" rather than "statistical" because it was calculated from the whole population rather than from a sample. Pennsylvania v. Local 542, International Union of Operating Engineers provides additional guidance as to the practical significance of a particular correlation coefficient. When an r or an r_s or an r_r is squared, the resulting value, r^2, is called the *coefficient of determination;* it measures the variation in one list that is "explained" or "taken into account" by the values or rankings in the other list. Thus, for an r that equals 0.87 in Kilroy v. Kilroy, r^2 equals $(.87)^2$, or 76. Thus 76% of the variation in duration of Kid's absences from school are "explained" by the length of Kid's visits to Mom. Correspondingly, 100% minus 76%, or 24%, of the variation is not explained. It is crucial that you note that this still does not prove that one event *caused* the other, no matter how large r^2 is. We can still only infer the extent to which the two variables change values in corresponding directions. To what extent does variation in ages of professors at Matchbox Law School, in the problem in Note 3 above, explain the variation in student evaluations? You can see from the Matchbox example that even a high correlation coefficient would not prove that increased age causes increased popularity. If we were to show a strong correlation, then we would suspect that there is something about age (increased experience, increased self-confidence, increased skill, a larger storehouse of amusing anecdotes) that "causes" popularity, not age itself.

CHAPTER VII

REGRESSION ANALYSIS

A. INTRODUCTION

The concept of regression is the natural climax to the study of descriptive and inferential statistics because the analysis of regression requires the use and understanding of all the statistical methods we have covered. *Regression analysis* is a method for examining the relationships among large numbers of variables. The concept of regression is probably most like that of the correlation coefficient. The correlation coefficient summarizes the relationship between two variables and indicates the degree to which changes in the value of one variable correspond to changes in the value of the other. Regression coefficients summarize the degree to which changes in the values of a number of variables all correspond to changes in the value of a single variable.

It should be apparent that such a technique has wide applicability. Calculating correlation coefficients involved a straightforward method of comparing the changes in values of two variables. Often, however, the value of a particular variable is due to changes in the values of more than one other variable. For instance, shoe size is determined not just by foot length but by foot width. The weather on a particular day is determined not just by the season but by high- and low-pressure areas, humidity, the jet stream, and numerous other factors. When more than one variable is used to "explain" the value of another variable, the tool of multiple regression is useful in indicating how much influence each of numerous "independent" variables has on the "dependent" variable. The summary statistics called *regression coefficients* describe the relationships between each of the *independent variables,* so-called because their values are independent of each other, and the *dependent variable,* so-called because its value depends on the values of the independent variables. As usual, these regression coefficients can be tested to determine their statistical significance.

While regression coefficients can be calculated by hand, they almost never are, outside of statistics courses. Computers calculate regression coefficients so handily that manual calculations only serve to acquaint the novice statistician with the theory of regression. We have already disclaimed any intent to explain the theoretical underpinnings of statistics in order to concentrate on the practical reality. The practical reality is that your statistical expert will get the computer to give you the regression coefficients and additional information sufficient for significance testing. What the attorney does need to know, however, is what the statistician is talking about when multiple regressions are offered as an element of proof.

B. FROM CORRELATION COEFFICIENTS TO BIVARIATE REGRESSION

The simplest case in regression analysis occurs when there are only two variables, in which case the regression is called *bivariate regression*. Bivariate regression has its roots in correlation. In Exhibit VI-7 on page 266, we described the relationship between remaining life expectancy and age as being negative in the sense that, as age increases, remaining life expectancy decreases. A perfect negative correlation, $r = -1$, suggests that each time one variable increases in value, the other decreases. The correlation coefficient, varying between -1 and $+1$, is a measure of the extent to which these two variables change in the same direction at the same time. But the correlation coefficient does not describe *how much* the value of one variable changes when the other does; this is the gap filled by the regression coefficient. It is symbolized by a capital B when we are referring to a population regression coefficient and by a lower-case b when we are talking about a sample.

1. Regression Coefficients: The Coleman Motor Case

The case of Coleman Motor Co. v. Chrysler Corp., 376 F. Supp. 546 (W.D. Pa. 1974), is illustrative of the nature and uses of the regression coefficient. Bivariate regression analysis was used to estimate damages suffered by Coleman Motor as a result of violations by Chrysler of §§1 and 2 of the Sherman Antitrust Act. Coleman Motor proved that by subsidizing some dealers, Chrysler had caused Cole-

B. From Correlation Coefficients to Bivariate Regression

man to lose sales and hence profits. The example we will use is adapted from the circumstances of that case.

You can see from the graph in Exhibit VII-2, on the following page, that profits and sales are not perfectly correlated. Generally, when sales rise, profits rise. But other factors, such as variations in expenses, affect the relationship. The regression coefficient estimates the *average* relationship between profits and sales in order to allow the court to project lost profits on the basis of lost sales, in this instance due to Chrysler's dealer-subsidy program.

To estimate the *average* relationship, we need to take into account two factors: (1) the extent to which the variables change in the same direction, and (2) a measure of *how much* deviation from this general direction is exhibited by the values of each of the variables as the value of the other changes. We have already measured correspondence in changes of two variables by the correlation coefficient. Deviations from a mean value, here the general direction of the correspondence, is measured as before, by the standard errors. The formula for the regression coefficient, b, which takes both factors into account, is:

$$b = r_s \times \frac{s.e._y}{s.e._x}$$

where r_s is the sample correlation coefficient given by:

$$r_s = \frac{\Sigma(Z_{x_i} \times Z_{y_i})}{n-1} = \frac{\Sigma\left[\frac{(x_i - m_x)}{s.e._x} \times \frac{(y_i - m_y)}{s.e._y}\right]}{n-1}$$

and $s.e._y$ and $s.e._x$ are the standard errors for the dependent and independent variables, respectively, given by:

$$s.e._y = \sqrt{\frac{\Sigma(y_i - m_y)^2}{n-1}} \quad \text{and} \quad s.e._x = \sqrt{\frac{\Sigma(x_i - m_x)^2}{n-1}}$$

In our case, profits is the dependent variable, y, because its value is affected by (explained by, accounted for by, determined by) sales, the independent variable, x. For the profit/sales data, r_s equals .876, $s.e._x$ equals 176 cars, and $s.e._y$ equals \$188,756, so b equals \$939 per car. The b tells us that the average profit per car sold was \$939. Now, if we can project lost sales, we can project lost profits.

2. The Slope of the Regression Line

The formula for b calculates the slope of a straight line through the observations graphed on the profit/sales graph in Exhibit VII-2. This line is not just any line, however. It is the *only* line that minimizes the distances between the data points and the line. To be

EXHIBIT VII-1.
Profit/Sales Table

Year	Sales	Profit ($)
1961	350	500,000
1962	200	250,000
1963	450	600,000
1964	730	800,000
1965	550	700,000
1966	700	650,000
1967	400	500,000
1968	450	350,000
1969	300	300,000

EXHIBIT VII-2.
Profit/Sales Graph

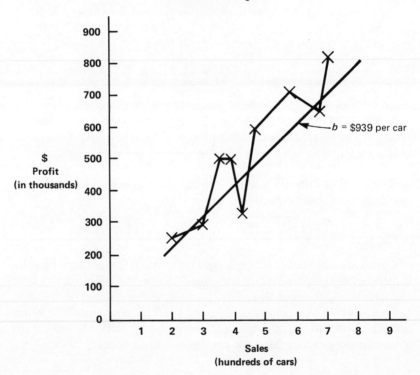

B. From Correlation Coefficients to Bivariate Regression

precise, it *minimizes the sum of the squared deviations* of the actual x and y pairs from the x and y pairs that are on the line drawn through them. The line indicated by $b = 939$ is called the *least squares line*. The mathematical technique by which this regression coefficient was calculated is called *ordinary least squares* (abbreviated *OLS*). If familiar with geometry, you will recognize b as the *slope* of the regression line.

3. The Intercept of the Regression Line

For the profit/sales case we note that if there are no sales, there will be no profit. But it will not always be true that a zero value for the independent variable implies a zero value for the dependent variable. Consider this example from an employment discrimination case. The plaintiff, a female bridge-builder, alleges that male bridge-builders with identical skills get higher pay. Assume that skill is determined by how many bridges a builder has previously constructed. The regression analysis will determine the average pay per bridge if our independent variable is the number of bridges each builder has constructed. A survey of bridge builders of all skill levels might yield the data in Exhibit VII-3, on page 298.

Our best OLS estimate of the regression coefficient associated with the number of bridges built is $12,545. This indicates that, on average, a worker's salary increases by $12,545 for each bridge built. It does not mean, however, that the salary of a bridge-builder will be two times $12,545 after two bridges. Why not? Because even a worker with no experience gets some pay. Notice, for instance, worker number 3 in the bridge-builder table. His salary is $20,000 with no experience. We clearly need another statistic to capture this unaccounted-for starting salary.

That new statistic is called the *intercept* and is abbreviated a. The intercept of a regression line is where that line hits or "intercepts" the ordinate, y, axis. The formula for the intercept is:

$$a = m_y - bm_x$$

where m_y and m_x are the means of x and y, and b is the regression coefficient.[1] In our bridge-builder example, the intercept is the best

---

1. There are a variety of ways of indicating that two values are to be multiplied together. Thus far, we have used the "times" sign ($\times$), and the word *times* to indicate that multiplication was required, e.g., 5×3 and $t \times s.e._m$ or 5 times 3 and t times $s.e._m$. Beginning with this section, a new convention is also used. When two symbols or a number and symbol appear next to each other in an equation, such as bx_i or $150m_x$, that means that the values should be multiplied together, that is, the particular value indicated by x_i should be multiplied by b, the regression coefficient, or the mean of x, m_x, should be multiplied by 150.

EXHIBIT VII-3.
Bridge-Builder Table

Worker Number	Number of Bridges Built	Salary ($)
1	2	45,000
2	1	30,000
3	0	20,000
4	2	35,000
5	3	60,000
6	1	32,000

EXHIBIT VII-4.
Bridge-Builder Graph

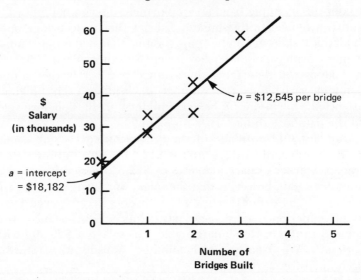

OLS estimate of the salary of an unskilled bridge-builder, which equals $18,182:

$$m_y - bm_x = \$37{,}000 - 12{,}545 \times 1.5 = \$18{,}182$$

The combination of the regression coefficient and the intercept allows us to predict the value of the dependent variable for any value of the independent variable. For instance, our best estimate of the salary (y) for a worker who has already built two bridges ($x = 2$), taking into account starting salary plus appropriate raises for skill increase, is represented by:

$$y = a + bx.$$

B. From Correlation Coefficients to Bivariate Regression

The predicted salary equals 18,182 plus 12,545 times two, or $43,272. We can compare our client's salary to this predicted level.

4. The Error Term in the Regression Line

You may have noticed that the regression line does not perfectly predict the values of y that we observed when we collected the data; no bridge-builder observed actually had a salary of $43,272. There is, then, some variation in salaries not accounted for by the slope, b, or intercept, a. This variation is obscured in the process of averaging the observed values. Remember, the regression line only predicts an average or expected value for the dependent variable. Recognizing that there is always some error left over after we use the calculated b and a to predict our observed y, the regression equation is often written as:

$$y = a + bx + u$$

where u is simply referred to as *the error term*. It would not be useful to our prediction to calculate an "average u," as we have calculated b and a, because the error term, u, varies with each observation; and since we assume that it is as often positive as negative, the average would be zero. It is still useful to recognize, however, that the regression line does not exactly predict the values of y that we in fact observed. And as we will see in the next section, the size of these errors determines the statistical significance of our regression coefficient. It makes intuitive sense that the larger are these errors, the less significant are the coefficients we use to predict values of the independent variable.

5. Significance Level and Bivariate Regression Analysis

As with other statistics that summarize sample data, regression coefficients and intercepts are used to test null hypotheses about the population from which the sample was drawn. In this regard the regression coefficients and intercepts are like sample means. After calculating a sample mean, we use a confidence interval to present a range of values within which we are fairly certain that the true mean, the population value, lies. We test regression results using a similar procedure.

Because regression coefficients describe the relationship between the dependent and independent variable, our null hypothesis is often,

"There is no relationship between these variables." Because b, the regression coefficient, indicates the "average" value of how much the dependent variable changes when the independent variable changes, b will be equal to zero if there is, in fact, no relationship between the variables. The test of the null hypothesis is, therefore, mathematically equivalent to a determination of whether the calculated b is significantly different from zero.

You can convince yourself that "b equals zero" is equivalent to "no relationship" by referring back to the formula for the regression coefficient:

$$b = r_s \times \frac{s.e._y}{s.e._x}$$

If b equals zero, then r_s must equal zero. The ratio of standard errors cannot equal zero, because no standard error will equal zero unless all the observations of the variable are the same, in which case it is constant and not variable at all. We know that the correlation coefficient, r_s, describes the degree of relationship between the variables. If r_s equals zero, there is no relationship between the variables, so b equals zero as well.

a. Calculating the t Value

To determine whether b is significantly different from zero, the t test is most frequently used. The formula for calculating t is rather complicated, but the t value is a statistic which, like the regression coefficient, is usually calculated by computers. The formula for the t statistic is:

$$t = \frac{b}{s.e._b}$$

where b is the regression coefficient and $s.e._b$ is the *standard error of the regression coefficient*. The only trick in this calculation is determining $s.e._b$, although you should remember that this is a value the statistician usually gets from the computer. The standard error of b is given by the formula:

$$s.e._b = \sqrt{\frac{s.e._e^2}{\Sigma(x_i - m_x)^2}}$$

where the x_i's are the observations of the independent variable, m_x is the sample mean for the independent variable, and $s.e._e^2$ is called the *residual variance,* which is the standard error of estimate squared or, somewhat less commonly, *the variance of the error.*

B. From Correlation Coefficients to Bivariate Regression

The residual variance may be defined as the variance in the observed values of the dependent variable that is not accounted for by the regression line. It is the variance of our error term, u. Remember that the regression line is the one line between the observed points that minimizes the squared deviations of the points from that line. The regression line rarely reduces these deviations to zero. The residual variance is one measure of the remaining deviations.

The formula for the residual variance of a bivariate regression is:

$$s.e._e^2 = \frac{\Sigma(y_i - y_i^*)^2}{n-2} = \frac{\Sigma u_i^2}{n-2}$$

where the y_i's are the actual observations of the dependent variable, n is the number of observations, and y_i^* is the estimated value of y_i. You can see that $y_i - y_i^*$ is equal to u, the error term. The estimated value of each y_i is calculated by plugging the value of each associated x_i into the regression equation, that is:

$$y_i^* = a + bx_i$$

Using the data for the bridge-builder case, we can calculate the t value for the regression coefficient by working backwards through the equations in this section. The observed values of x_i and y_i are shown in columns (2) and (3) below.

EXHIBIT VII-5.
Residual Variance Calculation for Bridge-Builder Data

(1)	(2)	(3)	(4)	(5)
Worker Number	x_i Bridges Built	y_i Salary	y_i^*	$u_i^2 = (y_i - y_i^*)^2$
1	2	$45,000	$43,272	2,985,984
2	1	$30,000	$30,727	528,529
3	0	$20,000	$18,182	3,305,124
4	2	$35,000	$43,272	68,425,984
5	3	$60,000	$55,817	17,497,489
6	1	$32,000	$30,727	1,620,529

From the regression equation, where y_i^* equals 18,182 plus 12,545x_i, we can calculate the estimated value of y_i for each worker. For example, for worker 1, x_i equals 2, so y_i^* equals 43,272 (18,182 + (12,545 × 2)). The y_i^*'s are shown in column (4). Column (5) shows the calculation $(y_i - y_i^*)^2$ needed to figure the residual variance, $s.e._e^2$.

The residual variance is therefore:

$$s.e._e^2 = \frac{\Sigma(y_i - y_i^*)^2}{n-2} = \frac{94,363,639}{6-2} = 23,590,910$$

Given this residual variance, we can calculate the standard error of the regression coefficient:

$$s.e._b = \frac{s.e._e^2}{\Sigma(x_i - m_x)^2} = \frac{23{,}590{,}910}{5.5} = 4{,}289{,}256 = 2071$$

where:

$$\begin{aligned}\Sigma(x_i - m_x)^2 &= (2 - 1.5)^2 + (1 - 1.5)^2 + (0 - 1.5)^2 \\ &\quad + (2 - 1.5)^2 + (3 - 1.5)^2 + (1 - 1.5)^2 \\ &= 0.25 + 0.25 + 2.25 + 0.25 + 2.25 + 0.25 \\ &= 5.5\end{aligned}$$

Given the standard error of the regression coefficient, we can proceed to calculate the t value for the coefficient as:

$$t = \frac{b}{s.e._b} = \frac{12{,}545}{2{,}071} = 6.06.$$

As with all test statistics, such as Z, χ^2, and r, we check the significance level of this t value against a table.

b. The t Test

There are two approaches to testing the null hypothesis. The first is through the use of a confidence interval. We can, for instance, estimate a 95% confidence interval for the true population regression coefficient, B, by the following interval around the estimated regression coefficient, b:

$$B = b \pm t_{.05} \times s.e._b$$

where $s.e._b$ is the standard error of the regression coefficient, b, calculated above, and $t_{.05}$ is the critical t value for a .05 significance level. The critical value is obtained from the t table, which we used for small samples in the Z confidence intervals and which is reproduced in Exhibit VII-6. A crucial fact to remember is that the degrees of freedom for bivariate regression is

$$d.f. = n - 2 \text{ for the } t \text{ test}$$

For six observations in our example, we therefore have four degrees of freedom. The critical value for $t_{.05}$ (a 95% confidence interval) is 2.776 for four degrees of freedom and for this two-tailed confidence interval. So the confidence interval is:

$$B = b \pm 2.776 \times s.e._b$$

B. From Correlation Coefficients to Bivariate Regression

and for the bridge builder example:

$$B = \$12{,}545 \pm 2.776 \times \$2{,}071 = \$12{,}545 \pm \$5{,}749$$

or

$$\$6{,}796 \leq B \leq \$18{,}294$$

We can be 95% certain that the true value is not zero since the interval does not encompass zero. We therefore reject the null hypothesis at a .05 significance level. This test of the null hypothesis is called a *two-tailed test* because it considers the possibility that the true population coefficient could be above *or* below the estimated regression coefficient.

With a *one-tailed test,* you can test the null hypothesis, "The true B is not greater than zero," if you believe, as in the bridge-building case, that the coefficient is likely to be positive. We believed it would be greater than zero in the bridge-building case because we expected salary to rise with skill level. To follow this approach, calculate the t value from:

$$t_{calculated} = \frac{b}{s.e._b}$$

If the calculated t is greater than the critical level for however many degrees of freedom you have ($d.f. = n - 2$) and whatever significance level you choose, (e.g., .05), then you may reject the null hypothesis.

For instance, the calculated t in our bridge-builder example is 6.06. At a significance level of .005, a 99.5% confidence level, the critical t value for a one-tailed test is 4.604 for four degrees of freedom. Since 6.06 is greater than 4.604, we can be 99.5% certain that the population B, the average effect of skill level on salary, is greater than zero.

Naturally we can do this for a null hypothesis that B is not less than zero as well. Imagine that we theorize that, as in the Coleman Motor Co. v. Chrysler Corp. case, increasing subsidies to franchised dealers will decrease profits to nonfranchised dealers. The dependent and independent variables are profits and subsidies, respectively, and we estimate an effect on profits of $\$.90$ decrease for every dollar of subsidy. The estimated regression coefficient in that case would be $b = -0.90$. If the standard error of that regression coefficient were 1.1, then t equals b divided by 1.1, or $-.818$. For a sample size of 20, there are 18 degrees of freedom, and the critical t value for a .025 significance level (a 97.5% confidence level) is 2.101. Ignore the minus sign on any t calculation. We see that $t_{.025}$ is greater than the calculated t, so we can refuse to reject the null hypothesis that B is not less than zero. We conclude that B is not significantly different from zero.

EXHIBIT VII-6.
Critical Values for Student's t

Degrees of Freedom	.45	.40	.35	.30	.25	.20	Significance Level One-Tailed Test .15	.10	.05	.025	.01	.005	.0005
	.90	.80	.70	.60	.50	.40	Two-Tailed Test .30	.20	.10	.05	.02	.01	.001
1	.158	.325	.510	.727	1.000	1.376	1.963	3.078	6.314	12.706	31.821	63.657	636.619
2	.142	.289	.445	.617	.816	1.061	1.386	1.886	2.920	4.303	6.965	9.925	31.598
3	.137	.277	.424	.584	.765	.978	1.250	1.638	2.353	3.182	4.541	5.841	12.924
4	.134	.271	.414	.569	.741	.941	1.190	1.533	2.132	2.776	3.747	4.604	8.610
5	.132	.267	.408	.559	.727	.920	1.156	1.476	2.015	2.571	3.365	4.032	6.869
6	.131	.265	.404	.553	.718	.906	1.134	1.440	1.943	2.447	3.143	3.707	5.959
7	.130	.263	.402	.549	.711	.896	1.119	1.415	1.895	2.365	2.998	3.499	5.408
8	.130	.262	.399	.546	.706	.889	1.108	1.397	1.860	2.306	2.896	3.355	5.041
9	.129	.261	.398	.543	.703	.883	1.100	1.383	1.833	2.262	2.821	3.250	4.781
10	.129	.260	.397	.542	.700	.879	1.093	1.372	1.812	2.228	2.764	3.169	4.587
11	.129	.260	.396	.540	.697	.876	1.088	1.363	1.796	2.201	2.718	3.106	4.437
12	.128	.259	.395	.539	.695	.873	1.083	1.356	1.782	2.179	2.681	3.055	4.318
13	.128	.259	.394	.538	.694	.870	1.079	1.350	1.771	2.160	2.650	3.012	4.221
14	.128	.258	.393	.537	.692	.868	1.076	1.345	1.761	2.145	2.624	2.977	4.140
15	.128	.258	.393	.536	.691	.866	1.074	1.341	1.753	2.131	2.602	2.947	4.073

16	.128	.258	.392	.535	.690	.865	1.071	1.337	1.746	2.120	2.583	2.921	4.015
17	.128	.257	.392	.534	.689	.863	1.069	1.333	1.740	2.110	2.567	2.898	3.965
18	.127	.257	.392	.534	.688	.862	1.067	1.330	1.734	2.101	2.552	2.878	3.922
19	.127	.257	.391	.533	.688	.861	1.066	1.328	1.729	2.093	2.539	2.861	3.883
20	.127	.257	.391	.533	.687	.860	1.064	1.325	1.725	2.086	2.528	2.845	3.850
21	.127	.257	.391	.532	.686	.859	1.063	1.323	1.721	2.080	2.518	2.831	3.819
22	.127	.256	.390	.532	.686	.858	1.061	1.321	1.717	2.074	2.508	2.819	3.792
23	.127	.256	.390	.532	.685	.858	1.060	1.319	1.714	2.069	2.500	2.807	3.767
24	.127	.256	.390	.531	.685	.857	1.059	1.318	1.711	2.064	2.492	2.797	3.745
25	.127	.256	.390	.531	.684	.856	1.058	1.316	1.708	2.060	2.485	2.787	3.725
26	.127	.256	.390	.531	.684	.856	1.058	1.315	1.706	2.056	2.479	2.779	3.707
27	.127	.256	.389	.531	.684	.855	1.057	1.314	1.703	2.052	2.473	2.771	3.690
28	.127	.256	.389	.530	.683	.855	1.056	1.313	1.701	2.048	2.467	2.763	3.674
29	.127	.256	.389	.530	.683	.854	1.055	1.311	1.699	2.045	2.462	2.756	3.659
30	.127	.256	.389	.530	.683	.854	1.055	1.310	1.697	2.042	2.457	2.750	3.646
40	.126	.255	.388	.529	.681	.851	1.050	1.303	1.684	2.021	2.423	2.704	3.551
60	.126	.254	.387	.527	.679	.848	1.046	1.296	1.671	2.000	2.390	2.660	3.460
120	.126	.254	.386	.526	.677	.845	1.041	1.289	1.658	1.980	2.358	2.617	3.373
∞	.126	.253	.385	.524	.674	.842	1.036	1.282	1.645	1.960	2.326	2.576	3.291

Source: R. Fisher and F. Yates, Statistical Tables for Biological, Agricultural and Medical Research 46 (Table III) (6th ed. 1963). Reprinted with permission.

Chapter VII. Regression Analysis

The one-tailed t test is very straightforward. Once the calculated t is found, just compare its absolute value to the critical t value for whatever confidence level you choose. Remember to use the one-tailed column of the t table. The rule for rejecting or refusing to reject the null hypothesis that b is not greater than zero or a hypothesis that b is not less than zero is:

If $|t_{calculated}| > t_{critical}$, reject the null.
If $|t_{calculated}| < t_{critical}$, don't reject the null.

This makes significance testing of regression coefficients very easy.

When regression results are reported by statisticians, they are usually reported with the calculated t ratios beneath the coefficient in the general form:

$$Y = a + bX$$
$$(t)$$

which, for the bridge-builder case, would be:

$$salary = 18{,}182 + 12{,}545 \times skill\ level$$
$$(6.06)$$

Occasionally the statistician will just report the standard error for the regression coefficient, $s.e._b$, which you simply divide into the coefficient, b, to calculate t. Also, either the sample size is reported or the statistician indicates the critical t values, so the reader can test the statistical significance.

Using a similar approach, two-tailed tests on the intercept, a, can be performed using the confidence interval:

$$A = a \pm t_{critical} \times s.e._a$$

where t critical is determined for whatever confidence level you choose and $n - 2$ degrees of freedom, and where $s.e._a$ is the standard error for the intercept. The standard error for the intercept is formulated as:

$$s.e._a = \sqrt{\frac{s.e._e^2}{n} + s.e._b \times m_x},$$

where $s.e._e^2$ is the residual variance, n is the sample size, $s.e._b$ is the standard error of the regression coefficient, and m_x is the mean of the values of x. The one-tailed formula for the calculated t is:

$$t_{calculated} = \frac{a}{s.e._a}$$

and we reject the null hypothesis that A is not significantly different from zero if t calculated is greater than t critical.

c. Null Hypotheses Related to Legal Maxima and Minima

Recall that when we first considered the Z test, we used the test:

$$Z_{calculated} = \frac{Observed - Expected}{Standard\ Deviation}$$

to determine the probability of a particular observed value's occurring by chance. Using the t statistic, the regression coefficient could be subjected to a similar test. Instead of the null hypothesis, "B is not greater than zero," we can test "B is not greater than the legal maximum." The procedure is to modify the calculated t formula,

$$t_{calculated} = \frac{b}{s.e._b}$$

which might be interpreted as:

$$t_{calculated} = \frac{b - 0}{s.e._b}$$

to read:

$$t_{calculated} = \frac{b - B'}{s.e._b}$$

where B' is the legal maximum. We then compare calculated t to the one-tailed critical t, as usual, and use the same old rules to reject or refuse to reject the null hypothesis. We can use the same approach to see whether the b is below a legal minimum.

SOUTH DAKOTA PUBLIC UTILITIES COMMISSION v. FEDERAL ENERGY REGULATORY COMMISSION
643 F.2d 504 (8th Cir. 1981)

STEPHENSON, Circuit Judge.

This is an appeal of an order entered by the Federal Energy Regulatory Commission (FERC) permitting an accelerated rate of depreciation for certain facilities owned by Northern Natural Gas (Northern). The South Dakota Public Utilities Commission (South Dakota) opposed Northern's proposed adjustments before the FERC and has appealed its decision. The FERC found that because of dwindling reserves of natural gas, Northern should be allowed to depreciate its equipment over a period shorter than the physical life of the equipment. That is, the FERC decided that the useful life of the pipeline systems would be shorter than the time that normal wear and tear would require abandonment.

South Dakota asserts primarily that the depreciation rates set by the FERC were premised upon baseless estimates of future reserves, that the FERC miscalculated the portion of the predicted future reserves Northern would be able to purchase and, therefore, the FERC decision was inconsistent with the standards imposed by the Natural Gas Act, 15 U.S.C. §§717, et seq., and as applied in Memphis Light, Gas and Water Division v. Federal Power Commission, 504 F.2d 225 (D.C. Cir. 1974). . . .

I. FACTUAL AND PROCEDURAL BACKGROUND

Northern Natural Gas is a major interstate transporter of natural gas with revenues exceeding a billion dollars per year. Its pipeline network moves natural gas from the producing areas of Texas, Oklahoma and Kansas northward to Nebraska, Iowa, South Dakota, Minnesota and Wisconsin. Northern also owns producing and gathering equipment offshore in the Gulf of Mexico plus an isolated system in Montana and Wyoming. For the purpose of computing depreciation, Northern's properties are divided into four components, two of which are important here. The first is referred to as the South End supply area. The South End links Northern's two major supply fields — the Hugoton-Anadarko and the Permian Basin — to the rest of the Northern system. The second major component is referred to as the Market Area and consists of the equipment north of the Kansas-Nebraska border.

The primary issues in the proceedings below were whether the FERC properly estimated the reserves of natural gas in the Hugoton-Anadarko and Permian Basin supply areas, and what share of the estimated reserves Northern would be able to acquire. The relationship of the supply of natural gas and depreciation rates is inversely proportional. The higher the estimates of natural gas supplies, the lower the depreciation rates should tend to be because it is more likely that the pipeline system will be a useful asset throughout its physical life. Conversely, the lower the estimated supplies the higher depreciation rates are called for because the pipeline system may become useless before it has physically deteriorated to the point where abandonment would be required. For example, in this case, under the FERC staff's estimates, Northern's facilities will be useful until approximately the year 2000. However, the physical life of the equipment will not end until about the year 2011. In these circumstances, the FERC concluded, an increased rate of depreciation was

B. From Correlation Coefficients to Bivariate Regression 309

appropriate. Therefore, the gravamen of this litigation, and the subject of nearly 7000 pages in this voluminous record, is how much natural gas is awaiting discovery in the Hugoton-Anadarko and Permian Basin fields, and how much of that supply will Northern be able to buy.

The FERC order approved settlement agreements in two related rate cases filed by Northern. The first, RP 76-89, was filed in April 1976. The second, RP 77-56, was filed about a year later, while the earlier case was still pending. Both were requests for general rate increases that eventually were narrowed to the single issue of proper rates of depreciation. . . .

III. THE RECORD

There are three basic elements to the FERC's decision in both RP 76-89 and RP 77-56 which are challenged by South Dakota. Two are studies or models designed to estimate the amount of natural gas in the Permian Basin and the Hugoton-Anadarko fields, and the third is an analysis of what portion or share of these estimated reserves Northern will be able to acquire. The first study used to predict natural gas supply is entitled the EHF Model. The study was conducted and supported at the two hearings by the staff's expert witness Edward H. Feinstein. The other study is based upon estimates compiled by the Potential Gas Committee (PGC), a diverse group made up of representatives from the natural gas industry, government and academia.

The following table represents the depreciation rates expressed in percentages which were supported by the various parties in RP 77-56. The settlement figures are the rates set by the FERC in its order. The Northern figures are its original proposal before negotiation. As noted above, Northern agreed to the settlement figures and supported those results during the hearing.

Area	Northern	FERC		SD	Settlement
		EHF	PGC		
South End	6.91	5.45	4.79	4.20	5.25
Market Area	4.67	3.96	3.10	2.00	3.75

The figures in RP 76-89 were slightly different. The settlement figure for the South End was 5.15 percent but was the same 3.75 for the Market Area. The PGC amounts were 4.99 percent and 3.33 percent for the South End and Market Area, respectively. The EHF predictions led to rates of 5.44 percent for the South End and 4.02

percent for the Market Area. South Dakota proposed the same 2.0 percent for the Market Area but had suggested that 4.35 percent would be appropriate for the South End.

The FERC suggests that the EHF and PGC studies, as rational predictions of an unknown quantity, form a "zone of reasonableness" for depreciation rates consistent with the standards expressed in Permian Basin [Area Rate Cases, 390 U.S. 747 (1967),] and *Memphis*. South Dakota attacks both of these studies along with the share analysis used by the FERC. However, as the table demonstrates, if the EHF and PGC represent legitimate estimates, the settlement rates are acceptable because they are within the range created by these two studies.

A. EHF MODEL

South Dakota argues that the "sole justification" for the settlement figures rest upon the EHF model. The depreciation rates which flowed from the EHF projections represent the high end of what the FERC maintains is a zone of reasonableness. This is because the EHF model produced the lowest reserves estimates of the two studies and consequently, the highest depreciation rates.

The EHF model is statistically based and predicts annual reserves based on a theory that relates drilling efforts to results. This study uses historical data to project future reserves. Feinstein, a petroleum engineer employed by the FERC, testified that the theory behind the EHF model is that for any finite depletable natural resource the large high-grade, easy-to-find deposits are discovered during the early years of the depletion cycle and that the mature years are marked by the discovery of smaller, scattered and lower grade deposits. He stated that both the Hugoton-Anadarko and Permian Basin fields had entered mature stages so this type of analysis would produce reliable results in this case.

Feinstein based his predictions on statistics from 1967-1976. He compared cumulative exploratory drilling footage to cumulative reserve additions. He extrapolated this historical data to predict the potential gas recoverable and the reserves discovered annually. This pattern was extended to the point where the productivity of efforts to results is negligible.[11] As part of this process, Feinstein calculated

11. In RP 76-89, (before ALJ Benkin), Feinstein did not use a least squares method to plot the extension of the historical data. At that hearing he testified that he had simply "eyeballed it on a least square basis." In the later proceeding, RP 77-56, Feinstein used data based upon the more precise least squares method. This procedure is defined as "a statistical method of fitting a line or plane to a set of observational points in such a way that the sum of the squares of the distances of the points from the line or plane is a minimum." Webster's Third New International Dictionary (Merriam-Webster 1971).

B. From Correlation Coefficients to Bivariate Regression 311

a figure referred to as "effectiveness of exploration" which was a comparison of new field drilling footage to new field discoveries. He next plotted the effectiveness of exploration data in relation to exploratory footage and time in separate graphs, and then compared cumulative reserve additions to cumulative exploratory footage. This information was compiled for each of the supply areas and was used to determine the respective depreciation rates.

[The following table is an example of the comparison between efforts and results made by the EHF model. This table details exploratory drilling and gas reserve statistics for the Permian Basin.

	Efforts		Results	
	New Field Annual	Footage Cumulative	Reserve Annual	Additions Cumulative*
Year	1000 ft.	1000 ft.	Bcf**	Bcf
		97,546		58,627
1967	3,598	101,144	3,307	61,931
1968	2,455	103,599	984	62,918
1969	2,122	105,721	1,107	64,025
1970	2,466	108,187	2,012	66,037
1971	1,765	109,952	1,871	67,908
1972	2,166	112,118	1,883	69,791
1973	2,331	114,449	1,567	71,358
1974	3,030	117,479	1,139	72,497
1975	4,310	121,789	791	73,288
1976	4,202	125,991	650	73,938

* Includes New Field Discoveries, New Reservoir Discoveries, Extensions and a certain adjustment for additions not reported in the year of occurrence.
** Billion cubic feet. (fn. 10)]

South Dakota primarily attacks the EHF model because it does not include "developmental drilling." Feinstein testified that his model considers "new field drilling" which he defined as efforts undertaken to discover new fields not directly related to those already in service. South Dakota argues that this approach ignores reserves added after the initial discovery of a field and as used in the Feinstein model represents a significant reduction in the amount of potential reserves. They argue, and Feinstein agreed, that developmental drilling would account for more activity than exploratory drilling in mature fields such as the Permian Basin and the Hugoton-Anadarko. Feinstein nevertheless maintained that this was not a serious omission. In support of his study, Feinstein constructed a second model based upon the same theory which separately calculated future additions from three categories: new fields, pre-1966 existing fields and existing fields more recently discovered. This second EHF model reached a result that differed by only three percent in total reserves from the first. The FERC also noted that a study conducted by Northern which took developmental drilling into account produced results very similar to the EHF model. Although it is not made clear in the record, at oral argument counsel for the FERC and Northern asserted that although developmental drilling is not included in the efforts side of the EHF relationship, it is a part of the results figures.

We cannot say that the FERC's conclusion that the EHF study was reliable is not supported by substantial record evidence. Both the South Dakota and Feinstein positions are plausible and in light of our standard of review we defer to the expert judgment of the FERC. The FERC's opinion demonstrates that the Commission has not ignored South Dakota's arguments but has given reasoned consideration to all the pertinent factors. We are not required to find more.

South Dakota claims further that the EHF study cannot be relied upon because the PGC estimates reserves three times larger than those predicted by the EHF model. This difference in estimates is so large, in South Dakota's view, that the EHF model cannot be used as substantial evidence. The FERC responds by suggesting that such seemingly inconsistent estimates are not uncommon. The Commission points to estimates for the United States ranging from 496 trillion cubic feet (Tcf) to 2250 Tcf. Again, we are constrained to affirm the FERC's conclusion. It is clear, as the evidence in this record suggests, that estimating natural gas reserves is not an exact science. In such areas where technical expertise is the basis for decision making and where the question is purely factual, courts must be mindful of their role. We think that the following quotation is pertinent:

> [I]n the end it was for the Commission, not us, to evaluate the respective justifications put forth on the record, and to choose between two divergent theories in setting the amount of the challenged factor. A conclusion on "conflicting engineering and economic issues is precisely that which the Commission exists to determine, so long as it cannot be said . . . that the judgment which it exercised had no basis in evidence and so was devoid of reason."

City of Cleveland v. Federal Power Commission, 525 F.2d 845, 849 n. 36 (D.C. Cir. 1976), (quoting United States ex rel. Chapman v. FPC, 345 U.S. 153, 171, 73 S. Ct. 609, 619, 97 L. Ed. 918 (1953)).

[South Dakota also claims that the model is so statistically unsupportable that its results are invalid. They rely upon the "t"-test of statistical significance. This test was conducted by the FERC staff following the hearing in RP 77-56. The "t"-test produces a signifi-

B. From Correlation Coefficients to Bivariate Regression 313

cance level which measures the validity of using the relationships between variables to support a hypothesis. According to South Dakota, the FERC staff determined significance levels of .66 and .87 for the EHF models concerning the Hugoton-Anadarko and Permian Basin areas, respectively. South Dakota maintains in its brief that a .90 level of significance is required before such a study can be considered to be reliable.

In its order denying rehearing, the Commission states that the "t"-test results were .80 and .90 for the two fields. The Commission also noted that if the figures for 1977 and 1978 are added to the data base, the "t"-test figure for the EHF model increases to over .97.

The "t"-test figures are not a part of the record and therefore it is not possible for us to fully evaluate these arguments. We do note, however, that reserve additions for 1977 support the estimates derived from the EHF model. According to the FERC opinion, the EHF model predicted reserve additions of 700 Bcf in the Permian Basin and 2079 Bcf in the Hugoton-Anadarko for 1977. Actual additions were later confirmed to be 730 Bcf and 2081 Bcf respectively. (fn.13)]

B. PGC MODEL

The PGC model is based upon estimates of natural gas reserves issued by the Potential Gas Committee in conjunction with the Potential Gas Agency at the Colorado School of Mines. The Potential Gas Agency is supported by the American Gas Association. It is one of the few studies which provide natural gas estimates by area. Most predictions only give figures for the entire United States. As noted above, the PGC includes representatives from energy, government and academic institutions.

The PGC divides its estimates into three categories: probable, possible and speculative. The FERC staff did not use the third category of the PGC estimates. They concluded that the speculative category was too uncertain to be included. South Dakota claims that the FERC has arbitrarily eliminated a large source of potential reserves for Northern noting that in the 1972 PGC report the speculative category constituted twenty-nine percent of the Hugoton-Anadarko estimate and eighteen percent of the Permian Basin's.

["Probable" is defined as "the most assured of the new supplies." The "possible" category is less assured supplies that will come from new field discoveries in previously productive formations. The "speculative" classification is described as the "most nebulous of new supplies." (fn. 14)]

We conclude that the FERC's view is supported by substantial evidence. . . .

Affirmed.

Notes and Problems

1. Projecting future values of variables, as was done in South Dakota Public Utilities Commission v. Federal Energy Regulatory Commission, is a common use of regression analysis. Because the statistician is estimating historical trends to project future values of the variables, the procedure is sometimes called *trend-line analysis*. In the *South Dakota* case, the parties wanted to determine how rapidly the natural gas in the field would be depleted. Two figures are relevant to this determination: how much gas has already been discovered and how much will be discovered in the future. As an example of trend-line analysis, focus on projecting future discoveries. We have, in the table on page 311, data describing annual addition to reserves resulting from exploration efforts. There you will find the annual additions to reserves under the heading *Reserve Annual Bcf*. The independent variable here is years (1967, 1968, . . . , 1976), and the dependent variable is billion cubic feet (BCF) of annual addition to reserves. We can graph this data and calculate a corresponding regression line. The resulting equation is:

Annual Addition to Reserves $= 317{,}293 - 160.16 \times Year$

In 1980, what is the projected new discovery? In what year do we expect the last gas to be discovered? What does the b mean in this equation? Is this b significantly different from zero? Form and test a two-tailed null hypothesis about this b. Then test a one-tailed null hypothesis regarding this b. Do you reject the null hypothesis? Estimate the probability of a type 1 error if you did reject the null hypothesis.

2. The graph and plotted regression line for this data looks like Exhibit VII-7. The intercept, $a = 317{,}293$, is not on our graph. The number 317,293 is the projected (backwards) billion cubic feet of gas we would have discovered in year zero, obviously an absurd result. Generally speaking, the further a projection is from the dates in the sample the more unreliable it is. As an advocate, you can use this fact to challenge your opponent's projections. When time, e.g., years, months, or days, is used as an independent variable, the bivariate regression is referred to as a *time-series analysis*.

3. If you wish to practice calculating a bivariate regression line, project the New Field exploration effort for 1978, based on historical efforts.

EXHIBIT VII-7.
New Annual Discovery Times-Series Graph

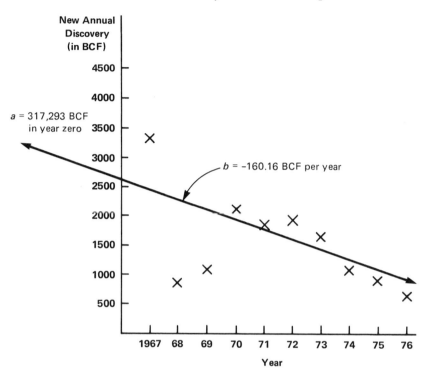

C. MULTIVARIATE REGRESSION

The basic concepts underlying bivariate regression extend to multivariate regression. The critical difference is that with multivariate or multiple regression techniques we can simultaneously identify the effects of more than one independent variable on a single dependent variable. The significance testing techniques used for bivariate analysis are used in exactly the same way for multiple regression.

Because the effects of numerous independent variables must be separated from one another in calculating the regression coefficient associated with each variable, the equations for the b's are very long and are usually taught using an advanced mathematical method called *matrix algebra*. Learning matrix algebra just for the purpose of statistical calculations is not worthwhile, particularly since the practicing attorney or her expert will use a computer for all calculations. Accordingly, we will discuss only the different forms of multiple regression and a new significance test, the F test, for summarizing the cumulative explanatory power of all the independent variables.

District Judge Higginbotham's decision in Vuyanich v. Republic National Bank, which follows, is an extraordinary elaboration of multiple regression techniques. When Judge Higginbotham presents the bivariate regression equation, he follows customary statistical practice by including the error term, u, as part of the equation:

$$y = a + bx + u$$

When more than one independent variable is used in a regression equation, each variable, x, is assigned a subscripted number, such as x_1, x_2, x_3, x_4, and so on. Thus the multiple regression line is represented by:

$$y = a + b_1x_1 + b_2x_2 + b_3x_3 + \ldots + u.$$

When discussing a particular observation of a given independent variable (for example, the sixth observation of the third variable) another subscript is added (e.g., $x_{3,6}$).

VUYANICH v. REPUBLIC NATIONAL BANK OF DALLAS
505 F. Supp. 224 (N.D. Tex. 1980)

HIGGINBOTHAM, District Judge. . . .

D. THE MATHEMATICS OF REGRESSION ANALYSIS

In today's society, among those who claim special insight denied the common run of men, the "esoteric language is mathematics; the special means of inspiration, the computer, the forbidden path of truth, science." B. Ackerman, et al., The Uncertain Search for Environmental Quality 1 (1974). Econometricians — who with multiple regression analysis can provide an important addition to the judicial toolkit necessary for reconstructing from bits and pieces of data the framework of past events — are no exception.

The practical use of multiple regression has grown markedly over the past 25 years due to the development of statistical methodology itself, increasing availability of statistical data, and most importantly, the development of the computer. Fisher, Multiple Regression in Legal Proceedings, 80 Colum. L. Rev. 702, 702 (1980). Regression

C. Multivariate Regression

analysis is increasingly being used in legal proceedings and commentary. Id.; Statistical Evidence on the Deterrent Effect of Capital Punishment: Editors' Introduction, 85 Yale L.J. 164, 167 n. 15 (1975) (Editor's Introduction). To record this court's understanding, correct or not, of the regression analyses presented by the experts in this case, and the limitations of those analyses — an understanding made necessary by technically complex trial challenges to their validity — we here describe how they are used, how they work, and when they do not. Cf. id. at 167 n. 15 & 169; Fisher, supra, at 702.

1. USES OF MULTIPLE REGRESSION

The two primary uses of multiple regression analysis can be illustrated through the following examples where such analyses have actually been used:

(i) For years after the disappearance of coal-burning locomotives, there was dispute on the preservation of the jobs of railroad firemen. One issue was whether the presence of a fireman on trains contributed to railroad safety.
(ii) Cable television systems (CATVs) have been involved in administrative proceedings where one issue is the magnitude of the effect of the entry and activity of CATVs upon the profits and growth of broadcast television stations. This issue presents such questions as the influence of CATVs on the viewing audience of particular stations and the effect of changes in a station's audience on its revenues. Of course some claim that such effects are small while others insist they are large.

Fisher, supra, at 703.

In the first case, multiple regression is being used to "test hypotheses"—does a particular variable[47] (presence or absence of

---

47. A variable is something that can take on different values. Since each variable can assume various values, it must be represented by a symbol instead of a specific number. For instance, we might represent price by P and imports by I. When we write $P = 3$ or $I = 18$, however, we are "freezing" these variables at specific values (in appropriately chosen units). A. Chiang, Fundamental Methods of Mathematical Economics 9 (2d ed. 1974).

Variables often appear in combination with "constants," such as in the expression $7P$ or $18.5I$. A "constant" is some number that does not change, and hence is the opposite of a variable. When a constant is joined to a variable, it is often referred to as the "coefficient" of that variable. We can let a symbol (for instance, "a") stand for a given constant; when symbols are used to designate constants, we often refer to them as "parameters." Id.

firemen) have *any* effect on some other variable (railroad safety). In the second case, multiple regression is being used for "parameter estimation"—there being little doubt that audience size affects television revenue, and the real question being *how much*. Fisher, supra, at 704.

Both the firemen and CATV cases above involve "conditional forecasting"—a prediction of what will happen to the "dependent variable"[49] (such as railroad safety) if an "independent variable"[50] (such as the number of firemen) is changed or, looking retrospectively, what would have happened to the dependent variable had the value of an independent variable been different. Fisher, supra.

Determining whether firemen do affect railroad safety faces two difficulties in the absence of multiple regression analyses. First, the factor whose influence one wishes to test or measure is usually not the only major factor affecting the dependent variable. Thus, for instance, the amount of traffic on the railroads affects accidents as well. If we could make controlled experiments, it would be easy to quantify the relationship. A controlled experiment here would involve varying number of firemen, traffic on railroads, and the other variables expected to affect the number of accidents one at a time, holding everything else constant and observing the resulting number of accidents. This would be difficult and costly. We are left then with analyzing nature's experiments. See id. at 705; cf. R. Wonnacott & T. Wonnacott, Econometrics 7 (2d ed. 1979). Second, even if the effects of other systematic factors can be accounted for, there typically remain elements of chance. Id. Falling objects follow a physical law, but the behavior of individuals does not. Cf. R. Wonnacott & T. Wonnacott, supra at 6.

2. ECONOMETRICS AND THE ORDINARY LEAST SQUARES FORM OF MULTIPLE REGRESSION ANALYSIS

Put in more formal terms, econometrics may be viewed as the science of model building, using quantitative tools used to construct and test mathematical representations of parts of the real world. R. Pindyck & D. Rubinfeld, Econometric Models and Economic Forecasts xi (1976). The fundamental underpinning of econometrics is the basic idea of relationships among economic variables. Relationships are grouped to form a model, the number of relationships included in an

---

49. Dependent variables are also referred to as "endogenous" variables. See G. Maddala, Econometrics 5 (1977).

50. Independent variables are also referred to as "exogenous variables" or "explanatory variables." Id.; Editors' Introduction, supra.

economic model depending on the objectives for which the model is constructed and the degree of explanation that is being sought.[51]

The models presented in this case involve, for the most part, one type of econometric modeling. They are "single-equation regression models." R. Pindyck & D. Rubinfeld, supra, at 1. Most of the single equation regression models are of a common variety: the behavior of the "endogenous" variable (a variable determined within the economic system under study) is assumed to be a linear function[52] of a set of "exogenous" variables (those determined outside the system),[53] and the variables are assumed to possess certain other properties such that the convenient "ordinary least squares" method of estimating the relationships among those variables can be used. R. Pindyck & D. Rubinfeld, supra, at 1,225; Editors' Introduction, supra; G. Maddala, supra n. 49, at 5; R. Wonnacott & T. Wonnacott, supra, at 334-35.

Multiple regression begins by specifying the major variables believed to affect the dependent variable. Fisher, supra, at 705. For instance, in our railroad example, we may wish to include as explanatory variables the number of firemen and the amount of railroad traffic, using as the dependent variable the number of railroad accidents. This involves using independent variables which reflect the important or systematic influences that may affect railroad safety. The "minor influences" are placed in a "random disturbance term," treating their effects as due to chance. Id. at 705-06. The relationship between the dependent variable and the independent variable of interest—for example, the relationship between the number of accidents and the number of firemen—is then estimated by culling the effects of the other major variables. Multiple regression is

---

51. For instance, the classic supply and demand model seeks to explain the sales of a commodity in a particular market in relation to price; it consists of three equations, namely, a demand equation, a supply equation, and a market adjustment equation. These equations will contain other variables in addition to the quantity and price of the commodity in question, such as disposable income in the demand equation and factor prices in the supply equation. The explanation achieved by the model is then conditional on the values of these other variables, variables which are not determined or explained by the model. J. Johnston, Econometric Methods 2 (2d ed. 1972).

52. We use the term "linear equation" or "linear function" for any relationship such as $Q = cA + fB + gC$ where $Q, A, B,$ and C are all variables and $c, f,$ and g are all constants. See W. Baumol, Economic Theory and Operations Analysis 14-15 (3d ed. 1972).

53. Certainly in a broad sense almost all variables are endogenous and the only exogenous variable one can think of are such things as weather. But in any particular econometric study, this is a matter of approximation. For instance, while studying the demand for gasoline by households, we can treat the quantity demanded as endogenous and income and price as exogenous, arguing that the household does not have control over these. G. Maddala, supra n. 49, at 5.

thus a substitute for controlled experimentation. Id. at 706. The results of multiple regressions—such as what we will call "coefficients" in the ordinary least square methodology—can be read as showing the effect of each independent variable on the dependent variable, holding the other independent variables constant. Moreover, relying on statistical inference, one can make statements about the probability that the effect described is due only to a chance fluctuation. Cf. id.

Central to the validity of any multiple regression model and resulting statistical inferences is the use of a proper procedure for determining what explanatory variables should be included and what mathematical form the equation should follow. The model devised must be based on theory, *prior to* looking at the data and running the model on the data. If one does the reverse, the usual tests of statistical inference do not apply. And proceeding in the direction of data to model is perceived as illegitimate.[54] Indeed it is important in reviewing the final numerical product of the regression studies that we recall the model's dependence upon this relatively intuitive step.

a. *Estimating Multiple Regression*

If the relationship of interest is to include only one independent variable ("x_1") to explain the behavior of a dependent variable ("Y"), and it is believed that the relationship of Y to x_1 is a straight line, then the relationship is expressed mathematically as:

$$(1) \quad Y = a + b_1 x_1$$

where a and b_1 are constants, Y is the dependent variable, and x_1 is the independent variable. Diagrammatically, the relationship is illustrated by the straight line in Figure 1. The econometrician thus seeks to determine the value of "a" (called the "intercept" or "constant") and the value of "b_1" (the "coefficient" or "slope" of x_1). Once he obtains these two numbers, for any value of x_1, he would know the exact value of Y. See G. Maddala, supra n. 49, at 74; J. Johnston, supra n. 51, at 122. Because there are random influences in life, it is unlikely that the relationship between Y and x_1 will be so exact. Instead, plotting values of Y against values of x_1 will likely produce a scatter of

---

54. See G. Maddala, supra n. 49, at 127; R. Wonnacott & T. Wonnacott, supra, at 88-90. Cf. Fisher, supra, at 714-15 ("[A] properly done study begins with a decent theoretical idea of what variables are likely to be important. . . . [A] study that casts about for a good-looking relationship by trying all sorts of possibilities is very likely to come up with relationships where none exist.")

C. Multivariate Regression

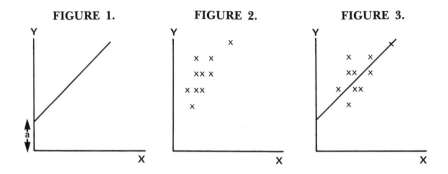

FIGURE 1. FIGURE 2. FIGURE 3.

points as in Figure 2. Thus the correct relationship is not described by equation (1) but instead by:

$$(2) \quad Y = a + b_1 x_1 + u$$

where u represents random influences and is called the "residual" or "error." See G. Maddala, supra n. 49, at 74. The econometrician attempts to cut through the noise generated by these random disturbances and extract the "signal"—that is, the line around which the points are scattered. He passes the line through the points so that it is as close as possible to the scatter of points, in the sense that the sum of the squared deviations between the predicted and actual Y values is minimized. Not inappropriately, this is called "least squares regression." G. Maddala, supra n. 49, at 75. See Figure 3.

In general, numerical estimates of a and b_1 are obtained by entering the scatter of points in a computer programmed to perform "least squares" calculations. These empirical estimates, based on the relatively crude "least squares" method of processing empirical observations, will be "good"—in that they are likely to be quite close to the true values—if and only if certain assumptions hold as to the nature of and relationship among the dependent variable, the independent variable(s), and the error term. See, e.g., R. Pindyck & D. Rubinfeld, supra, at 20-24, 55; R. Wonnacott & T. Wonnacott, supra, at 55-69; J. Johnston, supra n. 51, at 126; J. Kmenta, Elements of Econometrics 9-14, 161 (1971).

Regressions involving more than one explanatory variable (e.g. number of firemen as well as railroad traffic) are more frequently used and are known as "multiple regressions." Id.; R. Wonnacott & T. Wonnacott, supra, at 71. To illustrate, we may assume that there is a suspected relationship between Y, the dependent variable, and the explanatory variables, $x_1, x_2, x_3, \ldots x_k$, such that:

$$(3) \quad Y = a + b_1 x_1 + b_2 x_2 + b_3 x_3 + \ldots + b_k x_k + u$$

where $a, b_1, b_2, b_3, \ldots b_k$ are constants. As with a model with only one explanatory variable, a is called the intercept or constant; b_1, the coefficient of x_1; b_2, the coefficient of x_2; and so forth; and u, the "error term" or "residual." When empirical observations are placed in the computer programmed to do least squares manipulations (just as we input the "points" in the model with one explanatory variable), it will generate numerical estimates of $a, b_1, b_2, \ldots b_k$ based on how the dependent variable changes when the independent variables move in a variety of ways. Cf. Fisher, supra, at 712. These numerical estimates of $b_1, b_2, b_3 \ldots b_k$, are analogous to the b_1 in Figure 3. When only one explanatory variable is used, b_1 is the slope of the line: that is, it tells us how much Y will change for a unit change in the value of x_1. With a multiple regression, b_1 tells how much Y will change for a unit change in the value of x_1, holding the other explanatory variables constant; b_2 is the change in Y corresponding to a unit change in the value of x_2, holding the other explanatory variables constant; and so forth. Thus if the proper model for railroad safety were described by:

$$(4) \quad Y = a + b_1 x_1 + b_2 x_2 + u$$

where Y is the number of railroad accidents, x_1 is the number of firemen, and x_2 the miles of railroad traffic, we can "run" this model on the appropriate data, and obtain, for example, $Y = 50 + \frac{1}{2}x_1 + \frac{3}{4}x_2$. This equation tells us, among other things, that with each additional fireman, the number of railroad accidents is reduced by $\frac{1}{2}$ of an accident, while with each additional mile of railroad traffic, the number of railroad accidents is increased by $\frac{3}{4}$ of an accident.[57]

b. Statistical Inference

With certain assumptions about the particular probability distribution of the error term, one can go beyond estimating effects of

---

57. "Dummy variables" are sometimes used as explanatory variables. See R. Wonnacott & T. Wonnacott, supra, at 100-103 (use of dummy variable "1" for wartime and "0" for peacetime, in econometric modeling of sales of government bonds). "When some explaining factors assume only discrete values (such as "male" or "female") and others assume continuous values (such as age expressed in years, or, better, in months), a middle course between [analysis of variables] and regression can be effected by the use of multiple regression with 'dummy variables.' A dummy variable is given the values zero and one to correspond with the presence or absence of a particular attribute. For example, a regression could estimate an employee's wage by using the employee's years of work experience, years of education, and a dummy variable to reflect whether the employee were male. The weighting on the dummy variable would not be as readily interpretable as a weighting on a more usual type of explaining characteristic, but the question of the overall effect of the characteristic represented by the dummy variable could be answered." Note, Beyond the Prima Facie Case in Employment Discrimination Law: Statistical Proof and Rebuttal, 89 Harv. L. Rev. 387, 389 (1975) (footnote omitted).

independent variables on the dependent variable to gauging the certainty or accuracy of those effects. Fisher, supra, at 716; G. Maddala, supra n. 49, at 79. For instance, one can perform calculations that support statements about the range of values likely to contain the true coefficient of any of the explanatory variables.

In particular, one can determine the size of the interval on either side of the estimated coefficient which has a given probability (e.g., .95) of containing the true parameter. This range of variables is referred to as a "confidence interval." R. Pindyck & D. Rubinfeld, supra, at 31. For example, after performing the appropriate calculations, we might find that there is a 95% chance that the true coefficient of firemen in our railroad example is larger than $1/4$ and smaller than $3/4$. In other words, the 95% confidence interval for the firemen coefficient is $1/2 \pm 1/4$.

One can also perform calculations to test the hypothesis, at a given level of statistical significance,[58] that the true coefficient is actually zero; that is, that the independent variable to which it corresponds has no effect on the dependent variable. Fisher, supra, at 717. If what is known as the "t-statistic" for the particular coefficient is large enough, then we can reject the hypothesis that the true coefficient of the variable in question is equal to zero. See R. Pindyck & D. Rubinfeld, supra, at 31; R. Wonnacott & T. Wonnacott, supra, at 86. For example, if a 5% level of significance is used, a sufficiently large t-statistic for the coefficient indicates that the chances are less than one in 20 that the true coefficient is actually zero. Fisher, supra, at 717. The magnitude of the t-statistic necessary to reject such hypothesis varies with the desired level of significance. Thus the t-statistic would need to be larger with a 5% level of significance than with a 10% level of significance.

The magnitude of the t-statistic is also dependent on the particular type of hypothesis to be tested. When an analyst uses a "two-tailed hypothesis"—that is, where he wishes to determine whether there is *any* relationship, positive or negative, between the independent and

---

58. As this court explained in Cooper v. University of Texas, 482 F. Supp. 187, 194 (N.D. Tex 1979): "With a test of statistical significance we learn the probability that the observed value could have happened by chance, i.e., the probability that in a random sample of an appropriate test population the variable would exhibit a value as extreme as that observed. A test of statistical significance is thus itself based on the results of a hypothetical experiment. We suppose that an infinite number of samples of the same size are drawn from the test population. The probability of the observed value occurring by chance is equal to the proportion of the samples in which the value is at least extreme as the observed value. It has become a convention in social science to accept as statistically significant values which have a probability of occurring by chance 5% of the time or less. See generally N. Nie, C. Hull, J. Jenkins, K. Steinbrenner, & D. Bent, Statistical Package for the Social Sciences 222 (2d ed. 1975)."

the dependent variable — a t-statistic of approximately two means (in the case of large samples) that the chances are less than one in 20 that the true coefficient is actually zero. Cf. id. An explanatory variable is, however, usually included in the equation because of a prior theoretical reason for expecting it to affect the dependent variable in a specific direction. R. Wonnacott & T. Wonnacott, supra, at 86. For instance, we expect additional firemen to decrease, not to increase, the number of accidents. In such situations, we test a more specific hypothesis—a "one-tailed hypothesis"—not whether or not a particular coefficient is positive or negative as with the two-tailed test, but whether or not, in our railroad example, it is negative or zero. In this circumstance, the "one-tailed test" is one in which 5% would be the probability of observing some *negative* coefficient if the true value were zero. Cf. Fisher, supra, at 717 n. 26. The t-statistic required for significance at the 5% level on a one-tailed test is only approximately 1.6. Fisher, supra, at 717 n. 26.

Speaking purely from a statistical point of view, this does not mean that only results significant at the 5% level should be considered; less significant results may be suggestive. Id. at 718-19. Thus even if the t-statistic is small for a particular dependent variable, this does not mean there is no relationship between that dependent variable and the independent variable. R. Wonnacott & T. Wonnacott, supra, at 88. This is so because the size of the t-statistic corresponding to statistical significance is ultimately dependent on the level at which statistical significance is arbitrarily set. "The most commonly used significance level is five percent, but this is purely arbitrary." G. Maddala, supra n. 49, at 45. It follows that in many instances, reporting only whether a particular coefficient is significant or not should be avoided; it is more informative to report confidence intervals, test statistics, or other quantitative measures of significance. Id. at 45-46.

Moreover, significance tests and confidence intervals are controversial when the test data includes an entire universe of decisions. For example, in a promotion case, data on every promotion decision may be available, rather than a mere random sample of such decisions. When all such data are available, some statisticians argue that statistical tests are either meaningless or misleading, while others say that they can be useful. D. Baldus & J. Cole, [Statistical Proof of Discrimination (1980)] §9.32, at 316. See also Testimony of Dr. John Spaulding, Jr., at 1094-95; Johnson Post-Trial Brief at 4 n. 4.

Considering the arbitrary nature of the adoption of the 5% level of significance, it is not surprising that courts show flexibility in determining what level of significance to be required in a legal context. Cf., e.g., United States v. Georgia Power Co., 474 F.2d 906, 915 n. 11 (5th

Cir. 1973); D. Baldus & J. Cole, supra n. 55, §9.221, at 308 n. 36 & §9.41, at 318. Indeed, the Fifth Circuit has specifically stated that a 10% level of significance rather than the statisticians' more conventional and more stringent 5% level *"might* be acceptable" in the context of the validity of job tests under Title VII. Watkins v. Scott Paper Co., 530 F.2d 1159, 1187 n. 40 (5th Cir.), *cert. denied,* 429 U.S. 861, 97 S. Ct. 163, 50 L. Ed. 2d 139 (1976) (emphasis in original).[59]

Statistical tests more elaborate than t-statistics and confidence intervals for individual coefficients are also available. For instance, "R^2" is often calculated; this is a statistic which provides an overall index of how well Y can be explained by *all* the independent variables, that is, how well a multiple regression fits the data. R. Wonnacott & T. Wonnacott, supra, at 180-81. The higher the R^2, the greater the association between movements in the dependent and independent variables. Fisher, supra, at 720. There are problems, however, associated with the use of R^2. A high R^2 does not necessarily indicate model quality. See R. Pindyck & D. Rubinfeld, supra, at 58; G. Maddala, supra n. 49, at 122-24. Cf. Fisher, supra; D. Baldus & J. Cole, supra, §8.22, at 266-67. For instance, the addition of more explanatory variables to the regression equation can never lower R^2 and is likely to raise it. Thus one could increase R^2 by simply adding more variables, even though, because of "over-inclusion" and "multicollinearity" (terms we will later describe) it may be improper econometrically to do so. Cf. R. Pindyck & D. Rubinfeld, supra. . . .

E. ECONOMETRIC INDICATION OF DISCRIMINATION

In econometric terms, one way used (here and in the economic literature) to study differences among individuals' wages is to estimate a regression such as:

(7) $\quad Y = a + b_1 x_1 + b_2 x_2 + \ldots + b_k x_k + u$

where Y is the level (or natural logarithm) of earnings, income or wage rate, and $x_1, x_2, \ldots, x_k$ are observable productivity character-

---

59. We are discussing the concept of statistical significance, not "practical significance." "A measure of statistical significance . . . provides a basis for inferring whether there is, in fact, some disparity in results for the minority and majority groups. Practical significance, in contrast, refers to the magnitude of the disparity between the results for the two groups. This information is conveyed by the measure of disproportionate impact and provides a basis for assessing the level of injury to the minority group." D. Baldus & J. Cole, supra, §9.41, at 317 (footnote omitted). In Title VII cases, "[t]he degree of [d]iscrimination practiced by an employer is unimportant. . ."; "[d]iscriminations come in all sizes and all such discriminations are prohibited by the Act." Rowe v. General Motors Corp., 457 F.2d 348, 354 (5th Cir. 1972). See Hodgson v. American Bank of Commerce, 447 F.2d 416, 420 (5th Cir. 1971).

istics. Cf. Blinder, Wage Discrimination: Reduced Form and Structural Estimates, 8 J. Human Res. 436, 437 (1973). To illustrate, we could assume that the employee's wages are a function of his years of schooling and years of relevant experience, both factors being postulated by human capital theory to be productivity-enhancing. In such a model, Y would be earnings, x_1 would be the number of years of schooling, and x_2 would be the number of years of relevant experience. If the assumptions necessary for ordinary least squares hold, *and* if this model is a valid one, then by running the model on actual data for the firm, we would obtain actual numbers for a, b_1, and b_2. To determine the predicted earnings for any employee, we would plug his particular values of x_1 and x_2 into equation 7, using the estimates of a, b_1, and b_2.

Recalling from our discussion in section VI(B), supra, there are three ways the econometric models, if true mathematical representations of real-world behavior, may indicate that the employer is discriminatory: (1) use of non-job-related criteria with disparate impact; (2) unequal treatment of twins; and (3) improper treatment across twins.

The first type of indication of discrimination may occur, for instance, where the employer states that he behaves according to a certain model, and that model includes explanatory variables which do not meet the legal standard of business necessity. The process of mathematically representing an employer's behavior should at this stage be external to the legal framework, such as those concerned with the legality of the employer's use of various predictors. In determining whether the employer is violating Title VII through the use of non-job-related criteria with disparate impact, one must follow a three-step procedure:[65]

(i)   On purely econometric grounds, derive a mathematical model which best represents how the employer behaves. At this stage, it is not necessary for there to be *more* than business purpose to justify inclusion of a variable. Indeed, to require such a higher standard would be flawed methodologically. This is not to say that all relevant productivity-related variables should always be included: the problems which may be introduced by overinclusion and multicollinearity, described in section VI(D), supra, must always be considered.

---

65. Simply because a variable in an equation has a statistically significant coefficient associated with it does not prove it is job-related. All it may prove is that the employer has used the variable as a predictor. The law requires that predictors with disparate impact be job-related. Contra, Gwartney, Asher, Haworth, & Haworth, supra n. 40, at 655-58.

(ii) Examine the variables with disparate impact and see if they are adequately justified under the "business necessity" standard.
(iii) For those variables justified under (ii), see if those variables have been subject to group status-based manipulation by the employer.

Step (ii) above is in part a restatement of the job-relatedness requirement:

> Once the racially adverse impact of an examination is demonstrated by statistical or other evidence, the burden of proof shifts to the employer to prove that the exam is job-related. Employment *practices* having a racially adverse effect are subject to the same scrutiny: if a prima facie case of discriminatory practice is shown, it becomes the employer's burden to demonstrate the job performance validity of its practices. The employer's burden is not satisfied by establishing merely a rational basis for a test; the test must be validated. Validation, in general, requires a demonstration that "the qualifying tests are appropriate for the selection of qualified applicants for the job in question." Validation by any one of a number of methods of making such a showing, suffices to refute statistical evidence that a test has had a disproportionate racial impact. The same sort of validation will rebut the inferences drawn from statistical evidence of racial discrimination based on work practices.

Scott v. City of Anniston, 597 F.2d 897, 901-02 (5th Cir. 1979), *cert. denied*, 446 U.S. 917, 100 S. Ct. 1850, 64 L. Ed. 2d 271 (1980).

What is to be proven in job-relatedness is separate from the methods used in such proof. The employer must show that he is using the predictor to the same extent as would a race- (or sex-) blind employer.

Finally, predictors must be such that if "the legitimate ends of safety and efficiency could be served by a reasonably available alternative system of classification with less discriminatory effect, the classification cannot be continued." Note, 12 Ga. L. Rev. 104, 106 n. 12 (1977). More specifically, the Fifth Circuit has held that:

> [F]or a practice, which is not intentionally discriminatory or neutral but perpetuates consequences of past discrimination, to be justified by business necessity, the practice must "not only foster safety and efficiency, but must be essential to that goal . . . and there must not be a acceptable alternative that will accomplish that goal equally well with a lesser differential racial impact." Parson v. Kaiser Aluminum & Chemical Corp., 575 F.2d 1374 (5th Cir.), *cert. denied*, 441 U.S. 968, 99 S. Ct. 2417, 60 L. Ed. 2d 1073 (1979).

Swint v. Pullman-Standard, 624 F.2d 525, 536 (5th Cir. 1980).

328   Chapter VII. Regression Analysis

Step (iii) of our three-step analysis relates to use of predictors which are susceptible to manipulation by the employer. If a predictor which, while not valueless as a predictor of productivity, is subject to manipulation by a racist employer, the econometric results generated may well be biased in favor of a finding of nondiscrimination. Thus, for instance, in James v. Stockham Valves & Fittings Co., 559 F.2d 310 (5th Cir. 1977), *cert denied,* 434 U.S. 1034, 98 S. Ct. 767, 54 L. Ed. 2d 781 (1978), the defendant presented an earnings regression study which included a merit rating as one of its explanatory variables. The court observed, "If there is racial bias in the subjective evaluations of white supervisors, then that bias will be injected into . . . [the] earnings analysis." Id. at 332. Use of manipulable predictors in multiple regressions may cloak the employer's "personal biases in the mantle of a scientific judgment."

Since to include manipulable predictors as explanatory variables may be to bias the econometric results in favor of the employer, if despite such inclusion the plaintiffs are able to show non-productivity-related pay differentials, the results are all the more impressive. But in the defendant's hands, while models which include manipulable predictors among the explanatory variables are not valueless, especially if those predictors are shown not to have been manipulated, they may be suspect.

The second way in which econometrics may indicate discrimination occurs when the coefficients indicate that a white and a black are being rewarded differently for the possession of the same productivity characteristic(s): there is not equality of treatment of twins. Referring to the econometric model described in the beginning of this section (wherein x_1 is the number of years of schooling and x_2 is the number of years of relevant experience), we can see that there are at least two possible econometric indications of this sort of inequality:

1. First, one can run the same equation on data twice, the initial time running it on data solely for one group, and second time on data solely for the other group. After doing this, one compares the coefficients obtained for the two groups.

So if we obtain for males $Y = 10000 + 500x_1 + 350x_2$, and for females $Y = 9000 + 400x_1 + 300x_2$ at the same job, it is clear that females are being discriminated against. That is, any female regardless of her particular combination of schooling and experience is receiving less than a corresponding male: since the regression coefficients (a, b_1, and b_2) are each higher for males than for females, males are being rewarded more for each productivity-related characteristic. This is the simple situation, where it is clear that females are being discriminated against—where a (female) is less than a (male), *and* b_1

(female) is less than b_1 (male), b_2 (female) is less than b_2 (male), and so on.

A more difficult case would be that where males are rewarded more for one characteristic, while females are rewarded more for another: a "mixed" case. Here some men would be paid more than women with identical portfolios of productivity characteristics, while some women would be paid more than men with identical portfolios of productivity characteristics. For instance, suppose that $Y = 10000 + 500x_1 + 350x_2$ for males and $Y = 10000 + 400x_1 + 1000x_2$ for females.

(a) A woman with no schooling and two years of experience will be paid $10000 + 0 + 2000$ or $12,000; a similar man will be paid $10000 + 0 + 700$ or $10,700. Here a woman earns more than a man with the identical portfolio of productivity characteristics.

(b) A woman with three years of schooling and no experience will be paid $10000 + 1200$ or $11,200; a similar man will be paid $10000 + 1500 + 0$ or $11,500. Here a woman earns less than a man with the identical portfolio of productivity characteristics.

In such a "mixed" case, one might, for example, offer statistical evidence to show that as a result of different regression coefficients for the two groups, because of the empirical distribution patterns of productivity characteristics in the two groups one group is favored at the expense of the others.

One manner of comparing the regression coefficients for the two groups without actually running two separate regressions as described above is to use what are called "interaction" terms. By adding such terms to the regression equation to be estimated, one can produce estimates of the difference in the weights placed on each qualification. Thus, if we added an interaction term for race and experience, and one for race and schooling, the coefficient of the race-experience interaction term indicates the difference between the coefficient for experience for whites and blacks, and the coefficient of the race-schooling interaction term indicates the difference between the coefficients for schooling for whites and blacks. See D. Baldus & J. Cole, supra §8.123[2], at 262-64.

2. The other major way of detecting discrimination between two groups is to use "dummy variables." This method, in contrast to the more general method outlined above, can only be used if it is *assumed* that *all* the coefficients for the explanatory variables are exactly the same for the two groups being compared; that the only difference

with being female (or black) is represented by one overall effect (that is, the coefficient of the "dummy variable").

Thus, suppose that x_1 is the "dummy variable" for race ("1" is input for being black, and "0" for being white) and x_2 and x_3 are the two productivity predictors. If the multiple regression is run and we obtain $Y = \$5000 - 700x_1 + 300x_2 + 400x_3$, then this means that a black will earn \$700 a year less than an equivalent white. Cf. D. Baldus & J. Cole, supra, §8.02, at 242; R. Wonnacott & T. Wonnacott, supra, at 100-103.

The third way an employer may not be treating people solely according to productivity differences is what we have termed improper treatment across twins: (i) econometrically, this may mean that a particular race- (or sex-) correlative personal productivity characteristic may be rewarded by a racist (or sexist) employer more than it would be by a nonracist (or nonsexist) employer or (ii) the relative pay of various jobs may not correspond to their "worth." As for (i), suppose that a nonsexist employer would pay employees of both sexes according to the equation $Y = 5000 + 300x_1 + 400x_2$, and that men tend to possess higher amounts of x_1 and lower amounts of x_2. If a sexist employer pays instead according to the equation $Y = 5000 + 600x_1 + 400x_2$, then men are being paid more than is actually justified by productivity considerations. The weighting is improper.

In this section we have been discussing econometric detection of an employer's consideration of group status in addition to his consideration of productivity differences among individuals. In certain rare circumstances, such consideration of group status might not be violative of Title VII. Finally, we must emphasize that while we do examine certain indicators of discrimination — such as coefficients of dummy variables for group status — we do not mean to canonize particular indices of discrimination. To do so may encourage employers to focus unduly on improving the indices, despite all their shortcomings, rather than on ensuring real-world equality of opportunity. Moreover, if employers were to do so, the value of the indices as accurate indicators of discrimination would be depreciated. Cf. B. Ackerman, et al., supra, at 28-30 (dangers of using a simple number as proxy for an environmental problem). . . .

PRESSEISEN v. SWARTHMORE COLLEGE
442 F. Supp. 593 (E.D. Pa. 1977)

BEECHTLE, District Judge. . . .

IX. SALARY

Plaintiffs, in their attempt to establish a prima facie case that Swarthmore has a policy of discriminating against women with respect to salary, relied on statistical testimony and eleven individual cases of alleged discrimination. Defendants attempted to show that plaintiffs' statistics with respect to salary were unreliable and also introduced their own statistics with respect to salary.

Our discussion of the salary issue must, of course, begin with plaintiffs' Exhibit 43A. This exhibit shows that, as Dr. deCani admitted several times in his testimony, once you hold rank constant, there is no significant disparity in the average salaries of men and women. An examination of P-43A reveals that, except for academic years 1971-72 and 1972-73, at the Assistant Professor level, there is no significant statistical disparity between men and women in the same rank. In other words, any differences between men and women within the same rank could be due to chance alone. Plaintiffs contend, however, and Dr. deCani testified that, by comparing means and medians, the women for most years made less than men within the same rank. While that is true at the Assistant Professor level, at the Associate Professor level, mean-wise women made more than men four out of five years studied; median-wise, three out of the five years studied; and were equal in one out of the remaining two years. Also, at the Professor level, women had a higher salary median- and mean-wise in one out of the five years studied. Since plaintiffs were unable to show that Swarthmore had a policy of discriminating against women with respect to salary within rank, plaintiffs attempted to establish a prima facie case of salary discrimination by statistical analysis which excluded rank as a factor. In his multiple-regression analysis (discussed more fully infra), Dr. deCani excluded rank as a variable since, in his view, the inclusion of rank in the salary regression concealed the fact that women at Swarthmore took longer on the average to reach a given rank than did men. This testimony, of course, was based upon Exhibits P-45A through P-48A, which we have thoroughly discussed in the Promotion section. Swarthmore acknowledges that knowing whether women and men were promoted relatively equally with regard to time of promotion is *crucial* to the salary studies to make certain that there are no differences in the way men and women achieve rank. Accordingly, the precise issue before the Court is whether Dr. deCani was properly justified in excluding rank in his multiple-regression analysis. The answer is a simple one, and that is that he was not properly justified in excluding rank. The explanation for our conclusion has already been discussed in detail in the Promotion section. The Court has already found that Swarthmore does not have a policy of regularly and routinely discriminating against women with respect to promotions. Since P-43A shows that women are not discriminated against with respect to salary once you hold rank constant, the plaintiffs have failed to establish a prima facie case that Swarthmore discriminates against women with respect to salary. In addition, even if we were to assume that the eleven individual acts of discrimination were meritorious, the Court still would not find that plaintiffs have made out a prima facie case because all that you would be left with would be eleven individual examples, since there would be no statistical testimony to buttress.

However, even if Dr. deCani were correct in excluding rank from his multiple-regression analysis, we would still find that plaintiffs have not established by a preponderance of the evidence that Swarthmore discriminated against women with respect to salary.

Dr. deCani testified on the basis of a multiple-regression analysis he performed on the salaries for the full-time faculty at Swarthmore for the years 1971-72 through 1975-76, that sex appeared as a statistically significant factor in each of the five years in the study. [Hereinafter referred to as P-42.] In the multiple-regression analysis of salaries, Dr. deCani included the variables of sex, age, years since highest degree, years at Swarthmore, degree and division. Dr. deCani excluded rank as a variable in the regression analysis because, as mentioned above, the inclusion of rank in the salary regression, in his view, concealed the fact that women at Swarthmore College took longer on the average to reach a given rank than did men. Without going into the intricacies of a statistical definition of exactly what a multiple-regression analysis is, the plaintiffs' multiple-regression analysis of salaries at Swarthmore shows a correlation between the various factors mentioned above, *excluding rank,* which, when controlled for, shows that sex is a statistically significant factor in the average difference of salaries between men and women. The chances of a disparity as large, or larger, as that found by Dr. deCani in each of five years studied in P-42, which would be due solely to chance, is .01 for the first three years and .001 for the last two years, or less than one in one-hundred or one in one-thousand, respectively, using a conservative test of significance.

C. Multivariate Regression 333

From the above, plaintiffs contend that they have shown clearly that sex or a sex-dependent factor was statistically significant in effecting the salaries of men and women on the faculty of Swarthmore from 1971 to 1976. If defendants had not offered any evidence which sought to discredit Dr. deCani's multiple-regression analysis, and if this Court had concluded that Dr. deCani was justified in excluding rank from his analysis, then it is most likely that it would have been appropriate for the Court to find that plaintiffs made out a prima facie case with respect to salary; that is, excluding rank, the evidence would have shown that Swarthmore has regularly and routinely discriminated against women with respect to salary. However, for the reasons stated below, the Court finds that the defendants adequately rebutted such a would-be prima facie case by essentially showing that Dr. deCani's multiple-regression analysis was unreliable. Dr. Meier testified for Swarthmore that the multiple-regression analysis performed by Dr. deCani was unreliable for basically three reasons: First, there is a potential for bias that results from the problem of the inactives which we have already discussed in the Promotion section. Second, there is a potential bias introduced through the aggregation of junior and senior appointments. In other words, Dr. Meier testified that "in an observation study where different strata can be identified, strata that are dealt with differently, it is generally not appropriate to aggregate such strata prior to making an analysis." The Court notes that the different strata were identified by Mr. Pagliaro, the present Provost of the College, when he testified that, with respect to senior appointments, i.e., Associate Professors and full Professors, the practice of Swarthmore is to generally appoint them for five-year terms; that, with respect to junior appointments, i.e., Instructors and Assistant Professors, the practice has been to appoint them for one- to three-year terms. Dr. Meier went on to explain that the most obvious potential for bias with respect to the aggregation of different strata would arise if the pools from which such candidates are selected are different. He said: "[i]f, for example, it is the case — and I state this hypothetically — that the pool from which senior appointments are drawn is predominantly male and the pool from which junior appointments are drawn is mixed, the result of aggregation will be to mix the effect of seniority with the effect of sex in assessing the average length of initial appointment and make it impossible properly to sort out what policies may be due to seniority and what policies may be due to sex." Finally, Dr. Meier testified that, although the regression analysis allows for different intersects, it does not allow for possibly different slopes or different rates of change of salaries in different divisions. Dr. Meier went on to explain by illustrative example what he meant by the problem of the non-allowance for possibly different slopes:

> Let's suppose that one area, let me choose Physics, is predominantly a male area, that the situation for hiring in Physics and promotion is highly competitive, that is to say that the college competes for a scarce resource and must not only pay high initially but must raise rates rapidly; and contrast that to another area, let us suppose English, where there are a great many candidates, the competition that the college faces for recruitment is not so heavy and they might not raise people as rapidly in that area.
>
> If it were the case that in the departments such as Physics that were male-dominated and in departments such as English that were more mixed, then the differences between policies as regards Physics and English would show up in this analysis as affecting males and females, and that I would describe as a bias in the analysis as given.

The final criticism emanates from the testimony of Dr. Hollister, who the plaintiffs called in rebuttal. He testified that Dr. Iversen's regression analysis (D-56) was unreliable since it did not include such factors as scholarship, teaching ability, publications, some assessment of teaching ability, quality of degree, career interruptions, career continuity, quality of publications, administrative responsibility and some measure of committee work. As before, it is obvious that the criticisms of Dr. Hollister apply equally as well to the regression analysis of Dr. deCani. In fact, all parties concede that teaching, publications and research, and service to the community are factors in determining rank and salary at Swarthmore and that there is no way to measure accurately and reliably such data at Swarthmore to include it in a statistical analysis. The final criticism of Dr. deCani's regression analysis comes not from any testimony offered by any of the parties, but from an observation by the Court. Although there is no question that Dr. deCani is an eminently qualified statistician, we do not believe that his expertise extends into the discipline that governs what goes into the factors to be considered in assessing the amount of an individual's salary at a college. Although Dr. deCani is the chairman of the Statistics Department of the University of Pennsylvania, he was neither called nor qualified as an expert in that area of college administration concerned with the responsibility of the setting of salary. It seems to the Court that it was the plaintiffs' burden to call a witness especially qualified to testify with respect to this area. Thus, plaintiffs could have called an equivalent of the Provost of Swarthmore College from, for example, the University of Pennsylvania or a small liberal arts

C. Multivariate Regression 335

college like Swarthmore. He or she could have then testified as to what factors typically go into the setting of salaries of a college faculty member. Armed with that type of testimony, Dr. deCani perhaps could have made a more reliable analysis.

The defendants then offered their own evidence through the testimony of Dr. Iversen. He began by stating that the statistical analysis of faculty salaries cannot substitute for a critical analysis of the circumstances of the particular cases, especially at an institution such as Swarthmore where the number of faculty members is so small and the College's need so particularized for special, unique and diverse skills. With those limitations in mind, Dr. Iversen went ahead and did his own multiple-regression analysis. That analysis is contained in Exhibit D-56, which exhibit is an analysis of faculty salaries of full-time and regular part-time people for the years 1971-72 through the fall of 1976. The variables that Dr. Iversen used in his regression analysis differ somewhat from the variables used in Dr. deCani's. Dr. Iversen included rank, division, sex, years in given rank and degree. Exhibit D-56 shows that the sex variable does not yield a statistically significant disparity. In 1971, there was no measurable difference in the average salaries between men and women. In 1972, women on the average made $340 less than men. In 1973, women made on the average $292 less than men. In 1974, they made on the average $153 less than men. In 1975, women on the average made $303 less than men. In 1976, women on the average made $211 less than men. The P values for those years range from .37 through .66. P values are the means of measuring the probabilities that a given effect differs from zero by chance alone. Since a small value of P, i.e., less than .05 (for example, .04, .03, etc.), indicates an effect is statistically significant; and since all the P values in D-56 are not less than .05, the average differences mentioned above could be attributable to chance alone. In addition, Dr. Iversen conceded that his multiple-regression analysis does not take into account variables such as teaching, scholarship and the rest of the variables that Dr. Hollister testified were important to include in a regression analysis. Also, Dr. Iversen conceded that his regression analysis suffered from the same limitations as expressed by Dr. Meier. Finally, Dr. Iversen testified that he made an independent analysis of salaries within divisions, degrees and treating juniors and seniors in different categories, and using as his base point of departure Dr. deCani's regression analysis. He stated:

> I spent a considerable amount of time with [Dr. deCani's] equation but realized that it should be modified and I tried following the thought

that salaries move differently in the different divisions and for people with different degrees they may not get the same raises and it also struck me that the difference between junior and senior appointments would be important.

So I built these things into the analyses allowing for different slopes and found that the difference between men and women completely disappeared.

Plaintiffs did not at any time meet this testimony by Dr. Iversen.

It is true that Dr. deCani attempted to respond to the criticism regarding differences in divisions. Dr. deCani testified with respect to division one as follows:

You must remember the sample signs are very small. The maximum sample signs in division one was eight, which was 1973, '74, and '75. There are eight women. So you are not going to get anything with much significance; but it turns out that in 1973 I got a coefficient for women, which was significant at the five percent level, even with the sample size as small as eight.

There is one other thing. The sign of the coefficient tells you the direction of the disparity. That is the sign negative. That means the expected salary for women is less than the expected salary for a similarly situated male. In all five years the sign of this coefficient was negative.

Now if there were no real differences between men and women you would expect that sign to be positive half the time and negative half the time.

The probability of getting five out of five minuses — if you toss a coin, the probability of getting five out of five heads is one in 32, which is less than .05.

And so on that basis, on the basis of this very simple sign test, there is a significant disparity between the salaries of women and the salaries of similarly situated men in division one.

Dr. deCani did not run a similar test on divisions two and three.

When questioned about the above testimony, Dr. Meier testified that the sign test was an inappropriate calculation. He stated:

[t]he calculation that the probability, if there were no sex bias, of getting the same sign five times in a row assumes, of course, that the observations being referred to are independent. In this case in successive years one is dealing largely with the same set of people.

Thus, if you had a group of individuals and it happened that Group A, the men, were higher paid in 1966 than the women were, it would not be independent judgment that you might expect them also to be higher paid in 1967. These are not independent analyses and there really is no warrant for making the calculation one over two to the fifth power. There are one or two other minor points, but I think that is the major issue, and that calculation is simply inappropriate.

C. Multivariate Regression   337

Under cross-examination during his rebuttal testimony, Dr. de-Cani agreed with Dr. Meier's observation:

> Let me say the probability was calculated on the assumption that the successive years were independent. Dr. Meier's observation is quite correct. We are dealing with the same group of people in each year and, therefore, the observations in the successive years are not independent. The fact that the sign persisted means that the difference persisted also.

Thus, we can see that Dr. deCani did not adequately meet Dr. Meier's and Dr. Iversen's criticisms with respect to differences within divisions.

Plaintiffs in rebuttal then attempted to discredit Dr. Iversen's regression analysis. First, through the cross-examination of Dr. Iversen, plaintiffs show that Dr. Iversen included in his regression analysis the people that he excluded from his promotion studies. Dr. Iversen admitted that such exclusions and inclusions within the regression analysis had the potential for bias. Also, Dr. Iversen included regular part-time faculty in his regression analysis. He could not explain, however, the inflation factor that was consistently used to establish each regular part-time faculty member's full-time equivalent salary. Plaintiffs contend that, because of the possibility of over-estimation of full-time equivalent salaries for regular part-time faculty, the regression analysis is susceptible to considerable bias. In addition to the cross-examination of Dr. Iversen, plaintiffs presented another criticism through the testimony of Dr. Hollister. As mentioned earlier, Dr. Hollister testified that Dr. Iversen should have included more variables than he did in his rank analysis, such as publications and teaching experience. However, as we also stated earlier, those criticisms apply equally to Dr. deCani's regression analysis. Finally, Dr. deCani, while rejecting the potential for bias, during the trial reworked his regression analysis on male and female faculty at Swarthmore for the years 1970 through 1975. He attempted to eliminate the inactive bias by excluding from the new regression any people who were on the faculty prior to 1966. He also attempted to eliminate the baseline variable problem by excluding people who had not attained their highest degree by 1971, the first year of the regression study. This new regression study is contained in Exhibit P-642. The results of this reworked regression analysis were very similar to the results of the first regression analysis. In each of the five years studied, sex was statistically significant on both P-642 and P-42 and the percentage differences are about the same. Thus, Dr. deCani concluded that the inclusion or exclusion of inactives and baseline variables (years since highest degree) produce only a negligible bias. Although Dr. deCani attempted to meet Dr. Meier's criticisms with respect to the baseline variables and the inactives problem, Dr. deCani did not meet Dr. Meier's criticisms concerning the differences in division nor did he separate junior from senior appointments, nor did he include the Hollister variables, when he reworked his regression analysis.

The Court makes the following observations with respect to the above-described evidence. As we can see from Dr. deCani's regression analysis, to begin with P-42, if you exclude rank, which he did, sex is a statistically significant variable in each of the five years he studied. As mentioned at the outset, if that is where the testimony began and ended, then plaintiffs would have a very strong case. However, as set out above, Dr. deCani's regression analysis is suspect due to the inactives problem, the aggregation of junior and senior appointments, the failure to allow for different slopes within divisions, and the failure to include the Hollister variables. Moving over to Dr. Iversen's regression analysis (D-56), he found that sex was not a statistically significant variable for the years that he studied. Again, as set out above, Dr. Iversen's regression analysis is suspect due to some of the biases brought out by plaintiffs under cross-examination — namely, the inclusion in the regression analysis of persons excluded from his promotional studies, and the lack of explanation with respect to the inflation factor concerning regular part-time faculty members' full-time equivalent salary. Also, the fact that Dr. Iversen admitted that his analysis suffered from the same problems that Dr. Meier pointed out with respect to Dr. deCani's regression analysis. With respect to plaintiffs' Exhibit 642, which is Dr. deCani's reworked regression analysis, there is no question that Dr. deCani attempted to correct for the inactives problem and the baseline variable problem. However, Dr. deCani did not correct for the potential bias introduced by the problem of the aggregation of junior and senior appointments and the different slopes within the division. It seems to the Court that each side has done a superior job in challenging the other's regression analysis, but only a mediocre job in supporting their own. In essence, they have destroyed each other and the Court is, in effect, left with nothing. Again, it is interesting to note that plaintiffs failed to meet Dr. Meier's and Dr. Iversen's criticisms with respect to aggregation of junior and senior appointments and most especially with respect to different slopes within division. As mentioned above, plaintiffs never rebutted Dr. Iversen's testimony wherein Dr. Iversen stated that he reworked Dr. deCani's original regression allowing for different slopes, and found that the difference between men and women completely disappeared. Perhaps the only conclusion that one can reach, on the basis of all the above-described testimony, is that it is almost impossible, in the fact situation that the Court is presented with, to measure the differences in salaries between men and women by statistical analysis. Finally, if we simply conclude that in any multiple-regression analysis, by definition, there have to be biases, and if, in order to arrive at a result based on the analysis, we simply disregard those biases, what is the Court left with? We are left with Exhibits P-42 and P-642 versus Exhibit D-56. Essentially, the only difference between the two regression analyses is that Dr. deCani excluded rank from his analysis, whereas Dr. Iversen included rank in his analysis. Thus, we have come full circle back to the original problem—whether rank should be included or excluded in the multiple-regression analysis. We have already, in detail, gone through the reasons in the Promotion section why we believe that Swarthmore does not have a policy of discriminating against women with respect to promotion. Since the Court has found that rank, *in this particular case,* does not conceal more than it reveals about the effect of sex on salary, Dr. deCani improperly excluded rank from his regression analysis. By the same token, Dr. Iversen properly included rank in his regression analysis. Thus, whether the Court finds that Dr. deCani's regression analysis was at the outset improper because it excluded rank; or that the regression analysis is suspect because it does not take into account the criticisms as described in detail above; or whether we find that defendants adequately rebutted plaintiffs' prima facie case with respect to salary; or whether we find that the salary issue is inherently incapable of being analyzed through statistics, the conclusion is the same. Accordingly, the Court finds that Swarthmore does not regularly and routinely discriminate against women with respect to salary.

Notes and Problems

1. *Multicollinearity.* The statistical duel in Presseisen v. Swarthmore College is typical of judicial discussions of multiple-regression evidence. The question of which variables to include in a regression is always a critical one, and the inclusion or exclusion of the independent variable "rank" became the centerpiece of the debate. As Judge Higginbotham pointed out in Vuyanich v. Republic National Bank, the first step in doing a multiple regression is a theoretical one, that is, determining what independent variables are likely to affect the dependent variable. Because the purpose of the regression technique is to account for changes in the dependent variable, it is appropriate to include all measurable variables thought to have an influence on the dependent variable. Theoretically, one would expect faculty rank

to be relevant to salary determination. Why then did Dr. deCani, the plaintiffs' expert in *Presseisen,* exclude rank? His explanation was that inclusion of rank would conceal the fact that Swarthmore took longer on average to promote women than men to a higher rank. His claim was that sex and rank were correlated. The value of an independent variable must not be affected by the value of any other independent variable. Two or more independent variables that are correlated have values that are interdependent and by definition are not independent from one another. When two or more independent variables are correlated, the regression coefficients calculated by ordinary least squares may be misleading. This problem is referred to as *multicollinearity.*

Look more closely at the *Presseisen* example. The regression coefficient on each independent variable in the salary equation theoretically indicates the independent effect of changing the value of that variable by one unit while holding all other independent variables constant. The coefficient must have this meaning if we are to detect the separate influence of each variable. In *Presseisen,* we are trying to estimate the separate influence of the variable, "sex." If sex is unrelated to any other independent variable, then the coefficient on sex will show the difference between male and female salaries when every other explanatory influence is accounted for. But if sex affects rank and rank has some influence on salary, then all the effect of a change from male to female in the sex variable will not be captured in the coefficient on sex; some of the effect will be embodied in the associated change in rank. Therefore the regression coefficient on sex is misleading. Recall from the case that Dr. Iversen's regression, which included both sex and rank, showed a statistically insignificant coefficient on sex. This is typical of a problem of multicollinearity. The true effect of sex is obscured by the small regression coefficient and relatively large associated standard error on that coefficient. The combination of these two yields a low calculated t statistic.

It was in order to avoid this inaccuracy that Dr. deCani omitted rank. The court's discussion of the propriety of excluding rank was essentially one of determining whether rank and sex were correlated. The court concluded they were not and that therefore excluding the rank was improper. See the excerpt from *Presseisen* in Chapter VI (pages 262-263) that discussed this correlation. Sometimes, however, the relationship between two independent variables cannot be determined by calculating a correlation coefficient between the two. A bivariate regression of one variable on the other is suggested as a way of fully specifying the relationship. A most useful guide to the problems of detecting and treating problems of multicollinearity is Chapter 7 of Regression Analysis by Example, written by S. Chatterjee and B. Price (1977).

C. Multivariate Regression

While rank and sex might not have been correlated, there may have been other variables in Dr. deCani's equation that are correlated, e.g., age and years since highest degree. Are there others? Are there variables in Dr. deCani's equation, other than sex, that might be correlated with rank? Examine the variables in Dr. Iversen's equation. What possibilities for multicollinearity appear there?

2. *Categorical and Dummy Variables.* One of Dr. Meier's criticisms of Dr. deCani's equation in *Presseisen* is that the variable "division" did not allow for different slopes. Although the explanation of how teachers' salaries in different divisions or departments within a college might vary is clear, the import of this critique is not obvious.

To understand this problem, consider a college with three departments, Law, Economics, and Dance. Using these three categorical variables and following the approach we developed in our discussion of United States v. General Motors Corp. (see Note 2, page 256 above), we can assign a value of 1 or 0 to an individual teacher if she is or is not, respectively, on the Law faculty, a 1 or 0 if she is or is not on the Economics faculty, and a 1 or 0 if she is or is not on the Dance faculty. We have three categorical variables. A value of 1 indicates that a particular teacher fits in that category. Using three categorical variables in this way yields the type of data shown in Exhibit VII-8.

EXHIBIT VII-8.
Categorical Variables for College Faculty

	Department		
Teacher No.	Law	Economics	Dance
1	0	1	0
2	1	0	0
3	1	0	0
4	0	0	1
5	0	1	0

These categorical data can be included in a regression equation by the use of "dummy" variables. These variables are called *dummies* because they can only take the values zero and one. If we have these three divisions and two other variables, age and prior experience, we can determine the influence of these variables on salaries by setting up a regression equation that looks like this:

$$Salary = a + b_1\, L/E + b_2\, E/D + b_3\, Age + b_4\, Experience$$

where L/E takes on a value of 1 if the individual is a law teacher and 0 if an economics or dance teacher, and E/D takes on a value of 1 if the individual is an economics teacher and 0 if a dance or law teacher. Note that both variables have the value 0 if the person is a dance teacher. The coefficients b_1 and b_2 capture the effect on salary of being a law teacher and an economics teacher, respectively. The a captures the effect of being a dance teacher. The variables L/E and E/D are the dummy variables here. There is always one less dummy variable than there are categories — here, three categories and so two dummies.

The following hypothetical result indicates that law teachers make more than economics teachers, who in turn make more than dance teachers:

$$Salary = 13{,}000 \pm 10{,}000\ L/E + 5{,}000\ E/D + 300\ Age + 150\ Experience$$

To predict the salary of a 50-year-old economics professor with 12 years of experience, substitute 0 for L/E, 1 for E/D, 50 for Age, and 12 for $Experience$ to get:

$$Salary = 13{,}000 + 10{,}000(0) + 5{,}000(1) + 300(50) + 150(12)$$
$$= 13{,}000 + 5{,}000 + 15{,}000 + 1{,}800 = \$34{,}800$$

The comparable dance teacher makes only

$$Salary = 13{,}000 + 10{,}000(0) + 5{,}000(0) + 300(50) + 150(12)$$
$$= 13{,}000 + 15{,}000 + 1{,}800 = \$29{,}800$$

while an otherwise-identical law teacher makes $5,000 more than the economist.

This approach assumes that experience is valued at the same rate by each department and that there is simply a flat sum by which each department varies from others. This is probably the approach Dr. deCani used, though he used more independent variables. As an exercise, try to duplicate the full equation Dr. deCani tested.

Dr. Meier, defendant's expert, complained that this approach produced only an intercept for each department, not a slope. We have described the intercept as the starting point for the regression line. In this context, the intercept is the flat sum by which each department varies from others. Dr. Meier's complaint is that this amount may not in fact be a flat sum but may vary with some other factor, such as experience. He noted, for instance, that the hiring of physicists is highly competitive and that colleges might not only have to pay more initially but also give higher raises. In our hypothetical, a comparable situation might be that dancers' salaries start lower but increase faster than economists' as they get more experience. Dr. deCani's equation would not pick up this effect. His equation is *constrained* by the assumption that the effect of experience is independent of department. To pick up such differences among departments, one could calculate a separate regression equation for each department and test to see whether the coefficients on experience were significantly different from each other.

The important point for the attorney investigating the possible utility of multiple regression is that the effect on a dependent variable of an independent variable can be determined, even if it cannot be measured in the traditional sense.

"Sex," for instance, or "race" cannot be measured in the traditional sense, but using a dummy variable allows an advocate to quantify the effect of this variable on the dependent variable. Economists have traditionally used dummy variables to adjust for the effects of different seasons of the year on dependent variables like Gross National Product, crop harvest, or ski sales. If there are four seasons and the only other factor relevant to ski sales is the percentage of the population under age 26, what would a regression equation designed to predict ski sales look like? A torts lawyer trying to estimate forgone income might try to estimate the independent effects of occupational group, educational level, and previous experience on salary. How would you approach such a regression?

3. *Interpreting Regression Results.* To practice interpreting multiple regression results using dummy variables, consider the following data taken from an article titled The Determinants of Academic Lawyer's Salaries and Non-Institutional Professional Income, by John J. Siegfried and Charles E. Scott, 28 J. Legal Educ. 281, 287 (1977). The dependent variable is total salary, measured in thousands of dollars. The variables and their associated symbols are tabulated in Exhibit VII-9.

EXHIBIT VII-9.
Determinants of Academic Lawyer's Income

Variable	Measurement	Symbol
Total salary	thousands of dollars	TS
Base salary	thousands of dollars	(intercept)
Hours worked	hours per week	HR
Books and monographs	number	BM
Publications in last two years	number of articles	P
Research support	1 if yes, 0 if no	RS
Number of students	number per semester	S
Teaching award received	1 if yes, 0 if no	TA
Administrative position	1 if yes, 0 if no	AP
Public service consulting	1 if yes, 0 if no	PSC
Years since degree	number of years	YD
Recent job offer	1 if yes, 0 if no	RJO
Other professional position	1 if yes, 0 if no	OPP
Other professional noninstitutional income	thousands of dollars	OPI
Politically liberal	1 if yes, 0 if no	L
Major research university	1 if yes, 0 if no	RU

The regression results are as follows:

$$TS = 23.29 + 0.039HR + 0.113BM + 0.598P + 0.194RS$$
$$(1.115) \quad (0.92) \quad (4.09) \quad (0.26)$$
$$+ 0.010S - 0.527TA + 1.336AP - 0.479PSC + 0.364YD$$
$$(2.50) \quad (-0.55) \quad (1.59) \quad (-0.58) \quad (8.10)$$
$$+ 0.301RJO - 0.789OPP + 2.233NLD + 0.031OPI + 2.005L$$
$$(0.42) \quad (-1.07) \quad (2.96) \quad (1.11) \quad (2.63)$$
$$+ 2.363RU$$
$$(3.03)$$

The authors interviewed 137 law professors during the 1972-1973 academic year to get these results. The results represent a snapshot of what the distribution of salaries looked like across a sample of academic lawyers at one period in time. When time does not vary in a multiple regression, the approach is called *cross-sectional* analysis.

Recall that for bivariate regression, where there are two variables, there are $n - 2$ degrees of freedom. For regression analysis generally, where there may be many variables, the number of degrees of freedom for the t test is

$$d.f. = n - k - 1$$

where k is number of independent variables. Here we have 137 minus 16 equals 121 degrees of freedom.

A few clues will enable you to interpret these regression results. The measurement is a number, such as a number of hours, thousands of dollars, or 1 or 0. To find the $1000 increment to salary associated with a given variable, multiply the number in the measurement by the associated regression coefficient for that variable. Thus for political liberals, the dummy variable L equals 1, and liberals average 1 times 2.005 thousand dollars more each year in base salary than an otherwise-identical conservative. Individuals who teach 200 students per semester have an average salary $1600 higher than professors, otherwise similar, who teach 40 students per semester. Computing the increment to their salaries, we get 0.010 thousand dollars times 200 equals $2000, and .010 thousand dollars times 40 equals $400.

To test your understanding, answer the following questions:

(1) For which of the independent variables can we say that the effect is significantly different from zero at a 99% confidence level?
(2) On average, how much less does a professor without a nonlaw degree make than a similar professor with a nonlaw degree?

(3) Other things being equal, do law deans, who determine salaries, reward those who work longer hours?
(4) Should I expect a higher law salary if I go outside the university to get research support?
(5) What will happen to my salary as a result of publishing a book?
(6) Would my salary be increased more by writing an article than by writing a book next year?
(7) Would it pay me well to invest enough time in teaching to get an award?
(8) Could I be sure that taking an administrative position would increase my salary if everything else stays the same?
(9) What would happen to my salary if I obtained outside research support, took an administrative position, received a job offer from another law school, worked sixty more hours a week, and did public service consulting?

4. *Chewning v. Seamans.* Another use of dummy variables appears in Chewning v. Seamans, 20 E.P.D. ¶30,158 (D.D.C. 1979). In that employment-discrimination case, back-pay claims were attached to allegations that the Administrator of the Energy Research and Development Administration had systematically failed to promote the class of employees represented by plaintiffs. District Judge Flannery ruled that where the employer's records contained insufficient information to determine the amount of back-pay due, the master could use multiple regression analysis as a tool in making determinations. Dummy variables would play a key role in such an application, because critical to the amount of back-pay due each claimant would be the employee categories into which the claimant fell. Independent variables representing the categories would be included in the regression equation in an attempt to estimate back-pay due.

AGARWAL v. ARTHUR G. MCKEE & CO.
19 Fair Empl. Prac. Cas. (BNA) 503 (N.D. Cal. 1977)

ORRICK, District Judge:

This action involves individual and class claims for damages and injunctive relief under Title VII of the Civil Rights Act of 1964, 42 U.S.C. §2000e, et seq. Plaintiffs sought to establish that defendant's overall employment practices had the effect of preventing racial minorities from attaining positions within the top six of seventeen salary levels. . . .

Plaintiff Anand P. Agarwal was born in East India. He received a bachelor of science degree from Christ Church College in India and worked as a production engineer for an Indian employer for approximately six years. Plaintiff came to this country as a permanent resident in 1966. While working for the Roscoe Moss Company, plaintiff took graduate courses in engineering at UCLA and West Coast University and received a master of science in systems engineering from the latter in 1969. After graduation, plaintiff applied for and received a job at Arthur McKee & Company where he worked from May, 1969, to September, 1970.

Defendant, Arthur G. McKee & Company (McKee), is a Delaware corporation which specializes in the creation of production facilities required by the mining, steel, chemical, and petroleum industries. . . .

Defendant pays its employees with reference to a salary range within which a job classification falls. The salary range of a particular job classification may be ascertained from its number, higher-paying positions having generally lower numbers. The various annual salary ranges, as of January 5, 1976, were:

Level	Job Class Numbers	Salary Range
1	1100-1199	unavailable
2	1200-1299	unavailable
3	1300-1399	$36,240-$54,360
4	1400-1499	$31,464-$47,209
5	1500-1599	$27,480-$41,208
6	1600-1699	$24,156-$36,228
7	1700-1799	$21,360-$32,040
8	1800-1899	$19,704-$29,544
9	1900-1999	$17,628-$26,436
10-exempt	2000-2099	$15,660-$23,484
10-nonexempt	2500-2599	$15,660-$23,484
11-exempt	2100-2199	$13,992-$21,000
11-nonexempt	2600-2699	$13,992-$21,000
12	2700-2799	$12,588-$17,748
13	2800-2899	$10,872-$15,168
14	2900-2999	$ 9,408-$13,272
15	3000-3099	$ 7,848-$11,064
16	3100-3199	$ 6,672-$ 9,408
17	3200-3299	$ 5,592-$ 7,896

The parties presented a great deal of statistical information to support their representative contentions. The statistics derived from a variety of sources, including multiple regression analyses, parity studies, applicant flow records, nationwide census figures, standard metropolitan statistical area (SMSA) census figures, Equal Employment Opportunity Commission (EEOC) figures, and all-industry, construction industry, and mining industry employment figures. The Court found that each source had strengths and weaknesses in terms of its use in this particular case.

A multiple regression analysis is a computer-assisted method of assigning numerical weights, or "influence factors," to certain independent variables (e.g., age, race, education) in relation to a single dependent variable (e.g., salary). The analysis enables an observer to determine whether a particular independent variable has influenced a dependent variable and to estimate the extent of that influence. Furthermore, the analysis enables an observer to cumulate the various influence factors and to determine the degree to which the independent variables as a unit have influenced, or "explained," the dependent variable.

Plaintiff's regression analyses used the following independent variables: (a) minority status, (b) total years of education, (c) number of years since receipt of highest degree, (d) age of employee, (e) age of employee squared, (f) type of professional registration held by the employee, (g) years of prior experience, (h) years of experience at McKee, (i) years of experience at McKee squared, and (j) number of years of any break in service at McKee. Plaintiff's dependent variable for all regression analyses was salary.

The utility of a multiple regression analysis for evaluation of employment discrimination depends upon at least four considerations: (1) the "significance" of the influence factors discovered through the analysis, (2) the quantity of data used to evaluate a particular dependent variable, (3) the quality of that data, and (4) the ease with which independent variables may be converted to numerical equivalents.

The reliability of any given influence factor depends upon its statistical "significance." Statisticians measure significance through a variety of formulae which calculate the probability that an observed effect occurred by chance rather than through the influence of the variables under analysis. In general, statisticians reserve the term "significant" for results which would have recurred 90 or 95 times in 100 tests. Such results are termed significant at a 90 or 95% level. Alternatively, a statistician may declare that he has 90 or 95% "confidence" in his figures.

Plaintiff used information from the personnel files of approximately 700 class members to compile a "data base." The number of files was sufficiently large to dispel any inference that the results of the regression should be disregarded solely on the ground that the analysis did not include information from a representative cross-section of class members.

Plaintiff's regression analyses, however, contained a number of defects in terms of the quality of the data used. First, the personnel files lacked all of the information needed to assign a numerical value to each independent variable. Only 320 files contained all of the desired data. The total amount of missing information was not large but plaintiff's statistical expert could not state that the gaps had no effect on the overall quality of the regression analysis. Second, plaintiff's expert transferred the information in the personnel files to computer cards in a series of steps, each of which required subjective determinations of the need for certain data and created the possibility for errors in transfer. Third, plaintiff totally excluded certain kinds of information which could have had some bearing upon salary such as job level at McKee, prior salary, and past overseas assignment. The regression, therefore, did not purport to incorporate all variables which plaintiff had an opportunity to use. Fourth, plaintiff's expert did not attempt either a total or a random verification of the information contained in the personnel files. Plaintiff assumed the accuracy of the data source.

Plaintiff's regression analyses also contained a number of defects based upon the difficulty of assigning numerical values to independent variables having complex characteristics. First, plaintiff included only "years of experience" and did not code either the type or quality of that experience. Second, plaintiff included only "years of education," not the type or quality of that education. Third, plaintiff did not attempt to determine whether particular individuals had special abilities or characteristics which might have had a bearing upon salary level.

Finally, the regression analyses contained one deficiency which inhered in the regression technique, not in the quality of plaintiff's data or in the susceptibility of the independent variables to numerical expression. The study focused upon individuals, not positions. Plaintiff, therefore, treated all jobs as fungible and as requiring identical prerequisites. . . .

Examination of the starting salaries of minorities and nonminorities in specific jobs during each year does not reveal any pattern of bias against minorities. In the designer and draftsman positions, for example, the average starting salaries of persons hired into these positions in 1975 were as follows:

Title	Minority	Nonminority
2901 Draftsman	$ 988	960
2701 Senior Draftsman	1,150	1,075
2603 Designer	1,311	1,300
2501 Senior Designer	1,575	1,609

Given the relatively small numbers of employees involved in these comparisons, the average salaries of the minorities and nonminorities are very close. The evidence as a whole revealed that minorities and nonminorities are paid equally for equal work and that the company does not differentiate between the groups in their starting salaries. Employees hired into a specific job are started at the same general rate, and any deviations in the starting salary are based upon the person's background and experience. Defendant's statistical analyses of the salary differences of minorities and nonminorities within the individual job positions at WKE show that such differences are statistically insignificant.

Plaintiff's multiple regression analyses, on the other hand, revealed large differences between the salaries of minorities and nonminorities at WKE. As of December 31, 1975, the starting salaries of all employees, excluding clerical and service workers, average $1,530 while the starting salaries of minority employees average $1,164. Furthermore, the average salary of all WKE employees holding managerial, professional, and technical positions as of December 31, 1975, amounted to $1,816 while the average salary of all minority employees holding such positions amounted to $1,791. The monthly salary differential between minorities and nonminorities allegedly due to race alone amounted to $251.

As mentioned earlier, however, plaintiff's regression analyses contain a number of defects. Plaintiff failed to treat salary as a function of job position and salary grade. Furthermore, plaintiff treated all job positions as fungible, involving equal levels of knowledge, skill, and responsibility. Therefore, plaintiff's statistics do not refute defendant's contention that salary differences between minorities and nonminorities within each job position are not substantial.

In sum, the evidence does not show that defendant's employment practices with regard to salary have had a significant adverse impact upon racial minorities. . . .

JAMES v. STOCKHAM VALVES & FITTINGS CO. (I)
394 F. Supp. 434 (N.D. Ala. 1975)

GUIN, District Judge. . . .

[Black employees as a class and as individuals brought this action against their employer and union alleging racially discriminatory employment practices. The issue discussed in this excerpt is whether the employer offered equal opportunities in earnings to its employees without regard to race.— ED.]

(4) ADJUSTING FOR PRODUCTIVITY FACTORS BY REGRESSION ANALYSIS

Regression analysis is a statistical technique by which adjustment can be made simultaneously for more than three productivity characteristics. The technique allows one to place a dollar value on each variable reflecting its impact on earnings. In the case sub judice, adjustments were made by [defendant's expert] Dr. Gwartney simultaneously through the use of regression analysis techniques for each of these variables: years of schooling; seniority; skill level; outside craft experience; outside operative experience; absenteeism; merit rating; and achievement. Plaintiffs failed to introduce evidence of such adjustments for either the Stockham earnings data proffered by plaintiffs in this case, or the earnings data proffered by Stockham.

According to the uncontroverted testimony of Dr. Gwartney, it is better to adjust for *some* productivity factors than not to adjust for any; by adjusting for those productivity factors capable of measurement or reasonable estimate, it is possible to move toward a comparison of more similar employees in terms of their productivity characteristics.

There are different achievement levels for blacks and whites with equal years of schooling which are attributable to such things as differences in public expenditures, family background characteristics, home environment and community environment, and not to differences in intelligence or native ability.

Noting quantity of schooling alone is not an adequate productivity adjustment; quality of schooling must also be considered. The achievement level of whites exceeds that of blacks even when the two groups have equal years of schooling. At grade level 12 the black-white achievement differential is between 2.5 and 3.7 grade.

Two national studies indicate that when blacks and whites have an equal quantity of schooling, the achievement factor accounts for lower earnings for blacks measured at between 14.6% and 17.2% in one study and between 12.2% and 18.1% in the other.

Because of the educational achievement factor, blacks with the same quantity of schooling as whites would be expected to earn only about 85% as much as whites.

Outside craft experience (i.e., previous work experience) of Stockham employees was 2.7 years for whites and 1.2 years for blacks or an average of 1.7 years. Outside craft experience was tabulated in terms of years using the U.S. Census definition for craft jobs. This definition includes military or armed service time as craft skill and possibly accounts for what seems to be a high average of craft skill for both blacks and whites.

At the 96% confidence level, the mean of the range of estimates showed an unadjusted annual earnings differential of $448 in favor of whites. The differential was increased to $570 when seniority was considered (blacks had more seniority than whites). The addition of each of the other seven variables reduced the earnings differential and when all variables were considered simultaneously, the mean differential was −$299, with the range being between $16 to −$592. When adjustments were simultaneously made for seniority, years of schooling, skill level, outside craft experience, outside operative experience, absenteeism, merit rating and achievement, the annual earnings of whites at Stockham were estimated to be $299 less than the annual earnings of blacks.

After adjusting simultaneously for each of the eight factors, the hourly earnings of whites at Stockham exceeded those of blacks by 3.1 cents per hour with a range from +13.3 cents to −7.1 cents. The adjusted hourly wage rate of blacks was 99.2% of that of whites and the adjusted annual wage of blacks was 104.3% of that of whites.

(5) Conclusion

The application of the four methods separately and collectively to the earnings of Stockham's employees causes the Court to conclude that between 1965 and 1973, Stockham offered equal opportunities in earnings to its employees without regard to race.

JAMES v. STOCKHAM VALVES &
FITTINGS CO. (II)
559 F.2d 310 (5th Cir. 1977)

Wisdom, Circuit Judge. . . .

[Plaintiffs appealed the judgment of the district court.— Ed.]

Here the plaintiffs have produced evidence of gross disparities in job allocations at Stockham on the basis of race. All but two of the seniority departments were either predominantly white or predominantly black at Stockham at the time of trial in 1973. Only sixteen percent of the hourly jobs were integrated by that time. In 1973 the overwhelming majority of both incentive and non-incentive white workers were employed in jobs with the highest job classifications.

Blacks earn, on the average, $0.37 less per hour than whites, including overtime and incentive pay. Seventy percent of all black employees work in the monotonous, pressurized conditions of the incentive system, and 94 percent of all workers subject to the hot, dusty, dirty conditions of the foundry departments are black. The disparities revealed by the statistics on job allocations at Stockham are gross and the statistical evidence compelling; they establish a clear prima facie case of purposeful discrimination.

. . . Once the plaintiff in a Title VII case has presented a prima facie case of discrimination, the "onus of going forward with the evidence and the burden of persuasion" is on the defendant. The strength of the evidence presented in this case imposes on the defendant a heavy burden in attempting to counter the inference of systematic and purposeful discrimination. . . .

The defendant seeks to refute the plaintiffs' evidence that blacks earn an average of $0.37 less per hour than whites, including incentive and overtime pay, by means of the testimony of an expert witness. Dr. James Gwartney, an economist who testified for the defendant, conducted a study of the earnings of production and maintenance employees at Stockham in an attempt to determine the factors that explain earnings disparities between employees. He concluded that such productivity factors as education, skill, building experience, craft skill level, and absenteeism — not discrimination — explain the earnings differences between blacks and whites at Stockham. In studying the earnings opportunities at Stockham Dr. Gwartney considered four factors: (1) the earnings of employees at Stockham compared with the earnings of those in local, regional, and national labor markets and with earnings in other companies; (2) relative changes in the earnings of company employees over a long period of time; (3) relative changes in the earnings of company employees recently hired; and (4) application of the residual approach of scientifically adjusting earnings for productivity factors.

Dr. Gwartney's analysis does not meet the point that wage differences between blacks and whites at Stockham are explained by racially discriminatory job allocations. The first three factors are irrelevant to the question of discrimination at Stockham. The critical question is whether blacks at Stockham earn less than whites at Stockham, not whether blacks at Stockham earn more or less than blacks in various other geographic areas or in other companies. Those statistics will suggest only whether there is more or less discrimination in earnings opportunities for blacks in other settings as compared with Stockham. In addition, such statistics are totally irrelevant to the issue whether blacks are segregated by jobs and departments at Stockham and to the issue whether blacks must earn their wages under conditions less desirable than those of whites.

On its face, Dr. Gwartney's fourth factor deals with relevant and persuasive statistics on earnings disparities between blacks and whites. His regression analysis of productivity factors will not stand scrutiny.

Regression analysis is a statistical method that permits analysis of a group of variables simultaneously as part of an attempt to explain a particular phenomenon, such as earnings disparities between blacks and whites. The method attempts to isolate the effects of various factors on the phenomenon. Dr. Gwartney's analysis is based on the assumption that productivity factors, not discrimination, may explain the wage differences between Stockham's black and white employees. The productivity factors Dr. Gwartney employed were years of schooling, achievement, seniority, skill level, outside craft experience, outside operative experience, absenteeism, and merit ratings.

The rub comes with how these factors were defined in Dr. Gwartney's study. As the plaintiffs point out, the critical factors of "skill level" and "merit rating" were defined in such a way as to incorporate discrimination. "Skill level" was derived from an employee's job class; he had "skill" only if he worked in a job with a rating between JC 10 and 13. The systematic exclusion of blacks from promotion and training opportunities for such jobs, as is alleged here, will automatically produce no black employees with "skill level." A regression analysis defining "skill level" in that way thus may confirm the existence of employment discrimination practices that result in higher earnings for whites.

Dr. Gwartney used the merit ratings of Stockham supervisors, who are overwhelmingly white, for his "merit rating" factor; blacks average 71.3 in these ratings while whites average 79.3. If there is racial bias in the subjective evaluations of white supervisors, then that bias will be injected into Dr. Gwartney's earnings analysis.

Further, Dr. Gwartney included education as one of his productivity factors, even though education is not a job requirement at Stockham, because, according to the defendant, "an individual's educational level, regardless of race, impacts earnings." The fallacy in this conclusion stems from two facts: (1) as the defendant concedes, education is not a job requirement at Stockham, and (2) white employees at Stockham have more education than blacks. Thus, adjusting for education in a regression analysis of earnings where education is not related to job performance and where one race is more educationally disadvantaged than another, masks racial differences in earnings that may be explainable on the basis of discrimination. Certainly such differences cannot fairly be explained on the basis of a factor, such as education, concededly irrelevant to adequate job performance.

Significantly, although Dr. Gwartney asserted that his study proves that productivity factors and not discrimination explain the wage differences between black and white employees at Stockham, he concedes that he made no attempt to control or check for racial bias in his analysis. Our examination of his analytical approach compels us to conclude that the results of Dr. Gwartney's study in no way refute the plaintiffs' prima facie case of racial discrimination in job allocations at Stockham.

Stockham's attempt to refute the plaintiffs' evidence of racial job allocations by focusing on earnings differences misses the point. First, such an emphasis ignores the lopsided statistics on the number of all-black and all-white jobs at Stockham. Second, the defendant's focus on earnings avoids consideration of whether job segregation by itself, apart from any issue of economic harm, violates Title VII. This Court recently ruled on this issue in Swint v. Pullman-Standard, 539 F.2d at 89-90, in an opinion by Judge Clark:

> [A] Title VII plaintiff does not have to show economic loss to prove discrimination.
>
> . . . The key for this case is whether there was past discrimination. . . . Going further and requiring plaintiffs to prove that past assignment practices produced lower pay checks is contrary to law and precedent. . . .
>
> Title VII contains neither requirement nor implication that economic harm must be shown before a class can be found to have made out a prima facie case of racially discriminatory job assignment. Indeed, the statutory prohibitions of the enactment are explicitly broader than economic harm.

Thus, not only is the defendant's attempt to rebut the inferences of discrimination presented by the plaintiffs' evidence factually inadequate, it is also legally insufficient. . . .

Notes and Problems

1. Agarwal v. Arthur G. McKee & Co. and James v. Stockham Valves & Fittings Co. (I) and (II) illustrate the need to choose independent variables carefully. Aside from the court's criticism of potentially faulty data used by the plaintiff in *Agarwal,* the main problem was that the plaintiff failed to treat salary as a function of job position. By omitting job position as an independent variable, the variations in salary level were inadequately and inaccurately accounted for by variations in the ten independent variables that were used. In *James,* the defendant's expert witness sought to explain variations in wages by demonstrating that they were due to differences in the productivity of the workers. Circuit Judge Wisdom's criticism of Dr. Gwartney's multiple regression was that it failed to recognize that wage differences

were due to discriminatory job assignments. If blacks were unable, due to discrimination, to obtain higher-paying jobs, then the productivity factors offered by the defendant are irrelevant. The statistician's response to this criticism should be to perform a multiple regression analysis of the factors influencing job assignment. In such a regression equation, job assignment would be the dependent variable and various explanatory factors — both legitimate, such as an appropriate measure of skill level, and impermissible, such as race — would be the independent variables.

2. While the critiques of the multiple regression in these cases may seem arcane, technical, and confusing, they amount to nothing more than the kind of rebuttal to a legal argument with which lawyers are quite familiar. Take as an example Judge Wisdom's critique of Dr. Gwartney's multiple regression, which used productivity factors to explain wage differentials. Essentially, the plaintiffs complained of discrimination in the setting of wages. The defendant answered, "No, it is not discrimination but productivity factors that explain wage differences." Each party offered evidence to prove its assertion. Judge Wisdom exhibited more common sense than statistical expertise in recognizing that the legal focus should be on job assignments rather than productivity. An appreciation of the fancy statistical techniques is unnecessary to recognize whether a particular variable or, here, a particular set of productivity variables covering up the crucial issue. In evidentiary terms, the claim that productivity data is inapposite goes to the relevance of the multiple regression evidence offered by defendant. One might object that certainly productivity factors are relevant to job assignments and, therefore, the regression should be admitted into evidence. But, although the factors may be relevant, they should be admitted not for disproving or proving discriminatory wage differentials but rather for examining the determinants of job assignments itself. As far as we can tell, Dr. Gwartney did not test this multiple regression. What would the regression equation look like for a test designed to determine which factors affected job assignment?

3. What sources of multicollinearity appear in Dr. Gwartney's analysis of the productivity factors?

D. SIGNIFICANCE TESTING: R^2 AND F

Significance testing on single variables using a Z test or a t test indicates a range or confidence interval within which lies the population value of the variable that we are trying to estimate. For instance, a

95% confidence interval around a sample mean gives us a range within which the population mean lies. We have discussed the use of the t test to construct confidence intervals around estimated values of regression coefficients. And we have discussed using regression analysis to predict future values of the dependent variables. To determine whether the predicted future value of the dependent variable is accurate with a certain level of significance, we need another summary statistic, called the F *statistic*.

Computation of the F statistic is rather complicated. Fortunately, however, the F statistic is another number that is usually calculated by the computer. The F statistic is conceptually related to the r^2, which we have described as the percentage variation in one variable accounted for or explained by variation in another variable. In the multiple regression context, R^2 has the same meaning. The R^2, called the *multiple correlation coefficient* when applied to multiple regression, indicates approximately the percent of total variation in the dependent variable explained by all the independent variables together. It varies between zero and one and is used to test whether the independent variables are reliable predictors of the value of the dependent variable. R^2 is calculated by the formula:

$$R^2 = \frac{\Sigma(y_i^* - m_y)^2}{\Sigma(y_i - m_y)^2}$$

where y_i^* is the estimated value of y_i, and m_y is the mean of the y observations. The F statistic relates unexplained to explained variance by the formula:

$$F = \frac{R^2}{1 - R^2} \times \frac{n - k - 1}{k}$$

where n is the number of observations in the sample and k is the total number of independent variables. Manipulating the R^2 formula, we can also calculate F directly as:

$$F = \frac{\Sigma(y_i^* - m_y)^2}{\Sigma(y_i - m_y)^2 - \Sigma(y_i^* - m_y)^2} \times \frac{n - k - 1}{k}$$

These formulas for R^2 and F are applicable to bivariate or multivariate analysis.

The purpose of an F test is to establish whether the regression coefficients explain enough of the variation in y to give an accurate predicted value for the dependent variable. To use this test we must delve just a bit further into the mysterious realm of degrees of freedom. Because we have a ratio in the F statistic, we must calculate the degrees of freedom in both the top, *numerator,* and bottom, *denominator,* of the fraction. The rule is:

$$d.f._{numerator} = k$$
$$d.f._{denominator} = n - k - 1$$

where n is the number of observations of each variable, and k is the total number of independent variables.

For the study of the determinants of academic lawyers' salaries discussed on pages 343-345, the reported F statistic was 10.3. There, 137 lawyers were included in the study, and 15 independent variables were used. If, therefore, n equals 137 and k equals 15, then the degrees of freedom can be calculated as:

$$d.f._{numerator} = k = 15$$
$$d.f._{denominator} = n - k - 1 = 137 - 15 - 1 = 121$$

To find the significance level, use the F tables in Exhibit VII-10. Along the top of the first table, labelled *significance level = .05,* are degrees of freedom for the numerator. Along the side are degrees of freedom for the denominator. For our lawyers' salaries example, go along the top to 15 and then down to 120 (since 121 does not appear on the table, we use the next lower number). You will see the number 1.75. This number is the critical F value for a 5% significance level. On the following table, you will find the critical value for a 1% significance level, 2.19. Since the calculated F statistic, 10.3, is larger than either critical value, we can be more than 99% certain that the salary estimated by our regression equation is accurate. The null hypothesis is that these independent variables collectively have no relationship to the dependent variable. Another way of stating this null hypothesis is that all of the regression coefficients are simultaneously zero, so the equation has no predictive validity. We reject this null hypothesis with less than a 1% chance of a type 1 error.

As an exercise, determine the probability of a type 1 error if the F statistic, calculated from a regression with 27 independent variables and 90 observations, is equal to 2.10.

NORTHSHORE SCHOOL DISTRICT NO. 417
v. KINNEAR
530 P.2d 178 (Wash. 1974)

HALE, Chief Justice.
Education is a bulwark of this democracy. A system of free public schools, like a system of open courts, not only helps make life worth

EXHIBIT VII-10.
Critical Values for F (Significance Level = .05)

Degrees of Freedom in Denominator ($= n - k - 1$)	Degrees of Freedom in Numerator ($= k$)								
	1	2	3	4	5	6	7	8	9
1	161.4	199.5	215.7	224.6	230.2	234.0	236.8	238.9	240.5
2	18.51	19.00	19.16	19.25	19.30	19.33	19.35	19.37	19.38
3	10.13	9.55	9.28	9.12	9.01	8.94	8.89	8.85	8.81
4	7.71	6.94	6.59	6.39	6.26	6.16	6.09	6.04	6.00
5	6.61	5.79	5.41	5.19	5.05	4.95	4.88	4.82	4.77
6	5.99	5.14	4.76	4.53	4.39	4.28	4.21	4.15	4.10
7	5.59	4.74	4.35	4.12	3.97	3.87	3.79	3.73	3.68
8	5.32	4.46	4.07	3.84	3.69	3.58	3.50	3.44	3.39
9	5.12	4.26	3.86	3.63	3.48	3.37	3.29	3.23	3.18
10	4.96	4.10	3.71	3.48	3.33	3.22	3.14	3.07	3.02
11	4.84	3.98	3.59	3.36	3.20	3.09	3.01	2.95	2.90
12	4.75	3.89	3.49	3.26	3.11	3.00	2.91	2.85	2.80
13	4.67	3.81	3.41	3.18	3.03	2.92	2.83	2.77	2.71
14	4.60	3.74	3.34	3.11	2.96	2.85	2.76	2.70	2.65
15	4.54	3.68	3.29	3.06	2.90	2.79	2.71	2.64	2.59
16	4.49	3.63	3.24	3.01	2.85	2.74	2.66	2.59	2.54
17	4.45	3.59	3.20	2.96	2.81	2.70	2.61	2.55	2.49
18	4.41	3.55	3.16	2.93	2.77	2.66	2.58	2.51	2.46
19	4.38	3.52	3.13	2.90	2.74	2.63	2.54	2.48	2.42
20	4.35	3.49	3.10	2.87	2.71	2.60	2.51	2.45	2.39
21	4.32	3.47	3.07	2.84	2.68	2.57	2.49	2.42	2.37
22	4.30	3.44	3.05	2.82	2.66	2.55	2.46	2.40	2.34
23	4.28	3.42	3.03	2.80	2.64	2.53	2.44	2.37	2.32
24	4.26	3.40	3.01	2.78	2.62	2.51	2.42	2.36	2.30
25	4.24	3.39	2.99	2.76	2.60	2.49	2.40	2.34	2.28
26	4.23	3.37	2.98	2.74	2.59	2.47	2.39	2.32	2.27
27	4.21	3.35	2.96	2.73	2.57	2.46	2.37	2.31	2.25
28	4.20	3.34	2.95	2.71	2.56	2.45	2.36	2.29	2.24
29	4.18	3.33	2.93	2.70	2.55	2.43	2.35	2.28	2.22
30	4.17	3.32	2.92	2.69	2.53	2.42	2.33	2.27	2.21
40	4.08	3.23	2.84	2.61	2.45	2.34	2.25	2.18	2.12
60	4.00	3.15	2.76	2.53	2.37	2.25	2.17	2.10	2.04
120	3.92	3.07	2.68	2.45	2.29	2.17	2.09	2.02	1.96
∞	3.84	3.00	2.60	2.37	2.21	2.10	2.01	1.94	1.88

D. Significance Testing: R^2 and F

10	12	15	20	24	30	40	60	120	∞
241.9	243.9	245.9	248.0	249.1	250.1	251.1	252.2	253.3	254.3
19.40	19.41	19.43	19.45	19.45	19.46	19.47	19.48	19.49	19.50
8.79	8.74	8.70	8.66	8.64	8.62	8.59	8.57	8.55	8.53
5.96	5.91	5.86	5.80	5.77	5.75	5.72	5.69	5.66	5.63
4.74	4.68	4.62	4.56	4.53	4.50	4.46	4.43	4.40	4.36
4.06	4.00	3.94	3.87	3.84	3.81	3.77	3.74	3.70	3.67
3.64	3.57	3.51	3.44	3.41	3.38	3.34	3.30	3.27	3.23
3.35	3.28	3.22	3.15	3.12	3.08	3.04	3.01	2.97	2.93
3.14	3.07	3.01	2.94	2.90	2.86	2.83	2.79	2.75	2.71
2.98	2.91	2.85	2.77	2.74	2.70	2.66	2.62	2.58	2.54
2.85	2.79	2.72	2.65	2.61	2.57	2.53	2.49	2.45	2.40
2.75	2.69	2.62	2.54	2.51	2.47	2.43	2.38	2.34	2.30
2.67	2.60	2.53	2.46	2.42	2.38	2.34	2.30	2.25	2.21
2.60	2.53	2.46	2.39	2.35	2.31	2.27	2.22	2.18	2.13
2.54	2.48	2.40	2.33	2.29	2.25	2.20	2.16	2.11	2.07
2.49	2.42	2.35	2.28	2.24	2.19	2.15	2.11	2.06	2.01
2.45	2.38	2.31	2.23	2.19	2.15	2.10	2.06	2.01	1.96
2.41	2.34	2.27	2.19	2.15	2.11	2.06	2.02	1.97	1.92
2.38	2.31	2.23	2.16	2.11	2.07	2.03	1.98	1.93	1.88
2.35	2.28	2.20	2.12	2.08	2.04	1.99	1.95	1.90	1.84
2.32	2.25	2.18	2.10	2.05	2.01	1.96	1.92	1.87	1.81
2.30	2.23	2.15	2.07	2.03	1.98	1.94	1.89	1.84	1.78
2.27	2.20	2.13	2.05	2.01	1.96	1.91	1.86	1.81	1.76
2.25	2.18	2.11	2.03	1.98	1.94	1.89	1.84	1.79	1.73
2.24	2.16	2.09	2.01	1.96	1.92	1.87	1.82	1.77	1.71
2.22	2.15	2.07	1.99	1.95	1.90	1.85	1.80	1.75	1.69
2.20	2.13	2.06	1.97	1.93	1.88	1.84	1.79	1.73	1.67
2.19	2.12	2.04	1.96	1.91	1.87	1.82	1.77	1.71	1.65
2.18	2.10	2.03	1.94	1.90	1.85	1.81	1.75	1.70	1.64
2.16	2.09	2.01	1.93	1.89	1.84	1.79	1.74	1.68	1.62
2.08	2.00	1.92	1.84	1.79	1.74	1.69	1.64	1.58	1.51
1.99	1.92	1.84	1.75	1.70	1.65	1.59	1.53	1.47	1.39
1.91	1.83	1.75	1.66	1.61	1.55	1.50	1.43	1.35	1.25
1.83	1.75	1.67	1.57	1.52	1.46	1.39	1.32	1.22	1.00

(continued)

EXHIBIT VII-10.
Critical Values for F (continued) (Significance Level = .01)

Degrees of Freedom in Denominator ($= n - k - 1$)	Degrees of Freedom in Numerator ($= k$)								
	1	2	3	4	5	6	7	8	9
1	4052	4999.5	5403	5625	5764	5859	5928	5982	6022
2	98.50	99.00	99.17	99.25	99.30	99.33	99.36	99.37	99.39
3	34.12	30.82	29.46	28.71	28.24	27.91	27.67	27.49	27.35
4	21.20	18.00	16.69	15.98	15.52	15.21	14.98	14.80	14.66
5	16.26	13.27	12.06	11.39	10.97	10.67	10.46	10.29	10.16
6	13.75	10.92	9.78	9.15	8.75	8.47	8.26	8.10	7.98
7	12.25	9.55	8.45	7.85	7.46	7.19	6.99	6.84	6.72
8	11.26	8.65	7.59	7.01	6.63	6.37	6.18	6.03	5.91
9	10.56	8.02	6.99	6.42	6.06	5.80	5.61	5.47	5.35
10	10.04	7.56	6.55	5.99	5.64	5.39	5.20	5.06	4.94
11	9.65	7.21	6.22	5.67	5.32	5.07	4.89	4.74	4.63
12	9.33	6.93	5.95	5.41	5.06	4.82	4.64	4.50	4.39
13	9.07	6.70	5.74	5.21	4.86	4.62	4.44	4.30	4.19
14	8.86	6.51	5.56	5.04	4.69	4.46	4.28	4.14	4.03
15	8.68	6.36	5.42	4.89	4.56	4.32	4.14	4.00	3.89
16	8.53	6.23	5.29	4.77	4.44	4.20	4.03	3.89	3.78
17	8.40	6.11	5.18	4.67	4.34	4.10	3.93	3.79	3.68
18	8.29	6.01	5.09	4.58	4.25	4.01	3.84	3.71	3.60
19	8.18	5.93	5.01	4.50	4.17	3.94	3.77	3.63	3.52
20	8.10	5.85	4.94	4.43	4.10	3.87	3.70	3.56	3.46
21	8.02	5.78	4.87	4.37	4.04	3.81	3.64	3.51	3.40
22	7.95	5.72	4.82	4.31	3.99	3.76	3.59	3.45	3.35
23	7.88	5.66	4.76	4.26	3.94	3.71	3.54	3.41	3.30
24	7.82	5.61	4.72	4.22	3.90	3.67	3.50	3.36	3.26
25	7.77	5.57	4.68	4.18	3.85	3.63	3.46	3.32	3.22
26	7.72	5.53	4.64	4.14	3.82	3.59	3.42	3.29	3.18
27	7.68	5.49	4.60	4.11	3.78	3.56	3.39	3.26	3.15
28	7.64	5.45	4.57	4.07	3.75	3.53	3.36	3.23	3.12
29	7.60	5.42	4.54	4.04	3.73	3.50	3.33	3.20	3.09
30	7.56	5.39	4.51	4.02	3.70	3.47	3.30	3.17	3.07
40	7.31	5.18	4.31	3.83	3.51	3.29	3.12	2.99	2.89
60	7.08	4.98	4.13	3.65	3.34	3.12	2.95	2.82	2.72
120	6.85	4.79	3.95	3.48	3.17	2.96	2.79	2.66	2.56
∞	6.63	4.61	3.78	3.32	3.02	2.80	2.64	2.51	2.41

Source: E. Pearson and H. Hartley, eds., I Biometrika Tables for Statisticians 159 (Table 18) (1958). Reprinted with permission of the Biometrika Trustees.

D. Significance Testing: R^2 and F

10	12	15	20	24	30	40	60	120	∞
6056	6106	6157	6209	6235	6261	6287	6313	6339	6366
99.40	99.42	99.43	99.45	99.46	99.47	99.47	99.48	99.49	99.50
27.23	27.05	26.87	26.69	26.60	26.50	26.41	26.32	26.22	26.13
14.55	14.37	14.20	14.02	13.93	13.84	13.75	13.65	13.56	13.46
10.05	9.89	9.72	9.55	9.47	9.38	9.29	9.20	9.11	9.02
7.87	7.72	7.56	7.40	7.31	7.23	7.14	7.06	6.97	6.88
6.62	6.47	6.31	6.16	6.07	5.99	5.91	5.82	5.74	5.65
5.81	5.67	5.52	5.36	5.28	5.20	5.12	5.03	4.95	4.86
5.26	5.11	4.96	4.81	4.73	4.65	4.57	4.48	4.40	4.31
4.85	4.71	4.56	4.41	4.33	4.25	4.17	4.08	4.00	3.91
4.54	4.40	4.25	4.10	4.02	3.94	3.86	3.78	3.69	3.60
4.30	4.16	4.01	3.86	3.78	3.70	3.62	3.54	3.45	3.36
4.10	3.96	3.82	3.66	3.59	3.51	3.43	3.34	3.25	3.17
3.94	3.80	3.66	3.51	3.43	3.35	3.27	3.18	3.09	3.00
3.80	3.67	3.52	3.37	3.29	3.21	3.13	3.05	2.96	2.87
3.69	3.55	3.41	3.26	3.18	3.10	3.02	2.93	2.84	2.75
3.59	3.46	3.31	3.16	3.08	3.00	2.92	2.83	2.75	2.65
3.51	3.37	3.23	3.08	3.00	2.92	2.84	2.75	2.66	2.57
3.43	3.30	3.15	3.00	2.92	2.84	2.76	2.67	2.58	2.49
3.37	3.23	3.09	2.94	2.86	2.78	2.69	2.61	2.52	2.42
3.31	3.17	3.03	2.88	2.80	2.72	2.64	2.55	2.46	2.36
3.26	3.12	2.98	2.83	2.75	2.67	2.58	2.50	2.40	2.31
3.21	3.07	2.93	2.78	2.70	2.62	2.54	2.45	2.35	2.26
3.17	3.03	2.89	2.74	2.66	2.58	2.49	2.40	2.31	2.21
3.13	2.99	2.85	2.70	2.62	2.54	2.45	2.36	2.27	2.17
3.09	2.96	2.81	2.66	2.58	2.50	2.42	2.33	2.23	2.13
3.06	2.93	2.78	2.63	2.55	2.47	2.38	2.29	2.20	2.10
3.03	2.90	2.75	2.60	2.52	2.44	2.35	2.26	2.17	2.06
3.00	2.87	2.73	2.57	2.49	2.41	2.33	2.23	2.14	2.03
2.98	2.84	2.70	2.55	2.47	2.39	2.30	2.21	2.11	2.01
2.80	2.66	2.52	2.37	2.29	2.20	2.11	2.02	1.92	1.80
2.63	2.50	2.35	2.20	2.12	2.03	1.94	1.84	1.73	1.60
2.47	2.34	2.19	2.03	1.95	1.86	1.76	1.66	1.53	1.38
2.32	2.18	2.04	1.88	1.79	1.70	1.59	1.47	1.32	1.00

living but sustains our long-cherished ideas of individual liberty. Where the nation's constitution provides for a system of open courts, however, it makes no mention of free public schools. The people of this state found this oversight unacceptable in 1889 when they brought Washington Territory into the Union. Not only did they establish a judicial system, but at the same time they provided for a system of free public schools, imposing then and there a duty upon the state to make ample provision for the education of all children within its borders.

Since statehood, the legislature has structured a comprehensive system of public schools, enacting, re-enacting, amending and repealing a detailed code for the funding, operating and maintaining of that system which includes a code for the employment, certification, and retirement of teachers and school administrators. It is a system administered by a Superintendent of Public Instruction and a State Board of Education but puts direct responsibility and authority for actually operating the schools upon 320 separate school districts. The constitutionality of that system is now challenged.

Petitioners are 25 of the 320 school districts of this state, their directors, resident parents, taxpayers and children. They bring their original petition to this court for a writ of prohibition and mandamus to declare the state's system for funding its public schools unconstitutional and to prohibit state officers from collecting and disbursing public funds in support of it. So sweeping are the demands that, if their petition were upheld, the schools would have to be closed unless the legislature redesigned and restructured the statutes for the funding and operation of the public school system in consonance with the requirements of the decisional law which would be laid down by this court in sustaining the petition. For reasons now stated, we sustain the constitutionality of the laws creating, funding and maintaining the public schools and deny the petition. . . .

Petitioners' major claims of unconstitutionality, as indicated, are based on inequality in assessed valuation. Their claims do not arise, therefore, from asserted deficiencies in curriculum, but stem largely in this case from the idea that districts with lower assessed valuation per student raise less money per mill in special levies, or are less inclined to vote the extra millages in special elections. Thus, it is argued the school funding statutes and scheme are unconstitutional because local property taxes represent a substantial part of school revenues; that disparities in these revenues develop among the various districts because they reflect differences in assessed valuation per student; that these differences in valuation in turn are reflected in both the amounts derived from special millage elections and the voters' gen-

D. Significance Testing: R^2 and F 363

eral inclination in districts of low valuation to reject special millages. All of this, it is said, produces in turn unconstitutional inequality in educational opportunities available to children throughout the state. Differences arising from variances in local tax income per student, it is said, reflect unconstitutional differences in expenditures per student. Thus, the whole system, petitioners argue, is made unconstitutional because of claimed variances in educational opportunities stemming largely from differences in assessed valuation per student.

But the record does not bear out these claims of unconstitutional inequality of educational opportunity. There is no evidence whatever that one district or another provides unconstitutionally superior or unconstitutionally inferior opportunities; nor is there evidence as to which are the better or inferior of offending districts, if any, one way or another; nor is it denied that due to social, economic and demographic differences some districts will require substantially more money than others to provide approximately the same level of educational opportunities. Conclusions of fact resting largely as they do upon opinion and conjecture and drawn from the statistical data in the record, as will be shown, do not sustain the assertion of unconstitutional failure of the state to fulfill its duty.

Mr. Francis Flerchinger, from the Office of Superintendent of Public Instruction, with 11 years' experience in that work, testified that he had spent more than a year as a staff member of a committee studying the state school aid formula to determine whether there should be revisions in the school formula proposed. These studies, he said, had been continuous throughout the 11 years of his career, and it can be assumed as a matter of commonsense, that they have been going on at least for the last 40 or 50 years. The most current study in which he was then engaged, he said, had been initiated at the request of the state legislature.

Mr. Flerchinger testified that differentiations in assessed valuations had little or nothing to do with the quality of education supplied by the various districts. One criterion, he said, which is not controlling but significant is that which is called "basic expenditure per pupil." These basic expenditures

> are arrived at by subtracting from the total expenditures those items received by only some of the school districts including the expenditure for food services, transportation, all specially funded state and federal programs and those payments received for providing services to other school districts. The figure is then divided by the number of pupils.

He pointed out that the basic expenditures per pupil could and did vary to an extraordinary degree from a high of $4,517 per pupil down to $470 per pupil, but that the quality of education between the two districts might be about the same or the differences imperceptible. Thus, the Patterson District, with only 5 children in the entire district, had the highest basic expenditure per pupil, but could not be said to be providing superior educational opportunities than those of all of the other districts whose similar expenditures per pupil were much lower.

Referring to some of the charts, graphs and exhibits, Mr. Flerchinger said that the mean basic expenditure per pupil per district for all 320 districts in the state was $819, that the existing standard deviation from the $819 was $292, and that 158 of the largest school districts in the state contained 95 percent of the state's school pupils. Adverting to the mean expenditure per pupil for these 158 districts containing the 95 percent of the school population, he said it amounted to $667 with the standard deviation of only $84 as contrasted with $292 from the $819 mean basic expenditure for all 320 districts.

In Mr. Flerchinger's judgment, illustrated by reference to the material contained in the graphs, charts and other exhibits, he explained why there is a difference in the ratio of certificated staff per 1,000 students among the school districts in Washington. Here is his testimony on that point:

> Q. From your observation of the data contained in the study on public school financing and from your general experience with the public school system, do you have an opinion as to why the size of the enrollment in a district is the main factor which explains differences in the number of certificated staff per 1000 students?
> A. It has to do with basically the availability of students. Patterson, the smallest district here, has a staffing ratio of 200 certificated staff for 1,000 pupils. They don't have 1,000 pupils, they have 5, and they have 1 staff member. And you just concentrate on this end of the graph, the smallest districts, you can see the staff going down, because when they add that additional pupil, when they go from 5 to 6, that is a very significant change in the staffing per 1,000 pupils and in fact carries on throughout this. It is not a uniform effect because there is a mixing of districts in here, both elementary and secondary, and the staff ratios are different between elementary and secondary.

He thus concluded that one of the primary determinants of the differences in the basic expenditures for the school children of the state is mainly a result of the variations in the number of certificated staff members per 1,000 students. He carried this analysis one step farther by saying that 75 percent of all the variations in expenditure per pupil in Washington are accounted for by differences in pay and the differences in the number of certificated staff per 1,000. . . .

D. Significance Testing: R^2 and F

Differences in assessed valuation shown from the material before us have little to do with meeting state standards of education. Enrollment, rather than valuation, is the more significant key. Thus, for statistical purposes, the principal graphs and tables received in evidence are based on data from the 158 largest districts out of the total 320 districts in the state. Their significance is apparent when it is shown that these 158 districts contain 95 percent of the entire school enrollment. Among these 158 districts, there is, said Mr. Flerchinger, a standard deviation of .85, a point not fully explained in the testimony. In another context, the total expenditure per pupil per district would be $819, but the mean expenditure per pupil for the 158 of the large districts is $667. Thus, Mr. Flerchinger, using the data and charts received in evidence as a basis for his conclusions as earlier indicated, testified as follows:

Q. Can you compare the data in Respondents' Graph No. 35 with Respondents' Graph Nos. 28 and 29?
A. Yes.
Q. Have you?
A. Yes, we have. We performed a calculation called a step which is multiple regression analysis, which is a method of arriving at relating the interrelationships of the, in this case the three variables in such a manner you have eliminated the complications of the other variables on the main variable. In this case we are attempting to shall we say account for the basic expenditure per pupil and the program selected, the computer selected as the two primary factors the average pay for certificated staff and a staffing ratio per 1,000 pupils as accounting for 75 percent of the variation in the expenditure per pupil.
Q. So what you are saying is that 75 percent of all the variations in expenditure per pupil in Washington are accounted for by differences in pay and the differences in the number of certificated staff per 1,000?
A. That's correct.

He testified accordingly that the basic expenditure per pupil, as evidenced by graph No. 28 in evidence, would show little as to the differences in quality of education between large districts and small districts. When measuring the quality of the actual program conducted within the classroom there is, he said, "little that you can say from that graph."

Accordingly, he said, neither the teacher-student ratio nor the expenditures per pupil were adequate criteria for explaining or judging the quality of education a student is receiving in the various school districts of the state. Thus, alluding to a graph showing the relationship of expenditures per pupil to assessed valuation, he said that the higher assessed valuation per pupil in the larger districts is generally associated with a declining student enrollment. The size of the district, as measured by enrollment, he concluded, is the factor which most explains the difference between assessed valuation per pupil. He testified:

> If you divide assessed valuation . . . by the number of pupils in the district, the smaller districts are going to have the higher valuation. The other factor that comes in, of course, is the basic property within the district, the total valuation. But the diviser in that equation, that is the number of pupils, changes the other factors very significantly.

He concluded on this point:

> If you are taking the state as a whole, the smaller districts in expenditure per pupil bear an inordinate weight in the computation of the correlation coefficient of .85 which is the relationship between those two variables, and if you analyzed that smaller 162 and actually play around for a moment with the enrollment by adding one or two pupils here or there, you can significantly change that, and so therefore you have to conclude from that that the enrollment or the number of pupils available in the district as the diviser of these two factors has a very strong significance on the expenditure per pupil and the assessed valuation per pupil, and therefore it provides a spurious relationship when you do a statistical analysis using those two variables.

Variations in expenditures per pupil and assessed valuation per pupil are both mathematical functions of enrollment — a conclusion of ultimate fact inevitably to be drawn from the evidence in this case and as a matter of commonsense. . . .

The same lack of relationship between assessed valuation per student and quality of educational opportunity appears in an examination of another bar graph pertaining to and showing the assessed valuation per student among the following districts:

Clover Park	$10,784
Lake Washington	$14,061
Kent	$17,889
Everett	$25,338
Yakima	$15,559
Puyallup	$11,896
Northshore	$12,515
Central	$10,441
Bellingham	$17,109
Longview	$34,060

Thus, Northshore has a substantially higher assessed valuation per student than either Clover Park or Puyallup or Central and a modest amount less than Lake Washington, and one can draw no rational conclusion whatever from these differences as to the degree, if any, by which one district or the other, if any, fails to provide adequate educational opportunities for the children within its borders.

Petitioners here make much of the truism that it takes more mills for a district of low assessed valuation to raise the same amount of school money than it does in a district with a high assessed valuation. Thus, according to one bar graph, 11.1 mills will raise $392.47 per enrolled pupil in Seattle; 24.2 mills to raise $211.71 in Federal Way; 24.4 mills to raise $296.33 in Highline; 15.9 mills to raise $278.12 in Tacoma, while 28.7 mills will raise $358.92 in Northshore. But these graphs repudiate the very basis upon which petitioners would have us void the school statutes. Even this very limited comparison of but 5 of 320 districts shows that Northshore, by raising $358.92 from 28.7 mills is much better off financially under petitioners' theories than is Highline, for example, which raises only $296.33 from a 24.4 millage, and Federal Way which gets only $211.71 from 24.2 mills.

That the assessed valuation per pupil has little to do with the quality of education is also demonstrated in the graph showing the number of certificated personnel per 1,000 enrolled students: Northshore, one of petitioner districts, has 50.4 certificated personnel per 1,000 students whereas Shoreline district has only 39.6 certificated personnel per 1,000 students. Renton has 46.3 certificated personnel per 1,000 students; Federal Way, 46.9; Kent, 48.7; Clover Park, 49.3; Edmonds, 50; Auburn, 50.7; Highline, 51.6; Lake Washington, 52.5; Mercer Island, 54.9; Seattle, 55.8; Issaquah, 55.8; Bellevue, 55.8; Tacoma, 56.4; and South Central, 56.5. According to this, petitioner Northshore is much better off than Shoreline, Renton, Federal Way and Kent and about on a parity for certificated personnel per 1,000 students with Clover Park, Edmonds, Auburn and Highline.

These figures, taken largely from the Superintendent of Public Instruction's Office show conclusively that assessed valuation per pupil not only has little to do with the quality of education in the enumerated districts, but that no decision as to the equal protection of the laws nor the paramount duty to provide uniform education can be based upon it. The significance of assessed valuation per pupil is thus inconstant, tenuous, superficial and coincidental only.

Accordingly, petitioners' first claim of unconstitutionality, that children who live in school districts with low assessed valuation of property per pupil are denied equal protection of the laws contrary to the Fourteenth Amendment to the United States Constitution and Const. art. 1, §12, and, therefore, are victims of the state's failure to discharge its paramount duty to them, is not only not supported by the evidentiary data in the case, but is essentially disproved by it.

The record also fails to vindicate petitioners' position that differences in assessed value among the districts denies equal protection to the taxpayer. That it takes more millage to raise the same amount of dollars on low valued property than it does on high valued property is no more than a meaningless truism and can be answered with another truism that the lower the value of one's property the lower one's taxes, neither truism having anything to do with the equal protection clause of both constitutions so long as everybody in the taxing scheme pays the same rate. Differences in assessed valuation per pupil among the various districts do not to a constitutional degree substantially affect the amounts of revenue per pupil available nor the amount expended per pupil; nor the cost per pupil in providing about the same quality of education throughout the state. Disparities among the districts, it is shown in this record, arise not only from variances in revenue raised but in the necessary differences of money to be spent because: (1) Differences in appraisals of property for tax purposes by assessors may persist in the various counties. As between districts in different counties, a systematically high appraisal will produce more school revenue than a systematically low appraisal; (2) A lowering of the state's share has dropped from 59.2 percent of the total in 7 years to 49 percent, not because of a decrease in state appropriations but largely because the individual districts have put up proportionately more from local taxes and special levies; (3) All things are relative and, short of abolishing separate districts and converting the state into one school district, the disparities in tax revenues from the various areas of the state will persist; (4) Converting the entire state into a single district will not alter the differences in expenditures necessary to provide a substantially uniform system affording reasonably equal educational opportunities in the different areas of the state for the obvious reason that costs per child will vary due to the infinite differences in geography, climate, terrain, social and economic conditions, transportation and special services, and local choice as to extra curriculum and special services to be made available in consonance with the state's minimal requirements.

Where one district may offer a richer program in music and dramatic arts, another may go beyond the state's requirements in science or social studies, or physical education or agriculture, and others may emphasize more than one field of student activity beyond the college preparatory phases. One district may supply a more comprehensive remedial program for physical behavioral and emotional problems, and another may provide less than some experts may deem to be minimal. These are choices which inhere in the idea of viable local participation in establishing, operating and funding the common schools. If these differences are of constitutional dimension, there exists a remedy in equity to compel the particular school district or the state in a particular case to provide such services, but that is not the remedy these petitioners are seeking. . . .

Notes and Problems

1. The ability to explain variation in per pupil expenditure was critical to defenders of the Washington state school financing system examined in Northshore School District No. 417 v. Kinnear. Mr. Flerchinger, state employee and statistician, performed a multiple regression analysis using three variables: expenditures per pupil in each of Washington's 320 school districts, teachers' pay, and certified staff per thousand students. As an exercise, write out the regression equation for this analysis.

The only reported result of this analysis is that "the computer selected as the two primary factors the average pay for certificated staff and a staffing ratio per 1,000 pupils as accounting for 75 percent of the variation in expenditure per pupil." This means that the multiple correlation coefficient for the equation was .75. What is the evidentiary value of this assertion? For what purpose does the court use this evidence? Is that use logically justified? Might there be a correlation between teacher pay and educational quality?

2. If the multiple correlation coefficient is .75, what is the F statistic? What is the critical F value for this regression at a 99% confidence level? A 95% level? Are these independent variables accurate predictors of expenditures per pupil?

3. Mr. Flerchinger's testimony regarding the regression results is most curiously phrased from a statistical point of view. He testifies that the *computer* selected these two variables. Given his phrasing, it might be that he ran computer tests on independent variables randomly until he found a pair that gave the highest R^2. This egregiously violates the principle that the statistician should determine first what variables *theoretically* influence the dependent variable, and then do the computations. This principle avoids the potential for spurious results, results that show a correlation between the values of the variables that is due only to chance. There might, for instance, be a strong correlation between the apple crop in each school district and per pupil expenditure. The opinion reveals high per pupil expenditure in rural areas with few students. These areas probably grow the most apples. But this correlation is hardly of practical significance as an explanatory factor. The procedure of first constructing a "model" of what independent variables influence the dependent variable helps ensure that statistically significant results will also be practically signi-

ficant. It may be tempting to test the correlations first and then devise a theory to match, but such an approach, which the quantitative social sciences have been combating for years, is not logically sound, and your opponent has every incentive to find flaws in your reasoning.

4. The Environmental Protection Agency is required under Executive Order 12291 of February 17, 1981, 46 Fed. Reg. 13193, to apply the principles of cost-benefit analysis when promulgating new regulations or reviewing existing regulations. The following hypothetical is suggestive of the process by which the EPA projects future costs.

A review of regulations hypothetically in progress concerns reclassification of the Central New York Air Quality Control Region (AQCR) to require additional pollution-reduction efforts by industrial emitters of sulfur dioxide. Included in the cost analysis is an attempt to estimate the increase in enforcement costs due to this reclassification. It is expected that as emitters are required to pay more for pollution abatement, more of them will violate the regulations, and, therefore, more enforcement teams of inspectors, engineers, and attorneys will be needed. The following regression is used by the EPA to project increased enforcement costs in any area where there is a reclassification. The equation was obtained by evaluating increases in enforcement costs in each of the 70 AQCRs that were reclassified during the period from 1977 to 1980. The EPA is going to use this equation to estimate costs associated with this proposed reclassification. The dependent and independent variables are:

EC = additional enforcement costs in dollars
NP = number of polluters in the AQCR
$NPRC$ = additional pollution reduction costs to be borne by the polluters as a result of the reclassification.
GS = geographic size of the AQCR in square miles
SS = pre-reclassification size of the support staff in the regional EPA office measured in number of employees
UR = urban-rural character of the region. (The value 1 is assigned to this variable if more than 25% of the region's population lives in an urban area. Otherwise, the value 0 is assigned.)

The regression results are as follows:

$EC = 1200NP + .20NPRC + 93GS - 20,000SS - 593,500UR$
(t) (3.27) (1.99) (2.09) (-1.01) (-17.23)
$R^2 = .70$ $F = 12.30$

D. Significance Testing: R^2 and F

(a) The Central New York AQCR has 350 polluters in an area of 4,000 square miles, and 95% of the population lives in a rural area. The pre-reclassification regional EPA support staff of 20 persons estimates an increase in pollution-abatement costs of $12 million. What is the projected increase in enforcement costs, using the regression equation?

(b) Using the R^2 and F, evaluate the reliability of the estimate of the increase in enforcement costs calculated in (a).

(c) Which of the regression coefficients is (are) statistically insignificant? Which has the least practical significance?

Note: Corrected R^2

There is no substitute for thoughtful construction of the regression equation or model. The inclusion of irrelevant variables in a regression equation will almost certainly increase R^2 because of chance fluctuations in their values. Because of this possibility, and in order to discourage the statistician from including irrelevant variables merely in order to get statistically significant results, an adjusted or corrected R^2 may be computed. The formula for the corrected R^2 is:

$$R^2_{corrected} = R^2 - \frac{k}{n-k-1} \times (1-R^2)$$

where n is the number of observations of each variable and k is the total number of independent variables. The corrected R^2, rather than the R^2, is frequently reported for multiple regression studies. The unadjusted multiple correlation coefficient will never decrease as additional explanatory variables are added to a regression equation, but it is possible for the corrected coefficient to decrease if an additional independent variable produces too small a reduction in unexplained variance. Note that this correction will not eliminate all of the effect on R^2 of a highly correlated though irrelevant variable. Such an independent variable might contribute too much to the "explanation" of the variance in the dependent variable.

Note: In re Quaker Oats Co.

It is difficult to overemphasize the utility of significance tests on the predictive accuracy of entire regressions. In re Quaker Oats Co., 66 F.T.C. 1131 (1964), was an action charging a major producer of oat flour and other grain and cereal products with price discrimination and with selling below cost. The action was brought under §5 of the Federal Trade Commission Act. Evidence regarding Quaker's pricing policies and costing practices was critical to the government's case. The company kept no records from which actual costs could be calculated. Quaker's costing practice was to use multiple regression analysis to estimate crop conditions and prospective yields for various grains in different locations at different time periods. The estimates were then utilized to indicate the most profitable future price action for Quaker. The regression coefficients in these regressions indicated the relative importance to price of the various supply and demand factors included. A single factor might be more important in one part of the year than another. For instance, in the first quarter of the new crop year (July-September), the total supply of old crop oats will be important, while in the second quarter, new crop corn would be an important variable influencing the price of oats. Dummy variables representing different seasons must have been included in the regression equation to capture these effects.

To establish that Quaker's practice was to sell below cost, an understanding of the regression approach to prediction of optimal price is required. If Quaker followed the multiple regression projections exactly, then any anticompetitive behavior — the prevention of which is the purpose of the Clayton Act — would be incorporated into the independent variables in the regression equations. One can imagine that, for instance, not only the new crop corn but also the latest price set by the closest competitor might be included in the equation. The extent to which Quaker would be willing to follow the projections of its regression equations would depend on their predictive accuracy as measured by the R^2 and F tests. The closer R^2 came to a value of 1, and the larger the F for each equation, the more inclusive of all relevant variables the equation would be, the more reliable the estimate, the more willing Quaker would be to follow the projection, and the more important an appreciation of the regression technique would be to the Federal Trade Commission.

E. AUTOCORRELATION AND THE DURBIN-WATSON STATISTIC

We have assumed thus far that the best regression representation of a set of observations is a straight line. But often a straight line is a poor approximation of the relationship between two variables. Consider the data points in Exhibit VII-11, representing average hours worked each month for a year. During the early months of the year,

E. Autocorrelation and the Durbin-Watson Statistic

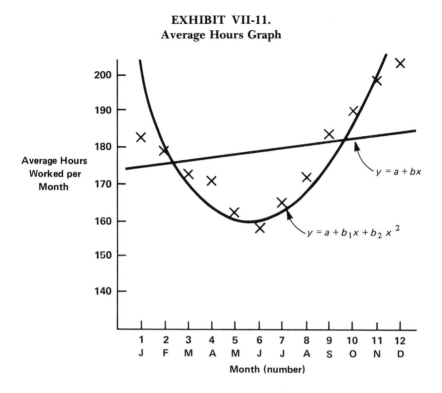

EXHIBIT VII-11.
Average Hours Graph

there is a great deal of work because clients are busy with new projects requiring legal expertise or new complaints resulting from seasonal activity. The pace slackens as summer approaches because clients, attorneys, and judges go on vacation. When Fall rolls around, the work increases to peak at year's end. A linear regression line, such as that depicted for these data points, would miss the significant changes in the dependent variable, average hours worked. As a result, the R^2 and F values would be very low and the regression results rejected. A *curved* regression line, such as the one drawn, would more accurately describe the variations in the dependent variable and give a better estimate of staffing requirements in different months, with a higher R^2 and F.

Attempting to describe a *curvilinear* relationship by a straight line is one type of misspecification that can occur in regression. This particular type of error in specifying the relationship between variables is called *autocorrelation*. Technically, autocorrelation means that the error terms, u, are correlated with one another; knowing the value of one helps you predict the value of the next one. The diagram in Exhibit VII-12 is called a *plot of the residuals*. It is merely a graph indicating, for each observed value of x, the error made in predicting the

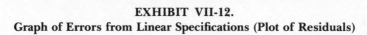

EXHIBIT VII-12.
Graph of Errors from Linear Specifications (Plot of Residuals)

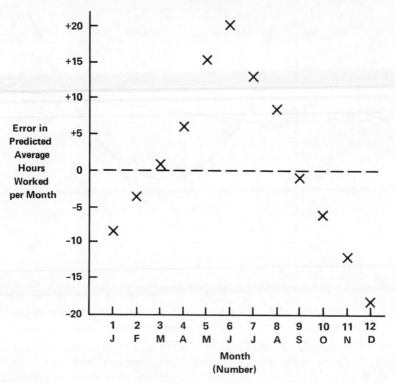

associated y value. Recall that the error is equal to the difference between the observed and predicted y values, that is, the vertical distance on the graph in Exhibit VII-11 between the regression line and the observed y value.

$$u_i = y_i^* - y_i$$

Exhibit VII-12 plots the u_i associated with each x_i. You can see that knowing the error for one month helps predict the error for the next month. For instance, the errors for all middle months are positive, for all months at each end of the year, negative. Moreover, from January to June, each succeeding month has a higher (closer to $+20$) error associated with it, and after June each succeeding error gets closer to -20. In a case where there is no autocorrelation, the errors are randomly distributed around the regression line.

A test statistic that indicates whether the errors are correlated with one another is called the *Durbin-Watson statistic, d*. For all practical purposes, the Durbin-Watson statistic, d, varies between zero and two,

E. Autocorrelation and the Durbin-Watson Statistic

though logically it varies between zero and four. If $d = 2$, there is no autocorrelation. The further d gets from the value of two, the greater is the probability that autocorrelation is present. A Durbin-Watson table, like the one in Exhibit VII-13, shows for different sample sizes, n, and numbers of independent variables, k, the deviation from two, where we reject the null hypothesis that there is no autocorrelation. The following is an example of how that table works.

EXHIBIT VII-13.
A Durbin-Watson Table

	$k = 1$	
n	d_L	d_U
.	.	.
.	.	.
.	.	.
25	1.29	1.45

Assume that for a linear regression, the computer calculated a $d = 1.25$. For one independent variable and a sample size of 25, the 95% confidence level table looks like Exhibit VII-13. The symbols d_L and d_U are upper and lower critical values of the Durbin-Watson statistic. If the calculated d is less than d_L, then we have autocorrelation. If d is greater than d_U, we do not have autocorrelation. If d is between d_L and d_U, then the Durbin-Watson test is inconclusive. Here, because the calculated $d = 1.25$, which is less than $d_L = 1.29$, we reject the null hypothesis that there is no autocorrelation at a 95% confidence level.

This very brief discussion of autocorrelation is necessary to explain a method that might help you gauge the accuracy of your regression coefficients and predictions when the R^2 and F for a particular regression are low, but you suspect that there really is a significant relationship between the dependent and independent variables. We have described the problem of autocorrelation as one of misspecification. We have specified that the variables change together and in the same direction (either increasing together or decreasing together) and at a regular constant rate. This is all implied by a linear (straight-line) specification. A nonlinear specification allows for relationships between variables that cannot be adequately described by a straight line. If we could only use linear specifications, we might look at the small b's and R^2 and F and conclude that there is no relationship between the variables. Looking at the Durbin-Watson statistic might explain why there is no relationship shown, and the plot of residuals might show what the nonlinear specification best describing the relationship is. From the plot of residuals shown above and from the average hours graph, we expect that a horseshoe-shaped curve might best describe the relationship between month and average hours worked.

F. NONLINEAR MULTIPLE REGRESSION

There are many shapes of curves that fit data. Each shape has a general formula describing it. The general formula describing straight lines is:

$$y = a + bx$$

The general formula describing the horseshoe-shaped curve in the prior example, involving average hours worked, is:

$$y = a + b_1 x + b_2 x^2$$

This multivariate equation has three variables, y, x, and x^2. The x^2 represents the value of whatever x is (in our example, the month), squared. Thus, the data for the three variables in the bivariate and multivariate cases are shown in Exhibit VII-14.

EXHIBIT VII-14.
Data for Multiple Regression

y Average Hours per Month	x Month	x^2 Month Squared
182	1	1
178	2	4
172	3	9
170	4	16
160	5	25
158	6	36
163	7	49
172	8	64
181	9	81
189	10	100
195	11	121
200	12	144

Bivariate: y, x
Multivariate: y, x, x^2

F. Nonlinear Multiple Regression

EXHIBIT VII-15.
Salary/Age Graph

The computer would calculate a separate regression coefficient for x and x^2. This would give the predictive accuracy we desire.

There is a general formula for almost every curve you might imagine; and if one of the curves fits your data better than a straight line, it can be used to predict the values of the dependent variable. The statistician uses the R^2 to test "goodness-of-fit" and selects the general formula that best fits the data, that is, the equation with the highest R^2 or corrected R^2. The tools and calculations for nonlinear multiple regressions are no different from those for linear regressions.

Look back at Agarwal v. Arthur G. McKee & Co. (page 345). Note that plaintiff's regression analysis used ten independent variables, two of which were "age of employee" and "age of employee squared." Plaintiff's expert hypothesized that the relationship between salary and age was not linear but curved, and he tested the equation that way. Why would this relationship be curved? One story might be that as age increases and as one becomes more skilled, salary increases quickly, but after a certain point, salary declines as the worker slows down or as skills become out of date. This hypothesis would be graphically represented by Exhibit VII-15. The complete regression calculated by plaintiff's expert in *Agarwal* had two such curved (also called *curvilinear*) variables. That regression equation appeared in full as follows:

$$Salary = a + b_1M + b_2E + b_3H + b_4A + b_5A^2 + b_6R + b_7PE + b_8EM + b_9EM^2 + b_{10}B$$

where M = minority status
    E = years of education
    H = years since receipt of highest degree
    A = age of employee
    R = type of professional registration held by employee
    PE = years of prior experience
    EM = years of employment at McKee, and
    B = number of years of any break in service at McKee

What theory probably underlies the nonlinear estimation on the variable EM?

As with the selection of independent variables, the selection of form of estimation, whether linear or nonlinear, should depend on the tester's theoretical understanding of the relationships between the variables. Choosing the form of estimation according to which method gives the highest R^2, however, does not introduce irrelevant explanatory terms into the equation in the way that random selection of independent variables might.

CHAPTER VIII

HEALTHY SKEPTICISM

A. SOURCES OF ERROR IN LEGAL STATISTICS

In the preceding chapters, we have discussed a variety of mathematical problems confronting the advocate who offers quantitative evidence. These include the problems of multicollinearity, autocorrelation, and the problems of small numbers of observations when using chi-square and Z tests. This chapter explores these and other mathematical problems in more detail. It also focuses attention on a variety of difficulties encountered in assembling and analyzing data that are more commonsense in nature but easy to overlook. As a review of sources of error, this chapter will serve as a warning to those preparing quantitative evidence as well as a guide to those attacking an opponent's statistical proof.

1. Errors in Gathering Data

Potential for error exists in the earliest stage of data preparation. Unlike scientific evidence, which is derived from controlled experiments, and like social science evidence, which measures phenomena in an uncontrolled environment, data used as statistical evidence often reflect measurements of nonreplicable events. The sales, cost, or profit data that describe business for the past years cannot be measured again, because the events they describe will not occur again. Nor can the accuracy of applicant-flow data for past years be improved, if methods for reporting data in past years were unsatisfactory. Because data used as proof are merely the by-product of other activities, and because social phenomena do not stand still as time passes and social environments change, attorneys offering evidence of past behavior or past outcomes face a variety of errors in the data with which they must deal.

One of the more obvious sources of errors is in designing samples. As we noted in our discussion of inferential statistics as proof, in Chapter III (see page 31), it is necessary to choose a sample that is representative of the population about which conclusions are to be made. The air quality one-half mile from a pollution source is unlikely to be typical of average quality in an air-quality-control region. Employment records for the month of June, when millions of college students enter the labor market, are unlikely to be representative of the whole year's experience. Opinions on the death penalty from people frequenting the local courthouse or church or law school are unlikely to mirror the opinions of a political subdivision from which jury venires are drawn.

The survey offered by the State of Oklahoma in Craig v. Boren (see pages 36-46 above) illustrates potential biases introduced by surveys. In *Craig,* the state surveyed drivers to determine the drinking and driving habits of Oklahoma youth. Cars may have been stopped at random, but consider the possibility of bias if the survey were done on Saturday night, as opposed to Wednesday afternoon. If social convention dictates that young males drive when they are out with young females, then the Saturday survey might incorrectly conclude that on average more drunk drivers are male than female. Justice Rehnquist relied on the conclusion that three-quarters of all drivers are male to support the rational basis for gender-based differential in drinking ages.

In United States v. Goff (see pages 93-95 above) the plaintiff surveyed food-stamp recipients to determine the percentage of poor persons who were registered to vote and, therefore, eligible for inclusion on jury venires. The conclusions from the sample may be incorrect if some poor people do not receive food stamps and have a different percentage registration than food-stamp recipients. It may have been easier to get a large sample of food-stamp recipients than to test poor people generally, but the attorney offering this evidence should be aware that a potential exists for incorrect results. One solution to this problem is to take a sample of poor persons who do not receive food stamps and use confidence intervals to compare the mean number registered to vote to that calculated for food-stamp recipients.

A cleverly designed survey will ask crucial questions in a variety of ways to increase the probability that favorable results are obtained. The employment data in Chinese for Affirmative Action v. FCC (page 163 above) came from the records of the radio station, rather than from a survey, but the data illustrate two different ways of asking the same question. Records were examined to determine how much of the KCBS work-force was Asian-American. They revealed that nine persons, or 3.6% of 249 employees during the license term, were Asian-Americans. Records also revealed that Asian-Americans worked only 94, or 3.1%, of the 2986 person/months worked by all employees of KCBS. A survey that asked only the number of employees of each race might miss the useful fact that the turnover rates for the two racial categories were different.

Conclusions from samples may also reflect data that are untrue. Answers to surveys may be deliberate or unintentional lies. A statistician who submitted an affidavit in United States v. General Motors Corp. (pages 249-255 above), where the factual issue was the number of trucks with defective wheels, concluded that affidavits obtained by the government from owners "might be inaccurate because the owners might not recall what it was they experienced and, additionally, because the government personnel assisting in the accumulation of affidavits might be biased thus tainting the information." The owners' misinformation is probably unintentional. The government employees' misinformation may be unintentional as well, though they may have interpreted ambiguous affidavits in a manner favorable to the government's case. Improperly trained data gatherers may also yield sloppily prepared information. The FTC considered the possibility of deliberately misleading information in In re Kroger Co. (pages 197-198 above), where prices increased on items before they appeared on Kroger's "Price Patrol" survey. Deliberate nonrandomness in surveys clearly gives incorrect conclusions. Falsification is easier where the methods of record-keeping are less well understood. It would be particularly difficult, for instance, to tell whether cost data or output projections submitted in an antitrust case such as In re Quaker Co. (pages 371-372 above) were falsified, because accounting practices of companies are often arcane and idiosyncratic.

Distortions may also result from inaccurate record-keeping. Applicant-flow data offered by the hotels alleged to have been discriminating in Gay v. Waiters' Local 30 (page 203 above) were not derived from a systematic and accurate list of applicants. Records were "sporatic and fragmentary." Race coding on the "walk-in log" was "speculative and unreliable" and "incomplete." In Agarwal v. Arthur G. McKee & Co. (pages 345-349 above), fewer than half of the personnel files contained the necessary data. The records had to be manually transferred to computer cards, which introduced the possibility of subjectivity and inaccuracy, and the plaintiff failed to test the accuracy of the computer records. Statistical sampling techniques like those used to test for defects in the shipments by A.B.G. Instrument and Engineering, Inc. to the United States government (see page 261 above) could be used to test the accuracy of these records.

Improper aggregation or lack of clear definition of categories used in gathering data might also lead to inappropriate conclusions. In Presseisen v. Swarthmore College (pages 331-339 above), for instance, plaintiff's expert, Dr. deCani, failed to differentiate between junior and senior faculty appointments. The court discussed the potential for bias resulting from this aggregation. Failure to disaggregate might "make it impossible properly to sort out what policies may be due to seniority and what policies may be due to sex." In Agarwal v. Arthur G. McKee & Co., the plaintiff treated all job positions as essentially identical and requiring equal levels of knowledge, skill, and responsibility. For that reason, the plaintiff failed to show that salary differences between minorities and nonminorities within each job position were substantial.

All of these sources of error may come together in the data accumulated in Census Bureau reports. Census Bureau computations made after the 1980 dicennial census were challenged by cities seeking to protest their federal funding. Significant errors in census data, particularly those under- and overstating population percentages of racial categories, may also have an important impact on discrimination cases by giving an inaccurate expected value for a particular category.

2. *Small Numbers Problems*

Many statistical tests depend on the assumption that the data analyzed is normally distributed. The Z test, for instance, measures the proportion of total observations, distributed in the assumed bell shape, that would lie beyond a given number of standard deviations from the mean. Other tests are mathematical alterations of the normal distribution. When a particular category makes up only a small proportion of the total population, we expect that the number of individuals of that category selected from the population into the group of suspect classification will be small. But if the group of suspect composition itself is small, we cannot expect a perfectly normal distribution of the numbers of the particular category selected into that group. That is to say, as the expected number of individuals in a particular category gets smaller, the likelihood that random selections would distribute themselves in a normal fashion decreases. Tests based on an assumption of normality may incorrectly estimate the probability that an observation could occur by chance.

A. **Sources of Error in Legal Statistics** 383

We have discussed two ways such an incorrect estimation can be corrected: the t statistic as a substitute for Z and the Yates-corrected chi-square test as a substitute of the uncorrected chi-square. In either case, the small-sample substitute gives a higher estimate of the probability that a particular event could occur by chance.

Consider a malpractice suit against Dr. X. To discover whether the percentage of her patients dying during open-heart surgery is significantly higher than that for other doctors in her hospital, we examine 30 recent cases. Of these patients, ten died. Seven of ten patients died at the hands of Dr. X. Three of 20 patients died at the hands of other doctors. If the average death rate is one-third (30 ÷ 10), does Dr. X's performance differ significantly from that of the other doctors? The chi-square test, uncorrected, and the Z test both indicate that we can be about 98% certain that the differences are not due to random factors. But because the number of observations is so small, it is reasonable to suspect some departure from normality. The expected number of deaths by Dr. X is about three, which is less than five, thus necessitating the Yates Correction. The total size of the group of suspect composition (dead patients) is less than 30, thus justifying reference to the t rather than the Z table. When adjusted for small size, the tests allow us to be only about 95% certain that the differences are not due to random factors. Note that the probability that the difference could be due to random factors has increased from 2% to 5%. By using the corrections, we adjust for departures from the normality assumption by measures which make us less certain of our outcomes. Note, however, that if we need only be 95% certain in order to reject the null hypothesis, we can do so in this case, even using the small-sample corrections. Because the tests of regression coefficients involve the t statistic and because the tests of significance correlation coefficients require adjustment for the number of pairs of observations, these other tools also have built-in adjustments for departures from normality caused by sample size.

What does this mean for the cases we have considered? In Inmates of the Nebraska Penal & Correctional Complex v. Greenholtz (pages 97-107 above), the court expressed "concern that the size of the Mexican-American group (18) on which the study was based might be too small to yield a reliable result." The court computed a disparity equal to less than two standard deviations from the mean. Using either a chi-square test or a Z test, we could calculate the probability that this disparity could occur by chance. Because the group of suspect composition from which we are sampling contains 535 members, which is greater than the small sample number of 30 for the t test, and because the expected number of Mexican-Americans (the smallest group) is 11 and so is greater than the small sample number of five for the chi-square test, we do not even need to apply a correction to the usual tests. The Z and chi-square tests are being used on a normal population, which presents no problems to our calculations.

It should be noted that statisticians are not in agreement as to the utility of the chi-square tests, even given the Yates Correction, when sample sizes are less than 30 or when the percentage of individuals in the minority group is relatively small. Some statisticians are skeptical enough of the accuracy of the chi-square test to recommend applying the Yates Correction whenever the minimum expected number is less than 500 and to use another test altogether when sample size drops below 30.

In EEOC v. United Virginia Bank/Seaboard National (pages 117-123 above), the court considered that one category of employee, containing only six members, was "probably too small a number for statistical comparison." In that case, the t test could be substituted for the Z test to calculate that, for six degrees of freedom, the probability of a random outcome -2.41 standard deviations from the mean is greater than 5%. This corroborates the intuitive conclusion of the court that, when adjusted for small sample size, the benchmark of the *Castaneda* two-to-three-standard-deviations test is not exceeded.

Judge Robinson's analysis of small-sample problems in Chinese Affirmative Action v. FCC is particularly useful. He recognizes that the value of a statistical showing depends not only on the size of the sample but also upon the size of the expected observation and the degree of disparity. "In short," he adds, "the court's conclusion that statistics for small groups are always unreliable is simply wrong." Judge Robinson also suggests that the sample size can be increased by including more years in the sample. This expansion must, of course, be legally justified. Irrelevant years or employees may not be included solely to improve statistical results.

B. SPECIFICATION ERRORS IN MULTIPLE REGRESSION

In Vuyanich v. Republic National Bank, Judge Higginbotham reviews the most important technical problems involved in using multiple regression. As described in the previous chapter, a regression line quantifies the relationship between dependent and independent variables. We have explored the fact that a regression will not exactly predict each value of the dependent variable we observed. Because the regression coefficient summarizes or averages the relationship between variables, there is some difference between the predicted values and the actual values. The size of this difference is reflected in the residual variance. This variance, denoted $s.e._e{}^2$, is the variance of predicted from actual values for different observations. The difference between each predicted value and the actual value we called the *error*, u. District Judge Higginbotham, in this short excerpt from *Vuyanich*, refers to this error, u, as the *random disturbance term*.

VUYANICH v. REPUBLIC NATIONAL BANK OF DALLAS
505 F. Supp. 224 (N.D. Tex. 1980)

HIGGINBOTHAM, District Judge. . . .

3. WHAT CAN GO WRONG?

A "perfect" model would explain completely the process under study. While such models are found in the physical sciences, they are rare in the social sciences. D. Baldus & J. Cole, supra, §8.21, at 264.[1] Indeed, it has been argued that no model in the social sciences ever meets the requirements for a perfect regression analysis. Id., §8.22, at 266. But this does not mean that because a model is subject to challenge, its results are valueless. Cf. Editors' Introduction, supra, at 169. Small departures from assumptions necessary for a perfect regression may have small deleterious effects. Cf. Fisher, supra, at 711.

The value of a regression would obviously be affected by problems in the underlying data or by mismeasurements of the explanatory variables. Cf., e.g., R. Pindyck & D. Rubinfeld, supra, at 194-202; Kmenta, supra at 336-41; G. Maddala, supra n.49 at 201-07; J. Johnston, supra, n.51, at 281-91; Beyond the Prima Facie Case, supra, n. 57, at 417-21. And, as discussed earlier, the model must be one justified by theory. We turn here to a few of the less intuitively obvious ways in which ordinary least squares can provide unreliable results important to this case.

---

1. Citations in this excerpt to classic statistical works have been retained for their utility as references. The full citations for the works can be found in the excerpt from this case reproduced in the previous chapter, pages 316-330.—ED.

In a perfect regression model using the ordinary least squares technique,[60] three major assumptions would hold:

> (a) that the effects of the random disturbance term are independent of the effects of the independent variable; (b) that the values of the random term for different observations are not systematically related and that the average squared size of the random effect has no systematic tendency to change over observations; and (c) that the sum of random effects embodied in the disturbance term is distributed normally, in the "bell curve" generally characteristic of the distribution of the sum of independent random effects.

Fisher, supra, at 708.

One way the first assumption may be violated is if some relevant explanatory variable has been left out of the analysis. This is one type of "misspecification" or "specification error." For instance, if yield of a crop is dependent on both amount of fertilizer and rainfall, and if in our regression we include only fertilizer, there will be improper omission of an explanatory variable. The problem caused by omission of variables is that the regression coefficient(s) of the included explanatory variable(s) (e.g., of fertilizer) would be "biased" (that is, not likely to be correct "on the average"),[61] *and* the usual tests of significance concerning the included regression coefficient(s) (such as calculation of a confidence interval) will be invalid. Kmenta, supra, at 392-95. See J. Johnston, supra n. 51, at 169. Certain statistical tests are available to suggest whether this sin of omission has occurred. See, e.g., Kmenta, supra, at 405.

However, in at least one circumstance, this problem may not be a serious one in cases where the issue is whether discrimination exists or does not exist. Where it is possible to use as proxy for the presence (or absence) of discrimination against a particular group a "dummy" or "group status" explanatory variable, such an omission will not threaten the validity of the group status coefficient (and hence, the validity of the model's suggestion of the existence or nonexistence of discrimination) unless the omitted variable is related to the group status variable. D. Baldus & J. Cole, supra, §8A.1, at 273. Thus, here,

---

60. When regression techniques more advanced than ordinary least squares are used, not all such assumptions may need to hold for valid results to be generated by the regression analyses.

61. For a simple explanation of bias, see Fisher, *supra,* at 709.

We do not discuss explicitly the associated concept of "efficiency." Because of the particular unbiased and "efficient" estimates of coefficients possible using the ordinary least squares technique, ordinary least squares is often described as the "best linear unbiased estimator." See R. Wonnacott & T. Wonnacott, supra, at 55-69; J. Johnston, supra n. 51, at 126; Kmenta, supra, at 9-14, 161; R. Pindyck & D. Rubinfeld, supra, at 20-24.

where the plaintiffs' model had compensation as its dependent variable, and various explanatory variables (including dummy variables for race and sex), Dr. Madden stated:

Q. Do you agree that your report would have been more valid if you measured all potential productivity?
A. Well, since my purpose was to analyze sex and race my report would have been more valid had there—to the extent that there are any omitted productivity variables that are correlated with race or sex. To the extent they're not correlated with race or sex [it] makes no difference whatsoever.

Another type of specification error occurs when one or more irrelevant variables are *included* in the model. This overinclusion by itself causes fewer problems than underinclusion. See R. Wonnacott & T. Wonnacott, supra, at 413; R. Pindyck & D. Rubinfeld, supra, at 190; Kmenta, supra, at 399; J. Johnston, supra n. 55, at 169. Overinclusion of variables, however, increases the risk of "multicollinearity."

Multicollinearity refers to a situation where due to the high (but not perfect) correlation of two or more variables (or combination of variables), it becomes difficult to disentangle their separate effects on the dependent variable. R. Pindyck & D. Rubinfeld, supra, at 67; G. Maddala, supra n. 49, at 183. Multicollinearity creates broad confidence intervals, and estimates of coefficients become sensitive to particular sets of sample data: a multicollinear model makes it difficult to establish that an individual explanatory variable influences the dependent variable. R. Wonnacott & T. Wonnacott, supra, at 353; J. Johnston, supra n. 51, at 160. Thus, even if two explanatory variables should be included in the regression, if multicollinearity is serious it may be necessary to drop one of them. This, in turn, may cause problems associated with omission of variables, but those problems might in certain circumstances be acceptable in the face of more serious problems of multicollinearity. See R. Pindyck & D. Rubinfeld, supra, at 68; G. Maddala, supra n. 49, at 190. Cf., e.g., Kmenta, supra, at 390-91 (an alternate solution to multicollinearity is acquisition of more data). There are some rough rules of thumb to judge whether multicollinearity is serious or not. G. Maddala, supra n. 49, at 186; Kmenta, supra, at 389-91. Thus, in a discrimination model, when too many qualification variables are included, the patterns of correlation among the explanatory variables will be such that the confidence interval of the group status coefficient is inflated. Cf. D. Baldus & J. Cole, supra, §8A.1, at 274. Hence, if multicollinearity exists, the probability will be increased that the net impact of group status will be judged statistically nonsignificant, even in cases where there are actual differences in the treatment. Id. at 275.

Another form of specification error occurs in the case where the analyst chooses to use a regression equation that is linear in the explanatory variables when the true regression model is nonlinear in the explanatory variables. For instance, the analyst may think the relationship is best described by, and thus runs the regression on:

$$(5) \quad Y = a + b_1 x_1 + u$$

while the true relationship is:

$$(6) \quad Y = a + b_1 x_1 + b_2 (x_1)^2 + b_3 (x_1)^3 + u$$

Specification of a linear model when the model is nonlinear — an error in the "form" of the specification — can lead to biased estimates. R. Pindyck & D. Rubinfeld, supra, at 190-91.

We have discussed the first major assumption underlying ordinary least squares with reference to specification errors. When the second assumption is violated, i.e., when the scatter or variance of the error terms about zero (the point of perfect prediction) is not approximately the same for all values of each independent variable, "heteroscedasticity" is said to exist. R. Pindyck & D. Rubinfeld, supra, at 17; R. Wonnacott & T. Wonnacott, supra, at 194-95; D. Baldus & J. Cole, supra, §8A.42, at 284. Heteroscedasticity can produce errors such as errors in confidence intervals. Id. at 285.

The third major assumption underlying ordinary least squares is that the error term follows the "normal distribution." With respect to this assumption, basic least squares regression models are "quite 'robust' in that they will tolerate substantial deviations without affecting the validity of the results." D. Baldus & J. Cole, supra, §8A.41, at 284. Nonnormality of errors can be detected, through the use of such techniques as the Kolmorgorov-Smirnov test. G. Maddala, supra, n. 49, at 306. . . .

[Judge Higginbotham used a wide variety of statistical tools, from Z test to multiple regression, to find that the defendant bank had discriminated against blacks in setting compensation but not in hiring and compensating women.]

1. Choice of Variables and the Error Term

It should not be surprising that the error term, u, would depend on our choice of independent variables. It does, after all, indicate our success in capturing the relevant factors influencing the dependent

B. Specification Errors in Multiple Regression   389

variable. The first assumption Judge Higginbotham discussed in *Vuyanich* was the independence of the distribution of error terms from the distribution of the independent variables. More simply, we want the random disturbance terms to be random. We don't want the size of the error term to depend on the particular independent variables chosen. There are three ways in which the error term will be affected by our choice of independent variables for the regression. We may omit relevant independent variables, include irrelevant independent variables, or include variables that are not independent of each other. Because the inclusion of irrelevant variables does not affect the estimations of the regression coefficients, b, in the absence of multicollinearity, we will focus our attention on the first and third ways in which the choice of variables affects the error term. The corrected R^2 (see page 371 above) adjusts R^2 to correct for some of the effects of including irrelevant variables in the regression.

2. Omitted Variables

Omission of relevant independent variables affects our estimates of the regression coefficients only if the omitted variable is correlated with one of the included variables. If we calculate the correlation between an omitted variable and the included variable, and if we find a statistically significant relationship between changes in the size of one and the size of the other, we know that the regression coefficient on the included variable is measuring not just the effect of the included variable but also some of the effect of the omitted variable. But our purpose in estimating a regression coefficient is to indicate the effect of the independent variable alone on the dependent variable. The existence of an omitted correlated variable means that the estimate includes the effect of both variables, not just the included variable.

When is the omitted-variable problem serious? Consider the two extreme values a correlation coefficient can take. As the calculated r approaches zero, the correlation between two variables is very small and will affect the regression coefficient very little because the effects of the included and omitted variables are unlikely to be related. When the calculated r approaches one or minus one, however, the regression coefficient on the included variable may be quite misleading because it is also picking up the effect of the omitted variable.

This is why Judge Higginbotham says that omitted variables may not present serious problems in discrimination cases. If we are trying to predict the influence of race on salary, for instance, we might choose salary as the dependent variable and race of worker as one of the independent variables. What happens to the regression coefficient for race if we omit age of worker as an independent variable? Probably nothing: age is almost certainly a relevant variable, but the effect of age on salary is not likely to be the same as the effect of race on salary. It is unlikely that age and race are highly correlated. If we omit the variable "number of promotions," however, we might bias the coefficient on race, because one's race might influence one's number of promotions. Race and promotability might be correlated.

In Presseisen v. Swarthmore College, Dr. Iversen, the defendant's expert, omitted such variables as scholarship and number of publications from his regression, which was designed to predict salary. Plaintiff's statistical expert on rebuttal, Dr. Hollister, testified that the defendant's regression was unreliable for that reason. When the omission of variables relevant to the salary determination affects the accuracy of the prediction of salary, it is reflected in a low R^2 for the entire regression. Because the omitted variables were unlikely to have been correlated with sex, however, his objection was irrelevant to the issue before the court. Because the omitted variables are unrelated to sex, the regression coefficient on the variable "sex" would be unbiased and an accurate measure of the influence of one's sex on the salary determination.

The obvious solution to the problem of omission of a relevant variable is to include it. But including it may give rise to the problem of multicollinearity. We have stated that the omitted variable gives misleading regression coefficients whenever the omitted variable is correlated to an included independent variable. We have also noted that inclusion of two correlated independent variables creates problems. If the two correlated independent variables are causally related, the regression coefficients are biased. Ordinary least squares cannot separate the causal contributions of such variables. When discussing the multicollinearity in the *Presseisen* regression (pages 339-341 above), we noted a potential causal connection between sex and rank. If these variables were causally related (sex influencing rank), then a regression that included sex but not rank would show an inaccurately high regression coefficient on sex, because part of the effect of sex on rank would be picked up. A regression that included both sex and rank would show an inaccurately low coefficient on sex, because the coefficient on rank has picked up part of the effect of sex.

It sounds like there is no solution to the problem of causally-related intercorrelated relevant independent variables. If you omit one, the coefficient is biased. If you include both, the results are biased. At this point, one option is to choose the alternative that results in the least bias. Your statistical expert will aid in this determination. Another option is to abandon the particular regression design altogether and choose another way to present this same evidence.[2]

When independent variables are correlated but not causally related, the regression coefficients are unbiased, but the standard errors of the regression coefficients, $s.e._b$, are very large. In our discussion of Northshore School District No. 417 v. Kinnear, we suggested a potentially high correlation between apple crop in each school district and per pupil expenditure (which tended to be higher in rural areas). The variables are correlated but not causally related. A regression with both of these as independent variables would show relatively large standard errors and correspondingly low t values for the regression coefficients. This might lead us erroneously to refuse to reject the null hypothesis that the regression coefficient is not significantly different from zero. It may be, of course, that even with the larger standard error, your significance test still allows you to reject the null hypothesis, but the problem with multicollinearity is an increased probability of a type 2 error, improper failure to reject the null. The problem may be avoided by respecification of your regression equation or modification of the OLS approach. Because the inclusion of irrelevant variables increases the chance of multicollinearity, the design of the regression equation is most important.

3. Heteroscedasticity

The second basic assumption of regression analysis is that the variance of the errors does not change systematically with the observations. This assumption is violated whenever knowing the magnitudes of the observations helps determine the variance of the actual from the predicted values. For instance, it may be that salaries of beginning law professors vary widely one from the other, depending on the kind of background and experience a new teacher brings to the job. Salaries for professors who have been around awhile are likely to be higher but show less variance as each approaches the top of the pay scale. If this is true, a regression designed to predict salary by looking at years of experience will predict better for older professors than young. A better prediction means that the error terms are smaller when predicting salaries for professors with more experi-

---

2. A regression method referred to as *two-stage least squares* has been suggested as one solution to this problem. This technique uses a regression estimate of the relationship between the two causally related, correlated, independent variables to aid in identification of the accurate, unbiased effects of each variable on the dependent variable. Discussions of two-stage least squares appear in most econometric texts. See, for instance, R. Wonnacott and T. Wonnacott, Econometrics, Chapter 9 (1970).

ence. Thus the size of error decreases with experience. This systematic relationship between the size of errors and the observations on level of experience is called *heteroscedasticity*. Heteroscedasticity does not affect the regression coefficient, but it does increase the standard error of that coefficient and, therefore, decrease the calculated t value. This affects the test of statistical significance of the regression coefficient. If there is heteroscedasticity, the standard error will not reflect the fact that errors are systematically greater for some predicted values than for others. The regression coefficient might in fact be statistically significant for some values of the independent variable — the ones that have smaller associated errors — even though the coefficient is not statistically significant for those at the other end of the range of the variable, which have large errors. The standard error for the regression coefficient averages these errors and may be large enough to cause you to say that the coefficient is not different from zero. Confidence intervals will also be less accurate.

There are several tests the statistical expert can perform to determine whether heteroscedasticity exists in a particular regression, but the systematic variance of errors with observations is detected primarily by examining the data and calculated errors. If you were, for instance, to rank the observations and rank the associated variance of errors for these observations, a Spearman rank correlation coefficient would indicate whether position on one list helped predict position on another list. A statistically significant Spearman's coefficient would indicate the presence of heteroscedasticity.

Solutions to the problem of heteroscedasticity involve either giving less weight to observations with higher variances or mathematically altering the data to eliminate the systematic variations between errors and observations. Each solution has its own attendant difficulties. The "weighted least squares" approach requires that the variances be known, which is not always the case. The transformation approach is basically a redesign of the regression equation, and the new regression design may be too great a departure from the original, desired model to give useful statistical results or to be convincing as proof.

C. CAUSATION AND INFERENCE

The adversary process is designed to decide ultimate issues of fact by juxtaposing opponents' explanations of the circumstances giving rise to the legal action. In a discrimination case, the circumstances may be an observed absence of minority employees. In antitrust, the circumstance requiring explanation might be a decline in sales, allegedly due to a competitor's trade practices. In products liability, the circumstance may be the occurrence of an automobile accident. In each case, some explanations would be sufficient to relieve the defendant of liability. Exculpatory explanations in these three examples might be a dearth of qualified minority workers in the relevant geographic area, a general business downturn affecting industrywide sales, and misuse of the vehicle that caused it to crash. Inculpatory explanations would be intentional discrimination by the employer, illegal tie-in sales by the competitor, and defective design by the manufacturer. Statistical evidence is used to support and attack both exculpatory and inculpatory explanations.

The value of statistical evidence depends on the quality of the data used to calculate the statistical results. Quality is determined by the relevance, materiality, and credibility of the magnitudes measured. Many of the statistical tools we have developed tell us how different one number is from another. The simplest such statistical method that most people learn is subtraction. Subtraction indicates the absolute difference between two numbers. We saw in Castaneda v. Partida (pages 82-89 above) and in United States v. Maskeny (pages 124-127 above) that absolute differences may impel legal conclusions. And in the environmental-law context, a large excess of pollutants emitted over permitted pollution may impel a conclusion that the emitter has violated some regulation promulgated under the Clean Air Act. But an inference drawn from an absolute difference is without value if the magnitudes being subtracted are inappropriate. Many of the arguments between statistical experts in these materials are based on the appropriateness of including or excluding certain data. Lawyers are at least as well equipped as statisticians to deal with the logical question of whether a piece of evidence is relevant proof of a factual issue. Each number going into a statistical calculation must be relevant and appropriate and can be challenged on that basis. This is true whether the calculation is an absolute disparity or a regression coefficient.

Once a statistic is calculated, even using the best available evidence, it is meaningless without an associated measuring stick. The measuring stick may indicate how large an absolute difference must be or how large a measured correspondence between two variables must be to have legal significance. Significance testing provides this measuring stick for statistical conclusions. The unit of measurement is the probability that such a statistic could occur by chance.

When all irrelevant factors causing a correspondence or absolute difference between two variables have been removed and the probability of random occurrence has been calculated, the statistic forms the basis for an inference, just as a piece of physical evidence does. For instance, police testify that a cloth money bag with the name of a recently robbed bank was found in the defendant's possession. This evidence creates an inference that the defendant was the robber, but the evidence does not prove it. A witness in a tort case testifies that one can be 99% certain that the increased fatality rate for patients of the defendant doctor is not due to chance. This evidence creates an inference of malpractice but, again, does not prove it. Likewise, evidence in a Title VII case indicates that if employees of a particular company were chosen without regard to race, then the racial make-up of the defendant's workforce would only occur by chance once in every five times a workforce were selected. This provides an inference that race is a factor in defendant's hiring decisions but does not prove it. The strength of the inference depends on the number of other circumstances that could account for the outcome. If identical money bags are for sale at nearby stores, if the doctor is forced to operate under battle conditions, if few members of racial minorities are qualified for the employer's jobs, then the inferences are weak. If no exculpatory explanation for the evidence is offered, however, the inference becomes stronger that an inculpatory explanation is appropriate. When statistics are used as proof, "chance" can also be an exculpatory explanation.

Even when we have an inference based on evidence for which there is no exculpatory explanation, we cannot say with certainty what *caused* the particular numbers or evidence to appear. Even multiple regression only points to a variety of possibly alternative explanations. It does not prove that changes in the values of the independent variables caused the value of the dependent variable to change. It only demonstrates, with a specified level of certainty, the extent to which changes in the values of the variables correspond to one another. There might be a high correlation between the number of clergymen in a town and the number of newborn children. Without more information, we are unwilling to say the former caused the latter. The more exculpatory explanations we can eliminate, however, the more likely we are to accept this explanation of causation.

Such is the relationship between inference and causation. Probative statistical evidence eliminates alternative explanations for the observed phenomenon. To provide the strongest inference, statistical evidence must be (1) based on accurate, relevant, and appropriate data and (2) of high statistical as well as practical significance. Such evidence will eliminate the greatest number of alternative explanations. When no significant exculpatory explanations remain, we become willing to take the additional logical step and to say that the prohibited activity must have caused the outcome we observe.

Another explanation of the role of inference and causation in statistical proof can be drawn from the dual requirements of accurate, relevant data and high statistical significance. The goal of the adversary process is to ferret out exculpatory explanations for observed outcomes. Tests of statistical significance can only eliminate chance as a significant explanation. The data must be selected to eliminate all other exculpatory explanations. With a high level of significance and appropriate data, only inculpatory explanations remain. From such careful, significant, statistical evidence, an inference of causation may be drawn: the prohibited activity caused the observed outcome.

APPENDIX A

STATISTICAL TABLES

Binomial Probabilities	398
Critical Values for χ^2	402
Z Values	404
Critical Values for Student's t	406
Critical Values for Correlation Coefficient r	408
Critical Values for Rank Correlation Coefficient r_r	409
Critical Values for F	410

Binomial Probabilities

n	x	.05	.10	.15	.20	.25	.30	.35	.40	.45	.50
1	0	.9500	.9000	.8500	.8000	.7500	.7000	.6500	.6000	.5500	.5000
	1	.0500	.1000	.1500	.2000	.2500	.3000	.3500	.4000	.4500	.5000
2	0	.9025	.8100	.7225	.6400	.5625	.4900	.4225	.3600	.3025	.2500
	1	.0950	.1800	.2550	.3200	.3750	.4200	.4550	.4800	.4950	.5000
	2	.0025	.0100	.0225	.0400	.0625	.0900	.1225	.1600	.2025	.2500
3	0	.8574	.7290	.6141	.5120	.4219	.3430	.2746	.2160	.1664	.1250
	1	.1354	.2430	.3251	.3840	.4219	.4410	.4436	.4320	.4084	.3750
	2	.0071	.0270	.0574	.0960	.1406	.1890	.2389	.2880	.3341	.3750
	3	.0001	.0010	.0034	.0080	.0156	.0270	.0429	.0640	.0911	.1250
4	0	.8145	.6561	.5220	.4096	.3164	.2401	.1785	.1296	.0915	.0625
	1	.1715	.2916	.3685	.4096	.4219	.4116	.3845	.3456	.2995	.2500
	2	.0135	.0486	.0975	.1536	.2109	.2646	.3105	.3456	.3675	.3750
	3	.0005	.0036	.0115	.0256	.0469	.0756	.1115	.1536	.2005	.2500
	4	.0000	.0001	.0005	.0016	.0039	.0081	.0150	.0256	.0410	.0625
5	0	.7738	.5905	.4437	.3277	.2373	.1681	.1160	.0778	.0503	.0312
	1	.2036	.3280	.3915	.4096	.3955	.3602	.3124	.2592	.2059	.1562
	2	.0214	.0729	.1382	.2048	.2637	.3087	.3364	.3456	.3369	.3125
	3	.0011	.0081	.0244	.0512	.0879	.1323	.1811	.2304	.2757	.3125
	4	.0000	.0004	.0022	.0064	.0146	.0284	.0488	.0768	.1128	.1562
	5	.0000	.0000	.0001	.0003	.0010	.0024	.0053	.0102	.0185	.0312

n	k											
6	0	.7351	.5314	.3771	.2621	.1780	.1176	.0754	.0467	.0277	.0156
	1	.2321	.3543	.3993	.3932	.3560	.3025	.2437	.1866	.1359	.0938
	2	.0305	.0984	.1762	.2458	.2966	.3241	.3280	.3110	.2780	.2344
	3	.0021	.0146	.0415	.0819	.1318	.1852	.2355	.2765	.3032	.3125
	4	.0001	.0012	.0055	.0154	.0330	.0595	.0951	.1382	.1861	.2344
	5	.0000	.0001	.0004	.0015	.0044	.0102	.0205	.0369	.0609	.0938
	6	.0000	.0000	.0000	.0001	.0002	.0007	.0018	.0041	.0083	.0156
7	0	.6983	.4783	.3206	.2097	.1335	.0824	.0490	.0280	.0152	.0078
	1	.2573	.3720	.3960	.3670	.3115	.2471	.1848	.1306	.0872	.0547
	2	.0406	.1240	.2097	.2753	.3115	.3177	.2985	.2613	.2140	.1641
	3	.0036	.0230	.0617	.1147	.1730	.2269	.2679	.2903	.2918	.2734
	4	.0002	.0026	.0109	.0287	.0577	.0972	.1442	.1935	.2388	.2734
	5	.0000	.0002	.0012	.0043	.0115	.0250	.0466	.0774	.1172	.1641
	6	.0000	.0000	.0001	.0004	.0013	.0036	.0084	.0172	.0320	.0547
	7	.0000	.0000	.0000	.0000	.0001	.0002	.0006	.0016	.0037	.0078
8	0	.6634	.4305	.2725	.1678	.1001	.0576	.0319	.0168	.0084	.0039
	1	.2793	.3826	.3847	.3355	.2670	.1977	.1373	.0896	.0548	.0312
	2	.0515	.1488	.2376	.2936	.3115	.2965	.2587	.2090	.1569	.1094
	3	.0054	.0331	.0839	.1468	.2076	.2541	.2786	.2787	.2568	.2188
	4	.0004	.0046	.0185	.0459	.0865	.1361	.1875	.2322	.2627	.2734
	5	.0000	.0004	.0026	.0092	.0231	.0467	.0808	.1239	.1719	.2188
	6	.0000	.0000	.0002	.0011	.0038	.0100	.0217	.0413	.0703	.1094
	7	.0000	.0000	.0000	.0001	.0004	.0012	.0033	.0079	.0164	.0312
	8	.0000	.0000	.0000	.0000	.0000	.0001	.0002	.0007	.0017	.0039

(continued)

Binomial Probabilities (*continued*)

n	x	.05	.10	.15	.20	.25	P .30	.35	.40	.45	.50
9	0	.6302	.3874	.2316	.1342	.0751	.0404	.0207	.0101	.0046	.0020
	1	.2985	.3874	.3679	.3020	.2253	.1556	.1004	.0605	.0339	.0176
	2	.0629	.1722	.2597	.3020	.3003	.2668	.2162	.1612	.1110	.0703
	3	.0077	.0446	.1069	.1762	.2336	.2668	.2716	.2508	.2119	.1641
	4	.0006	.0074	.0283	.0661	.1168	.1715	.2194	.2508	.2600	.2461
	5	.0000	.0008	.0050	.0165	.0389	.0735	.1181	.1672	.2128	.2461
	6	.0000	.0001	.0006	.0028	.0087	.0210	.0424	.0743	.1160	.1641
	7	.0000	.0000	.0000	.0003	.0012	.0039	.0098	.0212	.0407	.0703
	8	.0000	.0000	.0000	.0000	.0001	.0004	.0013	.0035	.0083	.0176
	9	.0000	.0000	.0000	.0000	.0000	.0000	.0001	.0003	.0008	.0020
10	0	.5987	.3487	.1969	.1074	.0563	.0282	.0135	.0060	.0025	.0010
	1	.3151	.3874	.3474	.2684	.1877	.1211	.0725	.0403	.0207	.0098
	2	.0746	.1937	.2759	.3020	.2816	.2335	.1757	.1209	.0763	.0439
	3	.0105	.0574	.1298	.2013	.2503	.2668	.2522	.2150	.1665	.1172
	4	.0010	.0112	.0401	.0881	.1460	.2001	.2377	.2508	.2384	.2051
	5	.0001	.0015	.0085	.0264	.0584	.1029	.1536	.2007	.2340	.2461
	6	.0000	.0001	.0012	.0055	.0162	.0368	.0689	.1115	.1596	.2051
	7	.0000	.0000	.0001	.0008	.0031	.0090	.0212	.0425	.0746	.1172
	8	.0000	.0000	.0000	.0001	.0004	.0014	.0043	.0106	.0229	.0439
	9	.0000	.0000	.0000	.0000	.0000	.0001	.0005	.0016	.0042	.0098
	10	.0000	.0000	.0000	.0000	.0000	.0000	.0000	.0001	.0003	.0010

	k	.05	.10	.15	.20	.25	.30	.35	.40	.45	.50
11	0	.5688	.3138	.1673	.0859	.0422	.0198	.0088	.0036	.0014	.0004
	1	.3293	.3835	.3248	.2362	.1549	.0932	.0518	.0266	.0125	.0055
	2	.0867	.2131	.2866	.2953	.2581	.1998	.1395	.0887	.0513	.0269
	3	.0137	.0710	.1517	.2215	.2581	.2568	.2254	.1774	.1259	.0806
	4	.0014	.0158	.0536	.1107	.1721	.2201	.2428	.2365	.2060	.1611
	5	.0001	.0025	.0132	.0388	.0803	.1321	.1830	.2207	.2360	.2256
	6	.0000	.0003	.0023	.0097	.0268	.0566	.0985	.1471	.1931	.2256
	7	.0000	.0000	.0003	.0017	.0064	.0173	.0379	.0701	.1128	.1611
	8	.0000	.0000	.0000	.0002	.0011	.0037	.0102	.0234	.0462	.0806
	9	.0000	.0000	.0000	.0000	.0001	.0005	.0018	.0052	.0126	.0269
	10	.0000	.0000	.0000	.0000	.0000	.0000	.0002	.0007	.0021	.0054
	11	.0000	.0000	.0000	.0000	.0000	.0000	.0000	.0000	.0002	.0005
12	0	.5404	.2824	.1422	.0687	.0317	.0138	.0057	.0022	.0008	.0002
	1	.3413	.3766	.3012	.2062	.1267	.0712	.0368	.0174	.0075	.0029
	2	.0988	.2301	.2924	.2835	.2323	.1678	.1088	.0639	.0339	.0161
	3	.0173	.0852	.1720	.2362	.2581	.2397	.1954	.1419	.0923	.0537
	4	.0021	.0213	.0683	.1329	.1936	.2311	.2367	.2128	.1700	.1208
	5	.0002	.0038	.0193	.0532	.1032	.1585	.2039	.2270	.2225	.1934
	6	.0000	.0005	.0040	.0155	.0401	.0792	.1281	.1766	.2124	.2256
	7	.0000	.0000	.0006	.0033	.0115	.0291	.0591	.1009	.1489	.1934
	8	.0000	.0000	.0001	.0005	.0024	.0078	.0199	.0420	.0762	.1208
	9	.0000	.0000	.0000	.0001	.0004	.0015	.0048	.0125	.0277	.0537
	10	.0000	.0000	.0000	.0000	.0000	.0002	.0008	.0025	.0068	.0161
	11	.0000	.0000	.0000	.0000	.0000	.0000	.0001	.0003	.0010	.0029
	12	.0000	.0000	.0000	.0000	.0000	.0000	.0000	.0000	.0001	.0002

Source: CRC Handbook of Tables for Probability and Statistics 183-184 (W. Beyer ed., 2d ed. 1968). Copyright © 1968 by The Chemical Rubber Co., CRC Press, Inc. Reprinted with permission.

Critical Values for χ^2

Degrees of Freedom	Significance Level (P)										
	.90	.80	.70	.50	.30	.20	.10	.05	.02	.01	.001
1	.0158	.0642	.148	.455	1.074	1.642	2.706	3.841	5.412	6.635	10.827
2	.211	.446	.713	1.386	2.408	3.219	4.605	5.991	7.824	9.210	13.815
3	.584	1.005	1.424	2.366	3.665	4.642	6.251	7.815	9.837	11.345	16.266
4	1.064	1.649	2.195	3.357	4.878	5.989	7.779	9.488	11.668	13.277	18.467
5	1.610	2.343	3.000	4.351	6.064	7.289	9.236	11.070	13.388	15.086	20.515
6	2.204	3.070	3.828	5.348	7.231	8.558	10.645	12.592	15.033	16.812	22.457
7	2.833	3.822	4.671	6.346	8.383	9.803	12.017	14.067	16.622	18.475	24.322
8	3.490	4.594	5.527	7.344	9.524	11.030	13.362	15.507	18.168	20.090	26.125
9	4.168	5.380	6.393	8.343	10.656	12.242	14.684	16.919	19.679	21.666	27.877
10	4.865	6.179	7.267	9.342	11.781	13.442	15.987	18.307	21.161	23.209	29.588
11	5.578	6.989	8.148	10.341	12.899	14.631	17.275	19.675	22.618	24.725	31.264
12	6.304	7.807	9.034	11.340	14.011	15.812	18.549	21.026	24.054	26.217	32.909
13	7.042	8.634	9.926	12.340	15.119	16.985	19.812	22.362	25.472	27.688	34.528
14	7.790	9.467	10.821	13.339	16.222	18.151	21.064	23.685	26.873	29.141	36.123
15	8.547	10.307	11.721	14.339	17.322	19.311	22.307	24.996	28.259	30.578	37.697

16	9.312	11.152	12.624	15.338	18.418	20.465	23.542	26.296	29.633	32.000	39.252
17	10.085	12.002	13.531	16.338	19.511	21.615	24.769	27.587	30.995	33.409	40.790
18	10.865	12.857	14.440	17.338	20.601	22.760	25.989	28.869	32.346	34.805	42.312
19	11.651	13.716	15.352	18.338	21.689	23.900	27.204	30.144	33.687	36.191	43.820
20	12.443	14.578	16.266	19.337	22.775	25.038	28.412	31.410	35.020	37.566	45.315
21	13.240	15.445	17.182	20.337	23.858	26.171	29.615	32.671	36.343	38.932	46.797
22	14.041	16.314	18.101	21.337	24.939	27.301	30.813	33.924	37.659	40.289	48.268
23	14.848	17.187	19.021	22.337	26.018	28.429	32.007	35.172	38.968	41.638	49.728
24	15.659	18.062	19.943	23.337	27.096	29.553	33.196	36.415	40.270	42.980	51.179
25	16.473	18.940	20.867	24.337	28.172	30.675	34.382	37.652	41.566	44.314	52.620
26	17.292	19.820	21.792	25.336	29.246	31.795	35.563	38.885	42.856	45.542	54.052
27	18.114	20.703	22.719	26.336	30.319	32.912	36.741	40.113	44.140	46.963	55.476
28	18.939	21.588	23.647	27.336	31.391	34.027	37.916	41.337	45.419	48.278	56.893
29	19.768	22.475	24.577	28.336	32.461	35.139	39.087	42.557	46.693	49.588	58.302
30	20.599	23.364	25.508	29.336	33.530	36.250	40.256	43.773	47.962	50.892	59.703

Source: R. Fisher and F. Yates, Statistical Tables for Biological, Agricultural and Medical Research 47 (Table IV) (6th ed. 1963). Reprinted with permission of Oliver and Boyd, Edinburgh.

Z Values

Z	0.00	0.01	0.02	0.03	0.04	0.05	0.06	0.07	0.08	0.09
0.0	.0000	.0040	.0080	.0120	.0160	.0199	.0239	.0279	.0319	.0359
0.1	.0398	.0438	.0478	.0517	.0557	.0596	.0636	.0675	.0714	.0753
0.2	.0793	.0832	.0871	.0910	.0948	.0987	.1026	.1064	.1103	.1141
0.3	.1179	.1217	.1255	.1293	.1331	.1368	.1406	.1443	.1480	.1517
0.4	.1554	.1591	.1628	.1664	.1700	.1736	.1772	.1808	.1844	.1879
0.5	.1915	.1950	.1985	.2019	.2054	.2088	.2123	.2157	.2190	.2224
0.6	.2257	.2291	.2324	.2357	.2389	.2422	.2454	.2486	.2517	.2549
0.7	.2580	.2611	.2642	.2673	.2704	.2734	.2764	.2794	.2823	.2852
0.8	.2881	.2910	.2939	.2967	.2995	.3023	.3051	.3078	.3106	.3133
0.9	.3159	.3186	.3212	.3238	.3264	.3289	.3315	.3340	.3365	.3389
1.0	.3413	.3438	.3461	.3485	.3508	.3531	.3554	.3577	.3599	.3621
1.1	.3643	.3665	.3686	.3708	.3729	.3749	.3770	.3790	.3810	.3830
1.2	.3849	.3869	.3888	.3907	.3925	.3944	.3962	.3980	.3997	.4015
1.3	.4032	.4049	.4066	.4082	.4099	.4115	.4131	.4147	.4162	.4177
1.4	.4192	.4207	.4222	.4236	.4251	.4265	.4279	.4292	.4306	.4319
1.5	.4332	.4345	.4357	.4370	.4382	.4394	.4406	.4418	.4429	.4441
1.6	.4452	.4463	.4474	.4484	.4495	.4505	.4515	.4525	.4535	.4545
1.7	.4554	.4564	.4573	.4582	.4591	.4599	.4608	.4616	.4625	.4633
1.8	.4641	.4649	.4656	.4664	.4671	.4678	.4686	.4693	.4699	.4706
1.9	.4713	.4719	.4726	.4732	.4738	.4744	.4750	.4756	.4761	.4767

2.0	.4773	.4778	.4783	.4788	.4793	.4798	.4803	.4808	.4812	.4817
2.1	.4821	.4826	.4830	.4834	.4838	.4842	.4846	.4850	.4854	.4857
2.2	.4861	.4864	.4868	.4871	.4875	.4878	.4881	.4884	.4887	.4890
2.3	.4893	.4896	.4898	.4901	.4904	.4906	.4909	.4911	.4913	.4916
2.4	.4918	.4920	.4922	.4925	.4927	.4929	.4931	.4932	.4934	.4936
2.5	.4938	.4940	.4941	.4943	.4945	.4946	.4948	.4949	.4951	.4952
2.6	.4953	.4955	.4956	.4957	.4959	.4960	.4961	.4962	.4963	.4964
2.7	.4965	.4966	.4967	.4968	.4969	.4970	.4971	.4972	.4973	.4974
2.8	.4974	.4975	.4976	.4977	.4977	.4978	.4979	.4979	.4980	.4981
2.9	.4981	.4982	.4983	.4983	.4984	.4984	.4985	.4985	.4986	.4986
3.0	.4987	.4987	.4987	.4988	.4988	.4989	.4989	.4989	.4989	.4990
3.1	.4990	.4991	.4991	.4991	.4992	.4992	.4992	.4992	.4993	.4993
3.2	.4993	.4993	.4994	.4994	.4994	.4994	.4994	.4995	.4995	.4995
3.3	.4995	.4995	.4996	.4996	.4996	.4996	.4996	.4996	.4996	.4997
3.4	.4997	.4997	.4997	.4997	.4997	.4997	.4997	.4997	.4997	.4998
3.5	.4998	.4998	.4998	.4998	.4998	.4998	.4998	.4998	.4998	.4998
3.6	.4998	.4998	.4999	.4999	.4999	.4999	.4999	.4999	.4999	.4999
3.7	.4999	.4999	.4999	.4999	.4999	.4999	.4999	.4999	.4999	.4999
3.8	.4999	.4999	.4999	.4999	.4999	.4999	.4999	.4999	.4999	.5000
3.9	.5000	.5000	.5000	.5000	.5000	.5000	.5000	.5000	.5000	.5000

Source: R. Fisher and F. Yates, Statistical Tables for Biological, Agricultural and Medical Research 45 (Table II) (6th ed. 1963), and E. Lovejoy, Statistics for Math Haters 239 (1975). Reprinted with permission of the authors and publishers.

Critical Values for Student's t

Degrees of Freedom	\.45	\.40	\.35	\.30	\.25	\.20	Significance Level One-Tailed Test .15	.10	.05	.025	.01	.005	.0005
	.90	.80	.70	.60	.50	.40	Two-Tailed Test .30	.20	.10	.05	.02	.01	.001
1	.158	.325	.510	.727	1.000	1.376	1.963	3.078	6.314	12.706	31.821	63.657	636.619
2	.142	.289	.445	.617	.816	1.061	1.386	1.886	2.920	4.303	6.965	9.925	31.598
3	.137	.277	.424	.584	.765	.978	1.250	1.638	2.353	3.182	4.541	5.841	12.924
4	.134	.271	.414	.569	.741	.941	1.190	1.533	2.132	2.776	3.747	4.604	8.610
5	.132	.267	.408	.559	.727	.920	1.156	1.476	2.015	2.571	3.365	4.032	6.869
6	.131	.265	.404	.553	.718	.906	1.134	1.440	1.943	2.447	3.143	3.707	5.959
7	.130	.263	.402	.549	.711	.896	1.119	1.415	1.895	2.365	2.998	3.499	5.408
8	.130	.262	.399	.546	.706	.889	1.108	1.397	1.860	2.306	2.896	3.355	5.041
9	.129	.261	.398	.543	.703	.883	1.100	1.383	1.833	2.262	2.821	3.250	4.781
10	.129	.260	.397	.542	.700	.879	1.093	1.372	1.812	2.228	2.764	3.169	4.587
11	.129	.260	.396	.540	.697	.876	1.088	1.363	1.796	2.201	2.718	3.106	4.437
12	.128	.259	.395	.539	.695	.873	1.083	1.356	1.782	2.179	2.681	3.055	4.318
13	.128	.259	.394	.538	.694	.870	1.079	1.350	1.771	2.160	2.650	3.012	4.221
14	.128	.258	.393	.537	.692	.868	1.076	1.345	1.761	2.145	2.624	2.977	4.140
15	.128	.258	.393	.536	.691	.866	1.074	1.341	1.753	2.131	2.602	2.947	4.073

16	.128	.258	.392	.535	.690	.865	1.071	1.337	1.746	2.120	2.583	2.921	4.015
17	.128	.257	.392	.534	.689	.863	1.069	1.333	1.740	2.110	2.567	2.898	3.965
18	.127	.257	.392	.534	.688	.862	1.067	1.330	1.734	2.101	2.552	2.878	3.922
19	.127	.257	.391	.533	.688	.861	1.066	1.328	1.729	2.093	2.539	2.861	3.883
20	.127	.257	.391	.533	.687	.860	1.064	1.325	1.725	2.086	2.528	2.845	3.850
21	.127	.257	.391	.532	.686	.859	1.063	1.323	1.721	2.080	2.518	2.831	3.819
22	.127	.256	.390	.532	.686	.858	1.061	1.321	1.717	2.074	2.508	2.819	3.792
23	.127	.256	.390	.532	.685	.858	1.060	1.319	1.714	2.069	2.500	2.807	3.767
24	.127	.256	.390	.531	.685	.857	1.059	1.318	1.711	2.064	2.492	2.797	3.745
25	.127	.256	.390	.531	.684	.856	1.058	1.316	1.708	2.060	2.485	2.787	3.725
26	.127	.256	.390	.531	.684	.856	1.058	1.315	1.706	2.056	2.479	2.779	3.707
27	.127	.256	.389	.531	.684	.855	1.057	1.314	1.703	2.052	2.473	2.771	3.690
28	.127	.256	.389	.530	.683	.855	1.056	1.313	1.701	2.048	2.467	2.763	3.674
29	.127	.256	.389	.530	.683	.854	1.055	1.311	1.699	2.045	2.462	2.756	3.659
30	.127	.256	.389	.530	.683	.854	1.055	1.310	1.697	2.042	2.457	2.750	3.646
40	.126	.255	.388	.529	.681	.851	1.050	1.303	1.684	2.021	2.423	2.704	3.551
60	.126	.254	.387	.527	.679	.848	1.046	1.296	1.671	2.000	2.390	2.660	3.460
120	.126	.254	.386	.526	.677	.845	1.041	1.289	1.658	1.980	2.358	2.617	3.373
∞	.126	.253	.385	.524	.674	.842	1.036	1.282	1.645	1.960	2.326	2.576	3.291

Source: R. Fisher and F. Yates, Statistical Tables for Biological, Agricultural and Medical Research 46 (Table III) (6th ed. 1963). Reprinted with permission.

Critical Values for the Correlation Coefficient r

Number of Pairs	Significance Level				
	.10	.05	.02	.01	.001
3	.98769	.99692	.999507	.999877	.9999988
4	.90000	.95000	.98000	.990000	.99900
5	.8054	.8783	.93433	.95873	.99116
6	.7293	.8114	.8822	.91720	.97406
7	.6694	.7545	.8329	.8745	.95074
8	.6215	.7067	.7887	.8343	.92493
9	.5822	.6664	.7498	.7977	.8982
10	.5494	.6319	.7155	.7646	.8721
11	.5214	.6021	.6851	.7348	.8471
12	.4973	.5760	.6581	.7079	.8233
13	.4762	.5529	.6339	.6835	.8010
14	.4575	.5324	.6120	.6614	.7800
15	.4409	.5139	.5923	.6411	.7603
16	.4259	.4973	.5742	.6226	.7420
17	.4124	.4821	.5577	.6055	.7246
18	.4000	.4683	.5425	.5897	.7084
19	.3887	.4555	.5285	.5751	.6932
20	.3783	.4438	.5155	.5614	.6787
21	.3687	.4329	.5034	.5487	.6652
22	.3598	.4227	.4921	.5368	.6524
27	.3233	.3809	.4451	.4869	.5974
32	.2960	.3494	.4093	.4487	.5541
37	.2746	.3246	.3810	.4182	.5189
42	.2573	.3044	.3578	.3932	.4896
47	.2428	.2875	.3384	.3721	.4648
52	.2306	.2732	.3218	.3541	.4433
62	.2108	.2500	.2948	.3248	.4078
72	.1954	.2319	.2737	.3017	.3799
82	.1829	.2172	.2565	.2830	.3568
92	.1726	.2050	.2422	.2673	.3375
102	.1638	.1946	.2301	.2540	.3211

Source: R. Fisher and F. Yates, Statistical Tables for Biological, Agricultural and Medical Research 63 (Table VII) (6th ed. 1963). Reprinted with permission.

Critical Values for the Rank Correlation Coefficient r_r

Number of Pairs	Significance Level			
	.10	.05	.02	.01
5	0.900	—	—	—
6	0.829	0.886	0.943	—
7	0.714	0.786	0.893	—
8	0.643	0.738	0.833	0.881
9	0.600	0.683	0.783	0.833
10	0.564	0.648	0.745	0.794
11	0.523	0.623	0.736	0.818
12	0.497	0.591	0.703	0.780
13	0.475	0.566	0.673	0.745
14	0.457	0.545	0.646	0.716
15	0.441	0.525	0.623	0.689
16	0.425	0.507	0.601	0.666
17	0.412	0.490	0.582	0.645
18	0.399	0.476	0.564	0.625
19	0.388	0.462	0.549	0.608
20	0.377	0.450	0.534	0.591
21	0.368	0.438	0.521	0.576
22	0.359	0.428	0.508	0.562
23	0.351	0.418	0.496	0.549
24	0.343	0.409	0.485	0.537
25	0.336	0.400	0.475	0.526
26	0.329	0.392	0.465	0.515
27	0.323	0.385	0.456	0.505
28	0.317	0.377	0.448	0.496
29	0.311	0.370	0.440	0.487
30	0.305	0.364	0.432	0.478

Source: N. Johnson and F. Leone, 1 Statistics and Experimental Design in Engineering and the Physical Sciences 547 (2d ed. 1977). Reprinted with permission.

Critical Values for F (Significance Level = .05)

Degrees of Freedom in Denominator ($= n - k - 1$)	\multicolumn{9}{c}{Degrees of Freedom in Numerator ($= k$)}								
	1	2	3	4	5	6	7	8	9
1	161.4	199.5	215.7	224.6	230.2	234.0	236.8	238.9	240.5
2	18.51	19.00	19.16	19.25	19.30	19.33	19.35	19.37	19.38
3	10.13	9.55	9.28	9.12	9.01	8.94	8.89	8.85	8.81
4	7.71	6.94	6.59	6.39	6.26	6.16	6.09	6.04	6.00
5	6.61	5.79	5.41	5.19	5.05	4.95	4.88	4.82	4.77
6	5.99	5.14	4.76	4.53	4.39	4.28	4.21	4.15	4.10
7	5.59	4.74	4.35	4.12	3.97	3.87	3.79	3.73	3.68
8	5.32	4.46	4.07	3.84	3.69	3.58	3.50	3.44	3.39
9	5.12	4.26	3.86	3.63	3.48	3.37	3.29	3.23	3.18
10	4.96	4.10	3.71	3.48	3.33	3.22	3.14	3.07	3.02
11	4.84	3.98	3.59	3.36	3.20	3.09	3.01	2.95	2.90
12	4.75	3.89	3.49	3.26	3.11	3.00	2.91	2.85	2.80
13	4.67	3.81	3.41	3.18	3.03	2.92	2.83	2.77	2.71
14	4.60	3.74	3.34	3.11	2.96	2.85	2.76	2.70	2.65
15	4.54	3.68	3.29	3.06	2.90	2.79	2.71	2.64	2.59
16	4.49	3.63	3.24	3.01	2.85	2.74	2.66	2.59	2.54
17	4.45	3.59	3.20	2.96	2.81	2.70	2.61	2.55	2.49
18	4.41	3.55	3.16	2.93	2.77	2.66	2.58	2.51	2.46
19	4.38	3.52	3.13	2.90	2.74	2.63	2.54	2.48	2.42
20	4.35	3.49	3.10	2.87	2.71	2.60	2.51	2.45	2.39
21	4.32	3.47	3.07	2.84	2.68	2.57	2.49	2.42	2.37
22	4.30	3.44	3.05	2.82	2.66	2.55	2.46	2.40	2.34
23	4.28	3.42	3.03	2.80	2.64	2.53	2.44	2.37	2.32
24	4.26	3.40	3.01	2.78	2.62	2.51	2.42	2.36	2.30
25	4.24	3.39	2.99	2.76	2.60	2.49	2.40	2.34	2.28
26	4.23	3.37	2.98	2.74	2.59	2.47	2.39	2.32	2.27
27	4.21	3.35	2.96	2.73	2.57	2.46	2.37	2.31	2.25
28	4.20	3.34	2.95	2.71	2.56	2.45	2.36	2.29	2.24
29	4.18	3.33	2.93	2.70	2.55	2.43	2.35	2.28	2.22
30	4.17	3.32	2.92	2.69	2.53	2.42	2.33	2.27	2.21
40	4.08	3.23	2.84	2.61	2.45	2.34	2.25	2.18	2.12
60	4.00	3.15	2.76	2.53	2.37	2.25	2.17	2.10	2.04
120	3.92	3.07	2.68	2.45	2.29	2.17	2.09	2.02	1.96
∞	3.84	3.00	2.60	2.37	2.21	2.10	2.01	1.94	1.88

10	12	15	20	24	30	40	60	120	∞
241.9	243.9	245.9	248.0	249.1	250.1	251.1	252.2	253.3	254.3
19.40	19.41	19.43	19.45	19.45	19.46	19.47	19.48	19.49	19.50
8.79	8.74	8.70	8.66	8.64	8.62	8.59	8.57	8.55	8.53
5.96	5.91	5.86	5.80	5.77	5.75	5.72	5.69	5.66	5.63
4.74	4.68	4.62	4.56	4.53	4.50	4.46	4.43	4.40	4.36
4.06	4.00	3.94	3.87	3.84	3.81	3.77	3.74	3.70	3.67
3.64	3.57	3.51	3.44	3.41	3.38	3.34	3.30	3.27	3.23
3.35	3.28	3.22	3.15	3.12	3.08	3.04	3.01	2.97	2.93
3.14	3.07	3.01	2.94	2.90	2.86	2.83	2.79	2.75	2.71
2.98	2.91	2.85	2.77	2.74	2.70	2.66	2.62	2.58	2.54
2.85	2.79	2.72	2.65	2.61	2.57	2.53	2.49	2.45	2.40
2.75	2.69	2.62	2.54	2.51	2.47	2.43	2.38	2.34	2.30
2.67	2.60	2.53	2.46	2.42	2.38	2.34	2.30	2.25	2.21
2.60	2.53	2.46	2.39	2.35	2.31	2.27	2.22	2.18	2.13
2.54	2.48	2.40	2.33	2.29	2.25	2.20	2.16	2.11	2.07
2.49	2.42	2.35	2.28	2.24	2.19	2.15	2.11	2.06	2.01
2.45	2.38	2.31	2.23	2.19	2.15	2.10	2.06	2.01	1.96
2.41	2.34	2.27	2.19	2.15	2.11	2.06	2.02	1.97	1.92
2.38	2.31	2.23	2.16	2.11	2.07	2.03	1.98	1.93	1.88
2.35	2.28	2.20	2.12	2.08	2.04	1.99	1.95	1.90	1.84
2.32	2.25	2.18	2.10	2.05	2.01	1.96	1.92	1.87	1.81
2.30	2.23	2.15	2.07	2.03	1.98	1.94	1.89	1.84	1.78
2.27	2.20	2.13	2.05	2.01	1.96	1.91	1.86	1.81	1.76
2.25	2.18	2.11	2.03	1.98	1.94	1.89	1.84	1.79	1.73
2.24	2.16	2.09	2.01	1.96	1.92	1.87	1.82	1.77	1.71
2.22	2.15	2.07	1.99	1.95	1.90	1.85	1.80	1.75	1.69
2.20	2.13	2.06	1.97	1.93	1.88	1.84	1.79	1.73	1.67
2.19	2.12	2.04	1.96	1.91	1.87	1.82	1.77	1.71	1.65
2.18	2.10	2.03	1.94	1.90	1.85	1.81	1.75	1.70	1.64
2.16	2.09	2.01	1.93	1.89	1.84	1.79	1.74	1.68	1.62
2.08	2.00	1.92	1.84	1.79	1.74	1.69	1.64	1.58	1.51
1.99	1.92	1.84	1.75	1.70	1.65	1.59	1.53	1.47	1.39
1.91	1.83	1.75	1.66	1.61	1.55	1.50	1.43	1.35	1.25
1.83	1.75	1.67	1.57	1.52	1.46	1.39	1.32	1.22	1.00

(*continued*)

Critical Values for F (*continued*) (Significance Level = .01)

Degrees of Freedom in Denominator ($= n - k - 1$)	\multicolumn{9}{c}{Degrees of Freedom in Numerator ($= k$)}								
	1	2	3	4	5	6	7	8	9
1	4052	4999.5	5403	5625	5764	5859	5928	5982	6022
2	98.50	99.00	99.17	99.25	99.30	99.33	99.36	99.37	99.39
3	34.12	30.82	29.46	28.71	28.24	27.91	27.67	27.49	27.35
4	21.20	18.00	16.69	15.98	15.52	15.21	14.98	14.80	14.66
5	16.26	13.27	12.06	11.39	10.97	10.67	10.46	10.29	10.16
6	13.75	10.92	9.78	9.15	8.75	8.47	8.26	8.10	7.98
7	12.25	9.55	8.45	7.85	7.46	7.19	6.99	6.84	6.72
8	11.26	8.65	7.59	7.01	6.63	6.37	6.18	6.03	5.91
9	10.56	8.02	6.99	6.42	6.06	5.80	5.61	5.47	5.35
10	10.04	7.56	6.55	5.99	5.64	5.39	5.20	5.06	4.94
11	9.65	7.21	6.22	5.67	5.32	5.07	4.89	4.74	4.63
12	9.33	6.93	5.95	5.41	5.06	4.82	4.64	4.50	4.39
13	9.07	6.70	5.74	5.21	4.86	4.62	4.44	4.30	4.19
14	8.86	6.51	5.56	5.04	4.69	4.46	4.28	4.14	4.03
15	8.68	6.36	5.42	4.89	4.56	4.32	4.14	4.00	3.89
16	8.53	6.23	5.29	4.77	4.44	4.20	4.03	3.89	3.78
17	8.40	6.11	5.18	4.67	4.34	4.10	3.93	3.79	3.68
18	8.29	6.01	5.09	4.58	4.25	4.01	3.84	3.71	3.60
19	8.18	5.93	5.01	4.50	4.17	3.94	3.77	3.63	3.52
20	8.10	5.85	4.94	4.43	4.10	3.87	3.70	3.56	3.46
21	8.02	5.78	4.87	4.37	4.04	3.81	3.64	3.51	3.40
22	7.95	5.72	4.82	4.31	3.99	3.76	3.59	3.45	3.35
23	7.88	5.66	4.76	4.26	3.94	3.71	3.54	3.41	3.30
24	7.82	5.61	4.72	4.22	3.90	3.67	3.50	3.36	3.26
25	7.77	5.57	4.68	4.18	3.85	3.63	3.46	3.32	3.22
26	7.72	5.53	4.64	4.14	3.82	3.59	3.42	3.29	3.18
27	7.68	5.49	4.60	4.11	3.78	3.56	3.39	3.26	3.15
28	7.64	5.45	4.57	4.07	3.75	3.53	3.36	3.23	3.12
29	7.60	5.42	4.54	4.04	3.73	3.50	3.33	3.20	3.09
30	7.56	5.39	4.51	4.02	3.70	3.47	3.30	3.17	3.07
40	7.31	5.18	4.31	3.83	3.51	3.29	3.12	2.99	2.89
60	7.08	4.98	4.13	3.65	3.34	3.12	2.95	2.82	2.72
120	6.85	4.79	3.95	3.48	3.17	2.96	2.79	2.66	2.56
∞	6.63	4.61	3.78	3.32	3.02	2.80	2.64	2.51	2.41

Source: E. Pearson and H. Hartley, eds., I Biometrika Tables for Statisticians 159 (Table 18) (1958). Reprinted with permission of the Biometrika Trustees.

10	12	15	20	24	30	40	60	120	∞
6056	6106	6157	6209	6235	6261	6287	6313	6339	6366
99.40	99.42	99.43	99.45	99.46	99.47	99.47	99.48	99.49	99.50
27.23	27.05	26.87	26.69	26.60	26.50	26.41	26.32	26.22	26.13
14.55	14.37	14.20	14.02	13.93	13.84	13.75	13.65	13.56	13.46
10.05	9.89	9.72	9.55	9.47	9.38	9.29	9.20	9.11	9.02
7.87	7.72	7.56	7.40	7.31	7.23	7.14	7.06	6.97	6.88
6.62	6.47	6.31	6.16	6.07	5.99	5.91	5.82	5.74	5.65
5.81	5.67	5.52	5.36	5.28	5.20	5.12	5.03	4.95	4.86
5.26	5.11	4.96	4.81	4.73	4.65	4.57	4.48	4.40	4.31
4.85	4.71	4.56	4.41	4.33	4.25	4.17	4.08	4.00	3.91
4.54	4.40	4.25	4.10	4.02	3.94	3.86	3.78	3.69	3.60
4.30	4.16	4.01	3.86	3.78	3.70	3.62	3.54	3.45	3.36
4.10	3.96	3.82	3.66	3.59	3.51	3.43	3.34	3.25	3.17
3.94	3.80	3.66	3.51	3.43	3.35	3.27	3.18	3.09	3.00
3.80	3.67	3.52	3.37	3.29	3.21	3.13	3.05	2.96	2.87
3.69	3.55	3.41	3.26	3.18	3.10	3.02	2.93	2.84	2.75
3.59	3.46	3.31	3.16	3.08	3.00	2.92	2.83	2.75	2.65
3.51	3.37	3.23	3.08	3.00	2.92	2.84	2.75	2.66	2.57
3.43	3.30	3.15	3.00	2.92	2.84	2.76	2.67	2.58	2.49
3.37	3.23	3.09	2.94	2.86	2.78	2.69	2.61	2.52	2.42
3.31	3.17	3.03	2.88	2.80	2.72	2.64	2.55	2.46	2.36
3.26	3.12	2.98	2.83	2.75	2.67	2.58	2.50	2.40	2.31
3.21	3.07	2.93	2.78	2.70	2.62	2.54	2.45	2.35	2.26
3.17	3.03	2.89	2.74	2.66	2.58	2.49	2.40	2.31	2.21
3.13	2.99	2.85	2.70	2.62	2.54	2.45	2.36	2.27	2.17
3.09	2.96	2.81	2.66	2.58	2.50	2.42	2.33	2.23	2.13
3.06	2.93	2.78	2.63	2.55	2.47	2.38	2.29	2.20	2.10
3.03	2.90	2.75	2.60	2.52	2.44	2.35	2.26	2.17	2.06
3.00	2.87	2.73	2.57	2.49	2.41	2.33	2.23	2.14	2.03
2.98	2.84	2.70	2.55	2.47	2.39	2.30	2.21	2.11	2.01
2.80	2.66	2.52	2.37	2.29	2.20	2.11	2.02	1.92	1.80
2.63	2.50	2.35	2.20	2.12	2.03	1.94	1.84	1.73	1.60
2.47	2.34	2.19	2.03	1.95	1.86	1.76	1.66	1.53	1.38
2.32	2.18	2.04	1.88	1.79	1.70	1.59	1.47	1.32	1.00

APPENDIX B

ANSWERS TO SELECTED PROBLEMS

Your answers may vary slightly due to rounding-off.

Page 71, Note 1

The probability of at least one job offer = 88.24%.

$p = .30$
$n = 6$
$x = 1$ or more (not zero)

Page 71, Note 2

(a) 2 cities.

$p = .30$
$n = 2$
$x = 1$ or more

(b) 6 cities.

$p = .30$
$n = 6$
$x = 2$ or more

Page 73, Note 4

Jury of 11 people.

$p = .20$
$n = 11$
$x = 1$ or more (not zero)

Page 73, Note 5

You should advise your client to settle. The probability of a conviction = 17.17%.

$p = .35$
$n = 5$ or more
$x = 9$

415

Page 73, Note 6

You should advise your client not to plead guilty. The probability of receiving the death penalty = .09%.

$p = .10$
$n = 5$ or more
$x = 9$

Page 74, Note 7

Take the $2,500 lump sum.

$p = .40$
$n = 5$
for $x = 1$, expected return = .2592 × $1000 = $ 259.20
      2, expected return = .3456 × $2000 = $ 691.20
      3, expected return = .2304 × $3000 = $ 691.20
      4, expected return = .0768 × $4000 = $ 307.20
      5, expected return = .0102 × $5000 = $ 51.00
Total expected return on contingency basis = $1999.80

Page 74, Note 8

4 states.

$p = .50$
$n = 4$
$x = 1$ or more

Page 92, Note 4

Yes.

$N = 600$
$P_1 = .15$
$P_2 = .85$
$Ex = .15 \times 600 = 90$
$Ob = 25$

$Ob - Ex = -65$

$SD = \sqrt{600 \times .15 \times .85} = 8.7$

$\dfrac{Ob - Ex}{SD} = \dfrac{-65}{8.7} = -7.5$ standard deviations

Page 92, Note 5

No, discrimination is in *favor* of males.

$N = 250$
$P_1 = .25$
$P_2 = .75$
$Ex = 62.5$
$Ob = 90$

$Ob - Ex = 27.5$

$SD = 6.8$

$\dfrac{Ob - Ex}{SD} = 4.0$ standard deviations

Page 95, Note 1

(a) The poor population is -201 standard deviations from the expected number.

$N = 1,000,000$
$P_1 = .1051$
$P_2 = .8949$
$Ex = .1051 \times 1,000,000 = 105,100$
$Ob = .0434 \times 1,000,000 = 43,400$
$SD = 307$

(b) The black population is -119 standard deviations from the expected number.

$N = 1,000,000$
$P_1 = .2633$
$P_2 = .7367$
$Ex = .2633 \times 1,000,000 = 263,300$
$Ob = .2106 \times 1,000,000 = 210,600$
$SD = 440$

Page 96, Note 2

(a) There is no discrimination against black people if $N = 23$.

$N = 23$
$P_1 = .2633$
$P_2 = .7367$
$Ex = 6.1$
$Ob = 4.8$
$SD = 2.1$

$\dfrac{Ob - Ex}{SD} = -0.62$ standard deviations

(b) There is no discrimination against poor people if $N = 23$.

$N = 23$
$P_1 = .1051$
$P_2 = .8949$
$Ex = 2.4$
$Ob = 1$
$SD = 1.5$

$\dfrac{Ob - Ex}{SD} = -0.93$ standard deviations

Appendix B. Answers to Selected Problems

Page 97, Note 4

(a) -247 standard deviations.

$N = 1,000,000$
$P_1 = .1251$
$P_2 = .8749$
$Ex = 125,100$
$Ob = 43,400$
$SD = 331$

(b) -1.2 standard deviations.

$N = 23$
$P_1 = .1251$
$P_2 = .8749$
$Ex = 2.9$
$Ob = 1.0$
$SD = 1.6$

Page 107, Note 3

-3.34 standard deviations.

$N = 535$
$P_1 = .02$
$P_2 = .98$
$Ex = 10.7$
$Ob = 0$
$SD = 3.2$

Page 108, Note 4

-1.1 standard deviations.

$N = 535$
$P_1 = .26$
$P_2 = .74$
$Ex = 139$
$Ob = 128$
$SD = 10$

Page 116, Note 3

-2 standard deviations.

$Ex\% = 5.1\%$
$Ob\% = 0.9\%$

$Ob\% - Ex\% = 0.9\% - 5.1\% = -4.2\%$

$SD = \sqrt{\dfrac{.051 \times .949}{109}} = 0.021 = 2.1\%$

Page 129, Note 2

(a) Observed % = 39.0% Number of standard deviations = −29
(b) Observed % = 28.4% Number of standard deviations = −37
(c) Observed % = 39.0% Number of standard deviations = −29
(d) Observed % = 47.4% Number of standard deviations = −23

Page 138, *Taylor* Problem

The variation in raises was greater in 1977 than in 1982.

$$C.V. (1977) = 1095/2500 = .438$$
$$C.V. (1982) = 1395/9220 = .151$$

Page 180, Note 3

If we reject, the probability of a type 1 error is between 2% and 5%.

Mean cancer rate = 20%

High-dosage *Low-dosage*
$E_j = 4.2$ $E_j = 2.8$
$O_j = 7$ $O_j = 0$

$d.f. = 1$
$\chi^2 = 4.67$

Page 180, Note 4

$\chi^2 = 5.3$

Page 181, Note 2

(a) Recalculating the problem in Note 3, page 180, the probability of a type 1 error is between 10% and 5%. Do not reject the null hypothesis.

Mean cancer rate = 20%

High-dosage *Low-dosage*
$E_j = 4.2$ $E_j = 2.8$
$O_j = 7$ $O_j = 0$

$d.f. = 1$
$\chi^2 = 3.2$

(b) Recalculating the problem in Note 4, page 180, the probability of a type 1 error is between 5% and 2%. Reject the null hypothesis.

Cancer rate for high dosage group = 13%

High-dosage *Low-dosage*
$E_j = 3$ $E_j = 4$
$O_j = 7$ $O_j = 4$

$d.f. = 1$
$\chi^2 = 4.15$

Page 182, Note 3

Do not reject null hypothesis that there is no relationship between elderliness and being accepted into the complex. The probability of a type 1 error would be between 10 and 20%.

Elderly
$E_j = 20 \times .20 = 4$
$O_j = 2$
$|O_j - E_j| - \frac{1}{2} = 2.5$

Not Elderly
$E_j = 20 \times .80 = 16$
$O_j = 18$
$|O_j - E_j| - \frac{1}{2} = 1.5$

$\chi^2 = 1.7$

Number of standard deviations = 1.1

Page 196, Note 1

The probability of a type 1 error is less than 1%. Reject the null hypothesis.

Pass rate for Caucasians = 61.3%

Caucasian
$E_j = 717.8$
$O_j = 717.8$

Black and Puerto Rican
$E_j = 174.7$
$O_j = 129.0$

$d.f. = 1$
$\chi^2 = 12.0$

Page 196, Note 3

The probability of a type 1 error is between 10% and 5%. Do not reject the null hypothesis.

$E_j = 10$ for each category
$d.f. = 5$
$\chi^2 = 9.4$

Page 196, Note 4

The probability of a type 1 error is between 2% and 1%. Refuse to reject the null hypothesis.

	Anglo	Black	Chicano	Oriental and Other
E_j	42	38	12	8
O_j	57	26	9	8

$d.f. = 3$
$\chi^2 = 9.9$

Page 197, Note 5

Number of SD for Black Principals:

Detroit	-3.5
Philadelphia	-3.4
Los Angeles	-3.9
Chicago	-8.0
New York	-10.2

Rank in order of randomness of selection using chi-square: Philadelphia, Detroit, Los Angeles, Chicago, New York.

Page 228, Note 1

(a) The last figure in the column should be 28, not 38.
(b) The court ignores the numbers after the decimal point.

1970	1.3
1970-71	3.9
1970-72	12.3
1970-73	17.0
1970-74	21.4
1970-75	28.4
1970-76	32.5
1970-77	36.5
1970-78	41.2
1970-79	43.3

Page 228, Note 2

There is a 20% chance of a type 1 error. Compare page 223, note 37.

$P_1^* = 8.3\% = 0.083$
$P_1 = 11.1\% = 0.111$
$P_2 = 88.9\% = 0.889$
$N = 206$
$Z = -1.28$

Page 230, *Alabama Nursing Home* Note

69.15% (mean cost plus one-half standard deviation = 50% + 19.15%).

Page 240.

Sample size for estimate within one ton at 90% = 3 days. Sample size for estimate within 500 pounds at 99% = 90 days.

(It is easier to convert 500 pounds to tons before doing this problem, but if you want to work it out in pounds, remember to find the pound equivalent of SD, rather than SD^2, and then square those pounds to find the sample variance, SD^2).

Page 248, Note 3

(a) .0300 to .0952 fibers per cc.

$Z = 2.33$
interval = $.0626 \pm .0326$

(b) .0396 to .0856 fibers per cc.

$Z = 1.645$
interval = $.0626 \pm .0230$

422 Appendix B. Answers to Selected Problems

Page 248, *Marathon Oil Co.* Note

We can be 99% certain that an exemplary plant will emit between 5.65 and 9.25 pounds per day.

$m = 7.45$

Page 256, Note 2

(a) At least 685 of the 2,361 owners reporting failures would submit affidavits.

(b) We can be 95% certain that between 144 and 1089 failures would be reported with 685 owners reporting.

$s.e.^2 = 1.2$
$s.e._m = .35$
interval $= .9 \pm .69$
interval $= .21$ to 1.59 failures/person

Page 260, Note 2

(a) At a 95% confidence level, between 24% and 96% will return affidavits.

$s.e._{proportional\ mean} = .16$
interval $= .6 \pm .36$

(b) At least 567 of the 2,361 owners reporting would submit affidavits.

(c) We can be 95% certain that between 62 and 958 failures would be reported by this minimum group.

$s.e.^2 = 1.2$
$s.e._m = .35$
interval $= .9 \pm .79$
interval $= .11$ to 1.69 failures/person

Page 260, Note 3

(a) Interval for recalculated Note 2 is 0.0024 to 0.1228 fibers per cc.

(b) Intervals for recalculated Note 3 with a 98% confidence level is 0 to 0.1601 fibers per cc., and 10% significance level is 0.0217 to 0.1035 fibers per cc.

Page 261, Note 1

(a) We can be 99% certain that between 2.75% and 9.95% will be defective.

$s.e._{proportional\ mean} = .014$
interval $= .0635 \pm 2.575 \times .014 = .0635 \pm .036$

(b) We can be 90% certain that between 4.05% and 8.58% will be defective.

interval $= .0635 \pm .023$

(c) We can be 66.3% certain that between 5.01% and 7.69% will be defective.

Appendix B. Answers to Selected Problems 423

(d) Assuming again that the other units in the required sample of 315 were all perfect, the answer is yes. The lower boundary of the interval, 0.0361, would be above 1%.

Page 261, Note 2

Yes; $0.0635 - 0.842 \times 0.014 = 0.0517$, or 5.17% defective.

Page 262, Note 3

(a) No; $1.9 - 1.476 \times 0.3 = 1.46$.
(b) No; $1.9 + 2.015 \times 0.3 = 2.50$.

Page 264, Note 1

The intervals do overlap.
The interval for males is 2.27 to 4.33.

$s.e._m = 0.49$ (using the reported standard deviation as the standard error of the sample).
interval $= 3.3 \pm 1.03$

The interval for females is 2.36 to 9.24.

$s.e._m = 1.52$
interval $= 5.8 \pm 3.44$

Page 264, Note 2

	Males	Females
Instructors	$M = 3.2 \pm 1.70$	$M = 5.2 \pm 3.71$
Associate Professors	$M = 7.6 \pm 1.41$	$M = 16.4 \pm 8.91$
Professors	$M = 13.9 \pm 1.78$	$M = 14.7 \pm 3.14$

Page 264, Note 3

(a) Yes; $16.4 - 1.796 \times 4.05 = 9.13$.
(b) No; $16.4 - 2.718 \times 4.05 = 5.39$.
(c) Yes; $16.4 - 1.363 \times 4.05 = 10.88$.

Page 283, Note 2

$m_x = 288.75 \quad m_y = 21.5$
$s.e._x = 475.75 \quad s.e._y = 25.68$
$\Sigma (Z_{x_i} \times Z_{y_i}) = -1.245 \quad r_s = -.415$

Not statistically significant

Page 284, *National Commission on Egg Nutrition* Note

The only finding that does not reach the 5% significance level is (i).

Page 289, Note 3

$r_s = .785$ Reject null hypothesis.
$r_r = .857$ Reject null hypothesis.

Page 291, Coefficient of Determination Note

Variation in age "explains" $(.785)^2 = 61.6\%$ of variation in evaluation.

Page 314, Note 1

(a) $176.2 BCF$ in 1980.
(b) $0 = 317{,}293 - 160.16(Y)$ $Y = 1981$
(c) $s.e._b = 72.5$, $t_{calculated} = -2.209$

Two-tailed test: Refuse to reject null hypothesis that B is not significantly different from zero at 95% confidence level because $-160.16 \pm 2.306 \times 72.5$ includes zero in the interval.

One-tailed test: Reject null hypothesis that B is not less than zero at 95% because, for a one-tailed test:

$$|t_{calculated}| > |t_{critical}|$$

(d)   Rejection in either case gives a 5% chance of a type 1 error.

Page 314, Note 3

New Field Annual Exploration $= -271{,}457 + 139.13$ (year). For 1978, projected exploration effort equals 3,742 thousand feet.

Page 343, Note 3

1. Two-tailed critical t for $120\ d.f. = 2.617$ for 99% confidence. Coefficients on P, YD, NLD, L, and RV are significantly greater than zero.
2. $2,233.
3. No, not significantly greater than zero.
4. No, not significantly greater than zero.
5. Probably nothing.
6. Yes.
7. No.
8. I can be between 90% and 95% sure.
9. If coefficients were significant, an increase of $3692, but they are not.

Page 357

Probability of type 1 error is just over 1%.

$d.f._{numerator} = k = 27$
$d.f._{denominator} = n - k - 1 = 90 - 27 - 1 = 62$
$F_{critical} = 2.12$ for 24, 60 degrees of freedom.

Appendix B. Answers to Selected Problems

Page 369, Note 2

$F = 476$, $F_{critical}$ at 99% confidence level = 4.79; $F_{critical}$ at 95% confidence level = 3.07, for 120 d.f. in the denominator.

$$F = \frac{R^2}{1-R^2} \times \frac{n-k-1}{k} = \frac{.75}{.25} \times \frac{317}{2} = 476$$

$d.f._{numerator} = k = 2$ independent variables

$d.f._{denominator} = n - k - 1 = 317$

Page 370, Note 4

(a) $2,792,000 (remember $UR = 0$).
(b) R^2 is high, F is statistically significant at .01 for 5, 64 d.f.
(c) $t_{critical} = 1.671$ at .05 for 64 d.f. and one-tail. Variable SS is not significantly less than zero.

TABLE OF CASES

Italic type is used to indicate location of principal cases.

A.B.G. Instruments & Engr., Inc. v. United States, 34, 261, 382
Agarwal v. Arthur G. McKee & Co., *345*, 377, 381, 382
Akins v. Texas, 129
Alabama Nursing Home Assn. v. Califano, 230

Ballew v. Georgia, 46, *58*, 145
B.F. Goodrich Co. v. Department of Transp., 138
Board of Educ. v. Califano, *108*
Boston Chapter, NAACP, Inc. v. Beecher, *272*
Brown v. Board of Educ., 46

Cassell v. Texas, 129
Castaneda v. Partida, *82*, 95, 116, 140, 144, 154, 172, 202, 229, 237, 393
Certified Color Manufacturers' Assn. v. Mathews, 34, *155*, 179
Chance v. Board of Examiners, *182*
Chewning v. Seamans, 345
Chinese for Affirmative Action v. FCC, *163*, 198, 380, 384
City of Chicago, United States v., *277*
Coca-Cola, Inc., In re, 34, 129
Coleman Motor Co. v. Chrysler Corp., 24, 25, 33, 294, 303
Craig v. Boren, *36*, 380

EEOC v. United Virginia Bank/Seaboard Natl., *117*, 384
Ethyl Corp. v. EPA, 33

Fatico, United States v., *146*
Forte-Fairbairn, Inc., In re, 34, *131*, 139

Gay v. Waiters' & Dairy Lunchmens' Union, Local 30, *203*, 381
General Motors Corp., United States v., 34, *249*, 341, 381
Goff, United States v., *93*, 380

Inmates of the Nebraska Penal & Correctional Complex v. Greenholtz, *97*, 383
ITT Continental Baking Co., In re, 34

James v. Stockham Valves & Fittings Co. (I), *349*
James v. Stockham Valves & Fittings Co. (II), *351*

Kroger Co., In re, 197, 381

Marathon Oil Co. v. EPA, 33, 248
Maskeny, United States v., *124*, 373

National Commn. on Egg Nutrition, In re, 283
National Lime Assn. v. EPA, 33
News Publishing Co. v. United States, 33
Northshore School Dist. No. 417 v. Kinnear, *357*, 391

427

Orchard View Farms v. Martin Marietta Aluminum Corp., 33

Pennsylvania v. Local 542, Intl. Union of Operating Engineers, 24, *26*, 31, *284*
Pennsylvania v. O'Neill, 21
Pfizer, Inc., In re, 34
Presseisen v. Swarthmore College, *262, 331,* 382, 390

Quaker Oats Co., In re, 371, 381

Reserve Mining Co. v. EPA, 33, *240*

South Dakota Public Utils. Commn. v. Federal Energy Regulatory Commn., 33, *307*
Sterling Drug, Inc., In re, 34
Swain v. Alabama, 128

Taylor v. Weaver Oil & Gas Corp., 138

United Fruit Co., In re, 34
United States v. City of Chicago, *277*
United States v. Fatico, 146
United States v. General Motors Corp., *249*, 341, 381
United States v. Goff, *93,* 380
United States v. Maskeny, 124, 373

Vuyanich v. Republic Natl. Bank of Dallas, *316,* 339, 384, *385*

SUBSTANTIVE LAW INDEX

This index is a guide to the areas of law considered in cases, notes, problems, and materials in this text. Following each area is a list of references and pages on which they appear. For principal cases, only the page on which the excerpt appears is noted.

Administrative Law

A.B.G. Instruments & Engr., Inc. v. United States, 261
Alabama Nursing Home Assn. v. Califano, 230
B.F. Goodrich Co. v. Department of Transp., 138
Board of Educ. v. Califano, 108
Certified Color Manufacturers' Assn. v. Mathews, 155
Chance v. Board of Examiners, 182
Chewning v. Seamans, 345
Chinese for Affirmative Action v. FCC, 163
EEOC v. United Virginia Bank/Seaboard Natl., 117
Ethyl Corp. v. EPA, 33
In re Coca-Cola Co., 129
In re Forte-Fairbairn, Inc., 131
In re Kroger Co., 197
In re National Commn. on Egg Nutrition, 283
Marathon Oil Co. v. EPA, 248
National Lime Assn. v. EPA, 33
Reserve Mining Co. v. EPA, 240
South Dakota Pub. Utils. Commn. v. Federal Energy Regulatory Commn., 307
United States v. City of Chicago, 277
United States v. General Motors Corp., 249

Antitrust (See also Trade Regulation)

Coleman Motor Co. v. Chrysler Corp., 24, 294
In re Quaker Oats Co., 371
In re Sterling Drug, Inc., 34
In re United Fruit Co., 34

Communication Law

Chinese for Affirmative Action v. FCC, 163

Constitutional Law

Akins v. Texas, 129
Ballew v. Georgia, 58
Board of Educ. v. Califano, 108
Brown v. Board of Educ., 46
Cassell v. Texas, 129
Castaneda v. Partida, 82
Chance v. Board of Examiners, 182
Craig v. Boren, 36
Gay v. Waiters' & Dairy Lunchmen's Union, Local 30, 203
Inmates of the Nebraska Penal & Correctional Complex v. Greenholtz, 97
Northshore School Dist. No. 417 v. Kinnear, 357

429

Pennsylvania v. O'Neill, 21
Swain v. Alabama, 128
United States v. Fatico, 146
United States v. Maskeny, 124

Consumer Protection

B.F. Goodrich Co. v. Department of Transp., 138
Certified Color Manufacturers' Assn. v. Mathews, 155
In re Coca-Cola Co., 129
In re ITT Continental Baking Co., 34
In re Kroger Co., 197
In re National Commn. on Egg Nutrition, 283
In re Pfizer, Inc., 34
United States v. General Motors Corp., 249

Contracts

A.B.G. Instruments & Engr. v. United States, 261
In re Forte-Fairbairn, Inc. 131

Criminal Law

Ballew v. Georgia, 58
Castaneda v. Partida, 82
United States v. Fatico, 146
United States v. Goff, 93
United States v. Maskeny, 124

Criminal Procedure

Ballew v. Georgia, 58
Casteneda v. Partida, 82
United States v. Fatico, 146

Damages

Agarwal v. Arthur G. McKee & Co., 345
Chewning v. Seamans, 345

Coleman Motor Co. v. Chrysler Corp., 24, 294
Orchard View Farms v. Martin Marietta Aluminum Corp., 33

Discrimination Law

Agarwal v. Arthur G. McKee & Co., 345
Akins v. Texas, 129
Board of Educ. v. Califano, 108
Boston Chapter, NAACP, Inc. v. Beecher, 272
Brown v. Board of Educ., 46
Cassell v. Texas, 129
Castaneda v. Partida, 82
Chance v. Board of Examiners, 182
Chinese for Affirmative Action v. FCC, 163
EEOC v. United Virginia Bank/Seaboard Natl., 117
Gay v. Waiters' & Dairy Lunchmen's Union, Local 30, 203
Inmates of the Nebraska Penal & Correctional Complex v. Greenholtz, 97
James v. Stockham Valves & Fittings Co., 349, 351
Pennsylvania v. Local Union 542, Intl. Union of Operating Engineers, 26, 284
Pennsylvania v. O'Neill, 21
Presseisen v. Swarthmore College, 262, 331
Swain v. Alabama, 128
Taylor v. Weaver Oil & Gas Corp., 138
United States v. City of Chicago, 277
United States v. Goff, 93
United States v. Maskeny, 124
Vuyanich v. Republic Natl. Bank of Dallas, 316, 385

Education Law

Board of Educ. v. Califano, 108
Brown v. Board of Educ., 46

Chance v. Board of Examiners, 182
Northshore School Dist. No. 417 v. Kinnear, 357

Employment

Agarwal v. Arthur G. McKee & Co., 345
Boston Chapter, NAACP, Inc. v. Beecher, 272
Chance v. Board of Examiners, 182
Chewning v. Seamens, 345
Chinese for Affirmative Action v. FCC, 163
EEOC v. United Virginia Bank/Seaboard Natl., 117
Gay v. Waiters' & Dairy Lunchmen's Union, Local 30, 203
James v. Stockham Valves & Fittings Co., 349, 351
Pennsylvania v. Local Union 542, Intl. Union of Operating Engineers, 26, 284
Pennsylvania v. O'Neill, 21
Taylor v. Weaver Oil & Gas Corp., 138
United States v. City of Chicago, 277
Vuyanich v. Republic Natl. Bank of Dallas, 316, 385

Energy Law

South Dakota Public Utils., Commn. v. Federal Energy Regulatory Commn., 307

Environmental Law

ABC Factory v. EPA, 233
Ethyl Corp. v. EPA, 33
Marathon Oil Co. v. EPA, 248
National Lime Assn. v. EPA, 33
Reserve Mining Co. v. EPA, 240

Equity

Board of Educ. v. Califano, 108
Agarwal v. Arthur G. McKee & Co., 345
Chance v. Board of Examiners, 182
Craig v. Boren, 36
Northshore School Dist. No. 417 v. Kinnear, 357
Pennsylvania v. O'Neill, 21
Reserve Mining Co. v. EPA, 240
United States v. City of Chicago, 277

Evidence

Agarwal v. Arthur G. McKee & Co., 345
Boston Chapter, NAACP, Inc. v. Beecher, 272
EEOC v. United Virginia Bank/Seaboard Natl., 117
Inmates of the Nebraska Penal & Correctional Complex v. Greenholt, 97
In re Coca-Cola Co., 129
In re Forte-Fairbairn, Inc., 131
In re Kroger Co., 197
James v. Stockham Valves & Fittings Co., 349, 351
Northshore School Dist. No. 417 v. Kinnear, 357
Pennsylvania v. Local Union 542, Intl. Union of Operating Engineers, 26, 284
Pennsylvania v. O'Neill, 21
Presseisen v. Swarthmore College, 262, 331
Reserve Mining Co. v. EPA, 240
South Dakota Public Utils. Commn. v. Federal Energy Regulatory Commn., 307
United States v. City of Chicago, 277
United States v. Fatico, 146

Family Law/Domestic Relations

Kilroy v. Kilroy, 265

Food and Drug Law

Certified Color Manufacturers' Assn. v. Mathews, 155
In re National Commn. on Egg Nutrition, 283

Juries

Akins v. Texas, 129
Ballew v. Georgia, 58
Cassell v. Texas, 129
Castaneda v. Partida, 82
Jury Selection Hypotheticals, 70
Swain v. Alabama, 128
United States v. Goff, 93
United States v. Maskeny, 124

Labor Law

Gay v. Waiters' & Dairy Lunchmen's Union, Local 30, 203
James v. Stockham Valves & Fittings Co., 349, 351
Pennsylvania v. Local Union 542, Intl. Union of Operating Engineers, 26, 284

Prisoners' Rights

Inmates of the Nebraska Penal & Correctional Complex v. Greenholtz, 97

Products Liability

United States v. General Motors Corp., 249

State and Local Governments

Chance v. Board of Examiners, 182
Pennsylvania v. O'Neill, 21
Northshore School Dist. No. 417, v. Kinnear, 357
United States v. City of Chicago, 277

Taxation

News Publishing Co. v. United States, 33

Trade Regulation

Coleman Motor Co. v. Chrysler Corp., 24, 294
In re Coca-Cola Co., 129
In re Forte-Fairbairn, Inc., 131
In re ITT Continental Baking Co., 34
In re Kroger Co., 197
In re National Commn. on Egg Nutrition, 283
In re Pfizer, Inc., 34
In re Quaker Oats Co., 371
In re Sterling Drug, Inc., 34
In re United Fruit Co., 34

Utilities Regulation

South Dakota Public Utils. Commn. v. Federal Energy Regulatory Commn., 307

Welfare

Alabama Nursing Home Assn. v. Califano, 230
United States v. Goff, 93

INDEX

Abscissa, 16
Absolute disparity, 91, 94, 127, 393. *See also* Disparity
Accuracy testing. *See* Significance testing
Autocorrelation, 372
Average, 75

Bias, 386
Binomial distribution
  Castaneda rule, 85, 172
  generally, 81, 85
  standard deviation, proportional formula, 82, 85, 92, 172
Binomial probability
  generally, 56, 65
  table, 66, 398
Bivariate regression. *See also* Regression analysis; Regression coefficient; Regression line
  bivariate regression line, 295, 297
  generally, 294, 340

Cardinal rankings, 288
Categorical variable, 256, 341. *See also* Dummy variables
Causation, 31, 93, 141, 162, 182, 223, n. 38, 229, 269, 291, 390, 392
Chi-square test
  chi-square value, 172
  degrees of freedom, 174
  formula, 172
  generally, 162, 169, 172, 190, 196
  probability of type 1 error, 175
  statistical significance, 159 n. 31, 175, 180
  table, 176, 402
  Yates correction, 181, 190, 383
Clustered sample, 255

Coefficient of determination, 286, 291
Coefficient of variation
  applied, 131, 133, 138
  formula, 132, 137
  generally, 137
Concurrent validation, 274, 278, 282
Confidence interval
  difference between means, 264
  formula, 260
  generally, 232, 235, 244, 246, 249, 254, 256, 261, 264, 351, 392
  one-tailed test, 202, 260, 261, 303, 324
  regression coefficients, 299, 323, 386. *See also* t test, for regression coefficients
  two-tailed test, 260
Confidence level, 249. *See also* Statistical significance
Content validation, 274, 281
Correlation coefficient. *See also* Rank correlation coefficient
  compared to rank correlation coefficient, 288, 290
  formula, 267, 270, 271, 295
  generally, 264, 274, 278, 282, 283, 366, 369
  negative correlation, 265, 283
  positive correlation, 265, 267, 275
  practical significance, 269, 272, 275, 281, 282, 284
  relationship to regression analysis, 293, 300
  statistical significance, 269, 272, 275, 277, 282, 284, 331, 347, 369, 371
  table of critical values for r, 271, 408
  to calculate
    for a population, 270
    for a sample, 271

433

Critical values
  r_s, 271
  r_r, 290
  t, 260, 261, 302
  χ^2, 175
  Z, 238, 248
Cross-sectional analysis, 344
Curvilinear relationships, 373, 376, 388

Data preparation, 379
Degrees of freedom
  bivariate regression, 302
  chi-square test, 174
  F test, 357
  multivariate regression, 344, 357
  t test, 260
Dependent variable, 293, 318. *See also* Regression analysis
Descriptive statistics
  generally, 3
  histograms, tables, graphs, 14
Direct relationship, 265. See also Positive correlation
Disparity, 92, 93, 99, 103, 107, 112, 116, 126, 168, 224
Dummy variable, 322 n. 57, 329, 341, 372, 386
Durbin-Watson statistic, 372

Error term. *See also* Residual variance
  autocorrelation, 374
  in regression analysis, 299, 321, 385, 388, 391
  plot of residuals, 373
Errors, types one and two. *See* Type 1 error; Type 2 error
Expected value, 80, 81, 85, 92, 96, 103, 105, 107, 112, 116, 121, 123, 168, 172, 180, 196, 219, 228, 229, 382
Explained variance. *See* Coefficient of determination

Frequency, 14, 139, 196
F statistic
  degrees of freedom, 357

  formula, 356
  generally, 316, 356, 369, 373
  statistical significance, 355, 356, 371
  table, 358, 410

Graphs
  bar graph, 9, 21
  examples of, 11, 19
  generally, 13
  line graphs distinguished from histograms, 14
  to read, 15, 16, 21
Group of suspect composition, 91, 96, 107, 115, 127, 229, 382

Heteroscedasticity, 388, 391
Histograms
  examples of, 7, 10
  generally, 14
  to read, 16

Independent variable, 293, 318, 348, 369, 372, 384, 386, 388. *See also* Regression analysis
Inferential statistics
  and causation, 31, 35
  generally, 31, 322. *See also* Statistical significance
  introduction, 3
Intercept
  formula for bivariate regression, 297
  generally, 297, 314, 320, 333, 342. *See also* Regression line
Inverse relationship, 265. *See also* Negative correlation

Least squares line. *See* Ordinary least squares; Regression analysis; Regression line

Matrix algebra, 315
Mean, 76, 231, 263, 267, 364
Mean standard error. *See* Standard error, of sample mean

Index 435

Measurement
  generally, 16
  ordinate axis on graphs, 16
  unit of measurement, 16
Median, 75, 263
Multicollinearity, 339, 355, 387
Multiple correlation coefficient
  corrected, 371, 377
  formula, 356
  generally, 325, 355, 369, 373, 377, 390
Multiple regression, 293, 315, 316. *See also* Regression analysis
Multivariate regression. *See* Multiple regression

Negative correlation, 265, 283. *See also* Correlation coefficient
Nonlinear multiple regression, 373, 376, 388
Normal distribution
  generally, 139, 198, 219, 382, 386
  statistical significance, 155. *See also* Z test
Normal distribution curve, 132
Normal table, 132, 200, 404
Null hypothesis, 144, 154, 162, 165, 170, 172, 179, 196, 199, 270, 289, 299, 307, 314, 351, 383. *See also* Significance testing

Observation, 80, 140, 198
Omitted variables, 386, 389
One-tailed test, 202, 260, 261, 303, 324. *See also* Confidence interval; Statistical significance
Ordinal rankings, 288
Ordinary least squares, 297, 310 n. 11, 318, 385. *See also* Regression line
Ordinate axis, 16

Percentage difference (compared to difference between percents), 128
Plotting residuals, 373
Point estimate, 232. *See also* Sample mean

Population, 231
Population mean
  distinguished from sample mean, 232
  formula, 76, 233
  generally, 231
Population regression coefficient, 294
Positive correlation, 265, 267, 275. *See also*, Correlation coefficient
Practical significance, 24, 143, 165 n. 57, 223, 269, 272, 275, 281, 282, 284, 369, 371, 394
Predictive validation of exam, 274
Predictive validity of regression results, 357

r. *See* Correlation coefficient
Random disturbance term. *See* Error term, in regression analysis
Rank correlation coefficient
  formula, 288
  generally, 284, 288
  heteroscedasticity, 392
  statistical significance, 285, 289
  table of critical values, 290, 409
Regression analysis. *See also* Bivariate regression; Multiple regression
  dependent variable, 293, 318
  error term, 299, 321, 385, 388, 391
  general form, 316, 320, 325, 384
  generally, 293, 316, 331, 345, 347, 350, 353, 365, 369
  independent variable, 293, 318, 348, 369, 372, 384, 386, 388
  regression coefficients. *See* Regression coefficients
Regression coefficients
  formula for bivariate regression, 295
  generally, 293, 316, 341
  statistical significance. *See* t test, for regression coefficients; F statistic

Regression coefficients—*Continued*
  population and sample regression coefficient distinguished, 302
Regression line
  generally, 316, 320, 325, 384
  intercept, 297, 314, 320, 333, 342
  slope of, 295, 320, 333, 341
Reliability testing. *See* Significance testing
Residual error. *See* Error term
Residual variance
  formula for bivariate regression, 301
  generally, 300, 385, 391
  variance of error, 300
r^2. *See* Coefficient of determination
R^2. *See* Multiple correlation coefficient
r table, 271, 408. *See also* Correlation coefficient
r_r. *See* Rank correlation coefficient
r_r table, 390, 409. *See also* Rank correlation coefficient
r_s. *See* Correlation coefficient

Samples, 4, 33, 34, 37, 44, 231, 244, 249, 253, 255, 261, 278, 283, 288, 380
Sample correlation coefficient. *See* Correlation coefficient
Sample mean. *See also* Confidence interval
  distinguished from population mean, 231
  formula, 232
  generally, 231, 299
  point estimate, 232
Sample size
  formula, 239
  how to calculate, 238
Sample standard deviation. *See* Standard error, of sample
Sample variance
  formula, 235, 257
  generally, 231
Sampling, 97, 255, 380
Sigma, 77
Significance level
  correlation coefficients, 272, 289

critical Z values for three common significance levels, 237
generally, 238
probability of type 1 error, 238
Significance testing. *See* Statistical significance
Simple random samples, 255
Slope of the regression line, 296, 320, 333, 341. *See also* Regression line
Small-number problems, 96, 101, 102, 120, 122, 123, 166 n. 73, 168, 180, 197, 257, 382
Spearman rank correlation coefficient. *See* Rank correlation coefficient
Specification errors, 384
Standard deviation
  binomial distribution, 81
  binomial formula, 82, 92, 103, 111, 172, 220
  Castaneda rule, 85, 92, 103, 108, 112, 115, 121, 172, 178, 198, 221, 237, 238, 384
  formula, 53, 80, 82, 85, 267
  generally, 56, 79, 85, 86, 91, 92, 93, 96, 103, 107, 111, 115, 121, 124, 126, 127, 130, 132, 137, 138, 139, 170, 198, 219, 228, 263, 279, 364
  proportionate formula, 111, 116
  statistical significance, 144. *See also* Z test
Standard error
  of dependent variable, 295
  of independent variable, 295
  of regression coefficient
    formula for bivariate regression, 301
    generally, 300, 302, 391
  of regression intercept, 306
  of sample
    formula, 235
    generally, 231
  of sample mean
    formula, 236, 257
    generally, 236, 246, 261
    proportionate formula, 256
Statistical notation, 76
Statistical significance. *See also* Practical significance

Castaneda rule as measure of significance, 143
Chi-square value as measure of significance, 159 n. 31, 169, 180
F statistic as measure of significance, 355, 371
generally, 143, 158, 161, 162, 165, 169, 180, 218, 263, 264, 275, 277, 285, 299, 323, 394
multiple regression, 299
r score as measure of significance, 269, 272, 275, 277, 282, 284, 331, 347, 369, 371
r_r score as measure of significance, 285
t statistic
   as measure of significance of regression coefficient, 300, 302, 312, 314, 323, 335, 340
   as measure of significance of regression intercept, 306
Z statistic as measure of significance, 223. *See also t* test
Statisticians, role of, 32
Stratified sample, 253, 256, 278
Student's *t* test. *See t* test
Subsample, 256
Summation, 77

Tables
   Appendix A, Statistical Tables, 397
   examples of, 14, 15, 18, 22, 23, 25, 27, 28
   generally, 13
   how to read, 15
Time-series analysis, 314
Trend-line analysis, 314
t test
   confidence interval of small samples, 257
   degrees of freedom
      for confidence intervals, 260
      for regression coefficients, 302, 344
      for regression coefficients, 300, 312, 314, 391
   generally, 257, 300, 312, 323, 383, 392

one-tailed test, 303, 314, 324
table, 258, 304, 406
t-calculated formula
   for regression coefficient, 300, 303, 307
   for regression intercept, 306
two-tailed test, 275, 303, 314, 323
Two-stage least squares, 391 n. 2
Two-tailed test. *See t* test
Type 1 error. *See also* Null hypothesis; Significance testing
confidence intervals, 238
generally, 61, 145, 155, 162, 165, 169, 173, 180, 203, 223 n. 38, 272
juries, 61, 145
probability of error, chi-square test, 203
probability of error, correlation coefficient, 272
F test, 357
t test, 314
Z test, 202, 223 n. 38, 238
Type 2 error. *See also* Null hypothesis
generally, 145, 391
juries, 61, 145

Unit of measurement. See *Measurement*

Variables
   categorical, 256, 341
   generally, 17
Variability
   range, 77
   variance, 77, 233
   squared deviation, 78
Variance
   explained variance. *See* Coefficient of determination
   population, 77
   sample, 233
Variance of the error. *See* Residual variance

Weighted least squares, 392

Yates correction, 181, 190, 383

Z test
  formula, 199, 221, 267, 307
  generally, 198, 219, 228, 382
  measure of statistical significance, 223
  proportional formula, 228
  probability of type 1 error, 202, 223 n. 38, 238
  relationship to t test, 257, 307
Z score, 198, 219, 267
Z scores corresponding to three common significance levels, 238
Z table
  discussed, 199, 236, 248
  table, 200, 404